Birkhäuser

Israel Kleiner

Excursions in the History of Mathematics

Birkhäuser

Israel Kleiner
Department of Mathematics
and Statistics
York University
Toronto, ON M3J 1P3
Canada
kleiner@rogers.com

ISBN 978-0-8176-8267-5 e-ISBN 978-0-8176-8268-2
DOI 10.1007/978-0-8176-8268-2
Springer New York Dordrecht Heidelberg London

Library of Congress Control Number: 2011940430

Mathematics Subject Classification (2010): 00-01, 00A30, 01Axx, 01A45, 01A50, 01A55, 01A60, 01A61, 01A99, 03-03, 11-03, 26-03

Printed on acid-free paper

www.birkhauser-science.com

With all my love to
Nava
Ronen, Leeor, Tania, Ayelet, Avy, Tamir
Tia, Jordana, Jake, Elise

Preface

The primary audience for this book, as I see it, is teachers of mathematics. The book may also be of interest to mathematicians desiring a historical viewpoint on a number of the subject's basic topics. And it may prove useful to those teaching or studying the subject's history.

The book comprises five parts. Thc first three (A–C) contain ten historical essays on important topics: number theory, calculus/analysis, and proof, respectively. (The choice of topics is dictated by my interests and is based on articles I have published over the past twenty-five years.) Part D deals with four historically oriented courses, and Part E provides biographies of five mathematicians who played major roles in the historical events related in Parts A–D.

Each of the first three parts – on number theory, calculus/analysis, and proof – begins with a survey of the respective subject (Chaps. 1, 4, and 7), which is followed in more depth by specialized themes. In number theory these themes deal with Fermat as the founder of modern number theory (Chap. 2) and with Fermat's Last Theorem from Fermat to Wiles (Chap. 3). In calculus/analysis, the special topics describe various aspects of the history of the function concept, which was intimately related to developments in calculus/analysis (Chaps. 5 and 6). The themes on proof discuss paradoxes (Chap. 8) and the principle of continuity (Chap. 9), and offer a historical perspective on a very interesting debate about proof initiated in a 1993 article by Jaffe and Quinn (Chap. 10).

The four chapters in Part D (Chaps. 11–14) describe courses showing how a teacher can benefit from the historical point of view. More specifically, each of Chaps. 11–14 describes a mathematics course inspired by history. Chapters 11 and 12 are about numbers as a source of ideas in teaching. Chapters 13 and 14 deal, respectively, with great quotations and with famous problems. Moreover, Chaps. 4 and 6 (on analysis and on functions) contain explicit suggestions for teachers, while such suggestions are implicit throughout the book.

Mathematics was discovered/invented by mathematicians. In each of the first 14 chapters the creators of the relevant mathematics are mentioned prominently, but because of space constraints are given shorter shrift than they deserve. I have therefore found it useful to set aside a chapter that will give a much fuller account of five mathematicians who have played important roles in the developments that I am recounting in the book. They are Dedekind, Euler, Gauss, Hilbert, and Weierstrass (Chap. 15). I hope these mini-biographies will prove to be instructive and inspiring. (My choice is limited to five by space considerations, but other mathematicians could justifiably have been picked.)

There is considerable repetition among the various chapters. This should make possible independent reading of each chapter. The book has many references, placed at the end of each of its fifteen chapters (in the case of Chap. 15, at the end of each of the five biographies). The references are mainly to secondary sources. These are, as a rule, easier to comprehend than primary sources, and more readily accessible. (Many of the secondary sources contain references to primary sources, which are often in German or French.)

I had two main goals in writing this book:

(a) To arouse mathematics teachers' interest in the history of mathematics.
(b) To encourage mathematics teachers with at least some knowledge of the history of mathematics to offer courses with a strong historical component.

Let me explain why I view these as important goals.

I come to the history of mathematics from the perspective of a mathematician rather than of a historian of mathematics. The two perspectives are, in general, not the same. My longstanding interest in the history of mathematics stems largely from trying to improve my teaching of mathematics.

Early in my teaching career I became dissatisfied with the exclusive focus on the formal theorem-proof mode of instruction. I admired the elegance of the logical structure of our subject, but over time I did not find it sufficient to sustain my enthusiasm in the classroom, perhaps because most of my students did not sustain theirs.

In due course I found that the history of mathematics helped boost my enthusiasm for teaching by providing me with perspective, insight, and motivation – surely important ingredients in the making of a good teacher. For example, when I taught calculus I was able to understand where the derivative came from, and how it evolved into the form we see in today's textbooks; and when I taught abstract algebra, I was able to understand how and why the concepts of ring and ideal came into being, and the source of Lagrange's theorem about the order of subgroups of finite groups.

Such examples could be multiplied endlessly. They supplied insight and added a new dimension to my appreciation of mathematics. I came to realize that while it is important to have technical knowledge of mathematical concepts, results, and theories, it is also important to know where they came from and why they were studied. The following quotation from the preface to C. H. Edwards' *The Historical Development of the Calculus* is apt:

> Although the study of the history of mathematics has an intrinsic appeal of its own, its chief *raison d'être* is surely the illumination of mathematics itself. For example, the gradual unfolding of the integral concept – from the volume computations of Archimedes to the intuitive integrals of Newton and Leibniz and finally the definitions of Cauchy, Riemann,

and Lebesgue – cannot fail to promote a more mature appreciation of modern theories of integration.

I hope to achieve my first goal – to arouse mathematics teachers' interest in the history of mathematics – by focusing in this book on two important areas, number theory and analysis, and on the fundamental notion of proof, a perennially hot topic of discussion among mathematics teachers (Chaps. 1–10). I trust that the five biographies will also capture the reader's interest (Chap. 15).

I have found *historical digressions* to be a useful device in teaching mathematics courses. For example, when introducing infinite sets I will give a brief history, starting with Zeno and culminating with Cantor, of how and why they came to be studied; when discussing Pythagorean triples in a first course in number theory, I will comment on Fermat's note in the margin of Diophantus' *Arithmetica* about the "marvelous proof" he had of what came to be known as Fermat's Last Theorem; and when appropriate I will briefly recount interesting stories of mathematicians (Archimedes and Galois come to mind). Such historical departures from standard teaching practice should convince students that mathematics is a human endeavor, that its history is interesting, and that it can give them some insight into the grandeur of the subject.

I have taught upper-level undergraduate mathematics courses with a strong historical orientation, dealing broadly with "mathematical culture," and a graduate course in the history of mathematics – a required course in an In-Service Master's Program for high school teachers of mathematics. (I was fortunate that my colleagues recognized – not without a "battle" – that these courses could be given in a *department of mathematics*, and that they formed a desirable component in the education of budding mathematicians.) Chapters 11–14, on numbers, great quotations, and famous problems, describe courses of the above types. Suitable material from Chaps. 1–10 can be used in these courses when appropriate. For example, the theme "Algebraic numbers and diophantine equations" in Sect. 11.3.4 will benefit from material in Chap. 3; the theme "Changing standards of rigor in the evolution of mathematics," Sect. 14.2.3, can usefully draw upon Chaps. 7 and 10; and Sect. 14.4, (e) and (g), will find Chap. 2 useful.

But a question presents itself: Why should we teach such courses in a mathematics department (or for that matter, why make historical digressions)? The answer depends on how we view the education of mathematics students. These courses (or digressions) may not make students into better researchers or theorem provers, but they *can* help make them "mathematically civilized."

The last phrase is the title of a note by Professor O. Shisha in the *Notices of the American Mathematical Society* (vol. 30, 1983, p. 603). In it he briefly discusses what it means for students to be mathematically civilized (or cultured). Among other desiderata, such students will have "good mathematical taste and judgment," and will know "how to express mathematical ideas, orally and in writing, correctly, rigorously, and clearly." We can encourage mathematical culture, according to Professor Shisha, by (among other things) "constantly pointing out in our courses the historical development of the subjects, their goals and relations with other subjects in and outside mathematics," and by "requiring students to take courses in the history of mathematics."

The implementation of the second important goal of this book – to encourage teachers to offer mathematics courses with a strong historical component (or courses in history with a strong mathematics component) – should result in mathematically cultured students. Such students might be able to discuss, for example, whether there are revolutions in mathematics (what is such a revolution anyway?), and what to make of Cantor's dictum that "the essence of mathematics lies in its freedom."

The following quotation, from an editorial about mathematics teaching by the then editors of *The Mathematical Intelligencer*, B. Chandler and H. Edwards, is a fitting conclusion to these comments (vol. 1, 1978, p. 125):

> Do let us try to teach the general public more of the sort of mathematics that they can use in everyday life, but let us not allow them to think – and certainly let us not slip into thinking – that this is an essential quality of mathematics.
> There is a great cultural tradition to be preserved and enhanced. Each new generation must learn the tradition anew. Let us take care not to educate a generation that will be deaf to the melodies that are the substance of our great mathematical culture.

I want to express heartfelt gratitude to my friend and colleague Hardy Grant for his kindness, support, and assistance over the past 40 years, and, in particular, for his help with this work. Of course all remaining errors (of omission or commission) are solely mine; I would be grateful if they were brought to my attention. Finally, I want to thank Tom Grasso, Katherine Ghezzi, and Jessica Belanger of Birkhäuser for their outstanding cooperation in seeing this book to completion.

Permissions

Grateful acknowledgment is hereby given for permission to reprint, with some changes, the following:

Numbers 1–6, credit for permission was granted by Springer Science + Business Media, LLC.

1. I. Kleiner, *A History of Abstract Algebra*, Birkhäuser, 2007.
2. I. Kleiner, The principle of continuity: a brief history, *Math. Intelligencer* 28: 4 (2006) 49–57.
3. I. Kleiner, History of the infinitely small and the infinitely large in calculus, *Educ. Studies in Math.* 48 (2001) 137–174.
4. I. Kleiner, From Fermat to Wiles: Fermat's Last Theorem becomes a theorem, *Elemente der Math.* 55 (2000) 19–37.
5. I. Kleiner & N. Movshovitz-Hadar, Proof: a many-splendored thing, *Math. Intelligencer* 19:3 (1997) 16–26.
6. I. Kleiner, Functions: historical and pedagogical aspects, *Science and Education* 2 (1993) 183–209.

Numbers 7–10, credit for permission was granted by The Mathematical Association of America.

7. I. Kleiner, Fermat: the founder of modern number theory, *Math. Mag.* 78 (2005) 3–14.
8. I. Kleiner, A history-of-mathematics course for teachers, based on great quotations. In *Vita Mathematica: Historical Research and Integration with Teaching*, ed. by R. Calinger, Math. Assoc. of Amer., 1996, pp. 261–268.
9. I. Kleiner & N. Movshovitz-Hadar, The role of paradoxes in the evolution of mathematics, *Amer. Math. Monthly* 101 (1994) 963–974.
10. I. Kleiner, Rigor and proof in mathematics: a historical perspective, *Math. Mag.* 64 (1991) 291–314.

Number 11, reprinted with permission from *Mathematics Teacher*, 1988, by the National Council of Teachers of Mathematics.

11. I. Kleiner, Thinking the unthinkable: the story of complex numbers (with a moral), *Math. Teacher* 81 (1988) 583–592.

Number 12, credit for permission was granted by Taylor & Francis.

12. I. Kleiner & S. Avital, Themes in the evolution of number systems, *Intern. Jour. of Math. Educ. in Science & Technology* 23 (1992) 445–461.

Number 13, credit for permission was granted by Cengage Learning/Nelson Education.

13. I. Kleiner, Number theory. In *History of Modern Science and Mathematics*, Vol. IV, ed. by B. Baigrie, Scribner's, 2002, pp. 1–21.

Number 14, credit for permission was granted by FLM copyright permission.

14. I. Kleiner, Famous problems in mathematics: an outline of a course, *For the learning of mathematics* 6 (1) (1986) 31–38.

Number 15, credit for permission was granted by Princeton University Press.

15. I. Kleiner, Weierstrass, Karl. In *The Princeton Companion to Mathematics*, ed. by W. T. Gowers, Princeton Univ. Press, 2008, pp. 770–771.

All the photographs of mathematicians that appear throughout the book are part of the public domain.

Contents

Part B Calculus/Analysis

Part A
Number Theory

Chapter 1
Highlights in the History of Number Theory: 1700 BC–2008

1.1 Early Roots to Fermat

Number theory, the study of the properties of the positive integers, which broadened in the 19th century to include other types of "integers," is one of the oldest branches of mathematics. It has fascinated both amateurs and mathematicians throughout the ages. The subject is tangible, the results are usually simple to state and to understand, and are often suggested by numerical examples. Nevertheless, they are frequently very difficult to prove. "It is just this," said Gauss, one of the greatest mathematicians of all time, "which gives number theory that magical charm that has made it the favorite science of the greatest mathematicians." To deal with the many difficult number-theoretic problems, mathematicians had to resort to – often to *invent* – advanced techniques, mainly from algebra, analysis, and geometry. This gave rise in the 19th and 20th centuries to distinct branches of number theory, such as algebraic number theory, analytic number theory, transcendental number theory, geometry of numbers, and arithmetic of algebraic curves. While number theory was considered for over three millennia to be one of the "purest" branches of mathematics, without any applications, it found important uses in the 20th century in such areas as cryptography, physics, biology, and graphic design. See [15].

The study of diophantine equations, so named after the Greek mathematician Diophantus (fl. c. 250 AD), has been a central theme in number theory. These are equations in two or more variables, with integer or rational coefficients, for which the solutions sought are integers or rational numbers. The earliest such equation, $x^2 + y^2 = z^2$, dates back to Babylonian times, about 1700 BC. This equation has been important throughout the history of number theory. Its *integer* solutions are called *Pythagorean triples*.

Records of Babylonian mathematics have been preserved on ancient clay tablets. One of the most renowned of these is named Plimpton 322. It consists of a table of fifteen rows of numbers that, according to most historians of mathematics, is a list of Pythagorean triples. There is no indication of how they were generated (*not* by trial), nor why (mathematics for fun?), but the listing suggests, as do other sources,

I. Kleiner, *Excursions in the History of Mathematics*,
DOI 10.1007/978-0-8176-8268-2_1,

that the Babylonians knew the Pythagorean theorem more than a millennium before the birth of Pythagoras (c. 570 BC), and that they studied number theory no later than algebra or geometry – nearly 4,000 years ago. See [12, 25].

Euclid's *Elements* (c. 300 BC) is known mainly for its axiomatic development of geometry. But three of its 13 Books (Books VII–IX) are devoted to number theory. Here Euclid introduces several fundamental number-theoretic concepts, such as divisibility, prime and composite integers, and the greatest common divisor (gcd) and least common multiple (lcm) of two integers. A number of basic results are also established.

The first two propositions of Book VII present the Euclidean algorithm for finding the gcd of two numbers. This is one of the central results in number theory. It is based on the important fact that if a and b are positive integers, there exist integers q and r such that $a = bq + r$, where $0 \leq r < b$. A very significant corollary of the Euclidean algorithm is that if d is the gcd of a and b then $d = ax + by$ for some integers x and y. Two other basic results in Book VII deal with primes: (1) every integer is divisible by some prime, and (2) if a prime divides the product of two integers, it must divide at least one of the integers.

Book IX resumes the study of primes (among other things). Proposition 20 proves that there are infinitely many primes, a far-from-obvious result. The beautiful, now-classic proof given by Euclid is used in textbooks to this day. One of the most important results in number theory (if not *the* most important) is undoubtedly the fundamental theorem of arithmetic (FTA) ("arithmetic" and "number theory" were at one time used interchangeably). It asserts that every integer $n > 1$ is a *unique* product of primes, so that (a) $n = p_1 p_2 \ldots p_s$, where the p_i are prime, and (b) if also $n = q_1 q_2 \ldots q_t$, with q_j prime, then $s = t$ and (after possibly rearranging the order of the q_j) $p_k = q_k$ for $k = 1, 2, \ldots, s$. There has been some debate among historians whether Proposition 14 of Book IX of the *Elements* is equivalent to the FTA. It states that "if a number be the least that is measured [divisible] by prime numbers, it will not be measured by any other prime number except those originally measuring it." In any case, the results (1) and (2) above readily yield a proof of the FTA.

Two final noteworthy topics in the *Elements* are "perfect numbers" and Pythagorean triples. The Pythagoreans of the 5th century BC believed that numbers (that is, positive integers) are the basis of all things. In particular, they assigned various attributes to specific numbers. For example, 1 was Godlike (although not considered a number, it was the generator of all numbers), 2 was feminine, 3 was masculine, and 5 was connected with marriage (since $5 = 2 + 3$). The number 6 was associated with perfection. Since 6 is the sum of all its *proper* divisors ($6 = 1 + 2 + 3$), any number with this property was called *perfect*. Another perfect number is 28 ($28 = 1 + 2 + 4 + 7 + 14$). Perhaps this number-mysticism motivated the Pythagoreans to initiate a serious study of properties of numbers, which later appeared in Books VII–IX of Euclid's *Elements*. In any case, the last result in Book IX, Proposition 36, deals with perfect numbers. Specifically, it shows that if $2^n - 1$ is prime for some integer n, then $2^{n-1}(2^n - 1)$ is perfect. Must all (even) perfect numbers be of this form? And for which n is $2^n - 1$ prime? These questions, suggested by the above result, will be discussed in subsequent sections.

Let us turn now to Pythagorean triples, integer solutions of $x^2 + y^2 = z^2$. Recall that the Babylonians determined 15 such triples. Euclid gave a formula that generates *all* of them (Book X, Lemma 1 preceding Proposition 29), namely $x = a^2 - b^2$, $y = 2ab$, $z = a^2 + b^2$, where a and b are arbitrary positive integers with $a > b$, although this is not how Euclid put it, since he had no algebraic notation. This implies in particular that there are infinitely many Pythagorean triples. See [12, 14, 25].

The other great Greek work in number theory is Diophantus' *Arithmetica* (c. 250 AD). It is divided into 13 "books," six of which have survived in Greek; four others were recently found in Arabic. The *Arithmetica* contains numerous problems, each of which gives rise to one or more equations, many of degree two or three. Although many of these sets of equations are "indeterminate," that is, have more than one integer or rational solution, Diophantus found in most cases a *single positive rational* solution. Here are three of his problems:

(a) To divide a given square into two squares (Book II, Problem 8). This requires solving $a^2 = x^2 + y^2$ for x and y, given a. Diophantus picked $a = 4$ and gave (16/5, 20/5) as the solution. There are, in fact, infinitely many solutions, which can be inferred from his method of solution; he said so explicitly in the body of Problem 19 of Book III. This problem would later motivate Fermat, one of the foremost mathematicians of the 17th century (see Sect. 2.4).

(b) To find two numbers such that their product added to either gives a cube (Book IV, Problem 26). This requires finding x, y, and z such that $xy + x = z^3$. Diophantus expressed y as a function of x; this led him to a cubic in x and z, which he proceeded to solve. The study of cubic Diophantine equations would be fundamental to mathematicians of the 18th and subsequent centuries (see Sects. 2.6.2, 3.8.1, and 3.8.3).

(c) Given two numbers, if, when some square is multiplied into one of the numbers and the other number is subtracted from the product, the result is a square, another square larger than the aforesaid square can always be found which has the same property (Book VI, Lemma to Problem 15). That is, if for fixed a and b, the equation $ax^2 - b = y^2$ has a solution, say $x = p$ and $y = q$, then it has another solution, $x = s$ and $y = t$, with $s > p$ and (necessarily) $t > q$. This problem dealt with the important idea of generating a new solution from a given one.

The solutions of many of the problems in the *Arithmetica* required great ingenuity – clever "tricks," as some would have it. However, Euler and others saw deep methods embedded in them. Recent historians have reconstructed Diophantus' work in the light of modern developments, suggesting that it contained the germ of an important geometric idea on the arithmetic of algebraic curves, the *tangent and secant method.* This entailed solving Diophantine equations by finding the intersection(s) of the curves determined by these equations with certain lines (tangent and secant lines) (see Sect. 3.8). The *Arithmetica* had great influence on the rise of modern number theory, especially on Fermat's work. See [3, 25], and Chap. 2.

While there was little if any progress in number theory in Europe during the Middle Ages, the Indian and Chinese civilizations were active in the field. The Indians especially were avid number theorists. For example, Brahmagupta (b. 598) solved the general linear diophantine equation $ax + by = c$ (a, b, and c fixed integers), and, for certain values of d, the Pell equation, $x^2 - dy^2 = 1$ (d a nonsquare positive integer). A solution of the general Pell equation was given by Bhaskara in the 12th century; this important equation is discussed more fully in Sects. 1.4.1, 2.6.3, and 2.6.4. By the fourth century AD, the Chinese had dealt with specific instances of what is today called the Chinese Remainder Theorem, the simultaneous solution of linear congruences, $x \equiv a_1 \pmod{m_1}, \ldots, x \equiv a_k \pmod{m_k}$, with the m_i relatively prime in pairs, although the Chinese did not use the congruence notation. The general problem was solved by Chin Chiu Shao in the 13th century. Both Indian and Chinese number-theoretic works were motivated to a large extent by problems in astronomy. See [11, 12, 24].

1.2 Fermat

Fermat was arguably the greatest mathematician of the first half of the 17th century. He made fundamental contributions to analytic geometry, calculus, probability, and number theory; the last was his mathematical passion. In fact, he founded number theory in its modern form.

Fermat's interest in the subject was aroused by Diophantus' *Arithmetica* (see [3]). The book had come to his attention in the 1630s through the excellent Latin translation, with extensive commentaries, by Bachet, a country gentleman of independent means who became very interested in number theory. Several of Fermat's important discoveries in the subject were given in the margins of this translation as commentaries on, and elaborations of, some of Diophantus' results. Most of his other results became known through his extensive correspondence with leading scientists of the day, principally Mersenne and Carcavi, who championed Fermat's work by disseminating it in the wider scientific community.

Fermat produced no formal publications in number theory, nor did he give any proofs (save one) in his letters or in the margins of the *Arithmetica*, although he did provide comments and hints. All his claims but one (see below) were later shown to be correct. One of his major aims was to interest his mathematical colleagues in number theory by proposing challenging problems (for which *he* had the solutions). As he put it:

> Questions of this kind [viz., number-theoretic] are not inferior to the more celebrated questions in geometry [other branches of mathematics] in respect of beauty, difficulty, or method of proof.

But his protestations were to no avail: mathematicians showed little serious interest in number theory until Euler came on the scene some 100 years later; see Sect. 2.1.

Several of Fermat's main results are given below. They turned out to anticipate major concerns of number theory.

1.2.1 Fermat's Little Theorem

This theorem says that for any integer a and prime p, $a^p - a$ is divisible by p; nowadays it is expressed in terms of congruences as $a^p \equiv a \pmod{p}$. An equivalent, useful way of putting it is that $a^{p-1} - 1$ is divisible by p, provided that a is *not* divisible by p. This is one of Fermat's most important results, and it found significant applications in cryptography in the second half of the 20th century [24].

Fermat is thought to have become interested in this problem through Euclid's result on perfect numbers, which raised the question of primes of the form $2^n - 1$ (see above). First, Fermat showed that for $2^n - 1$ to be prime, n must be prime (this is easy), and then studied conditions for $2^n - 1$ to have divisors. This led him to the special case $a = 2$ of Fermat's Little Theorem (that is, that $2^{p-1} - 1$ is divisible by p, if p is an odd prime), and thence to arbitrary a.

Numbers of the form $2^p - 1$ (p prime) are called Mersenne numbers (since they were studied by Mersenne) and denoted by M_p. It is not the case that M_p is prime for every prime p, for example, $M_{11} = 23 \times 89$. Mersenne claimed that for $p = 2, 3, 5, 7, 13, 17, 19, 31, 67, 127$, and 257, M_p *is* prime; these are called *Mersenne primes*. He was wrong about $p = 67$ and 257 (keep in mind that the corresponding M_p are huge numbers), and he missed $p = 61, 89$, and 107 (among those p less than 257). By the end of 2000, 38 Mersenne primes were known; the 38th, for $p = 6{,}972{,}593$, was found June 1, 1999 by one of the 12,000 participants in the Great Internet Mersenne Prime Search (see www.mersenne.org); it is the first Mersenne prime to have more than a million decimal digits. Eight more Mersenne primes were found in this century; the latest two are $2^{43112609} - 1$ (about 12.9 million digits, August 2008) and $2^{37156667} - 1$ (c. 11.1 million digits, September 2008). Are there infinitely many Mersenne primes? This is still an open question, about 350 years after it was posed, though probabilistic arguments suggest that the answer is yes. See [10, 24, 27], and Sect. 2.3.

1.2.2 Sums of Two Squares

Diophantus remarked that the product of two integers, each of which is a sum of two squares, is again a sum of two squares, that is, that $(a^2 + b^2)(c^2 + d^2) = (ac + bd)^2 + (ad - bc)^2$, though he did not state it in this generality. This appears to have prompted Fermat to ask which integers are sums of two squares. Since every integer is a product of primes, the above identity reduces the problem to asking which *primes* are sums of two squares. Now, all (odd) primes are either of the form $4n + 1$ or $4n + 3$, and it is easy to see that no prime (in fact, no integer) of the form $4n + 3$ can be a sum of two squares. Fermat claimed to have shown that every prime of the form $4n + 1$ *is* a sum of two squares, and that it is a *unique* such sum. It is then not difficult to characterize those integers that are sums of two squares. About 200 years later, Jacobi gave expressions for the number of ways in which an integer can be written as a sum of two squares (for example, $65 = 8^2 + 1^1 = 7^2 + 4^2$). See [1, 11, 24, 25] and Sect. 2.4.

Fermat also proved the following related theorems: every prime of the form $8n + 1$ or $8n + 3$ can be written as $x^2 + 2y^2$, and every prime of the form $3n + 1$ can written as $x^2 + 3y^2$. These were no idle results. For example, the latter was used by Euler in his proof of Fermat's Last Theorem (FLT) for $n = 3$ (see below). Moreover, these results raised the question of representation of primes in the form $x^2 + ny^2$ for general n (Fermat had difficulty already with $n = 5$). This was an issue with important ramifications, which evolved in the following two centuries into the question of representation of integers by "binary quadratic forms," $ax^2 + bxy + cy^2$, one of the central problems in number theory (see below). See [24] and Sect. 2.4.

1.2.3 Fermat's Last Theorem

In the margin of Problem 8 of Book II of Diophantus' *Arithmetica*, which asked for the representation of a given square as a sum of two squares (see above), Fermat said that, unlike that result,

> It is impossible to separate a cube into two cubes or a fourth power into two fourth powers or, in general, any power greater than the second into powers of like degree. I have discovered a truly marvelous demonstration, which this margin is too narrow to contain [7, p. 2].

Fermat was claiming that the equation $z^n = x^n + y^n$ has no (nonzero) integer solutions if $n > 2$. This has come to be known as FLT. Many mathematicians in the know doubt whether Fermat had a proof of this result. In later correspondence on this problem, he referred only to proofs of the theorem for $n = 3$ and 4 (see Sect. 2.5). This "theorem" was perhaps the most outstanding unsolved problem for 360 years. The Princeton mathematician Andrew Wiles gave a proof in 1994 (a detailed discussion is given in Chap. 3).

As mentioned, Fermat gave only one proof in number theory, and that was the case $n = 4$ of FLT, namely that $x^4 + y^4 = z^4$ has no nonzero integer solutions (it is easier than the case $n = 3$). This he did in the context of a problem on Pythagorean triples. What he showed was that "the area of a right-angled triangle whose sides have rational length cannot be a square of a rational number." Here he was responding to a problem raised by Bachet, based on one in Book VI of Diophantus' *Arithmetica*, that of finding a right-angled triangle whose area equals a given number.

It can be shown that the above problem is equivalent to showing that the area of a right-angled triangle with *integer* sides cannot be a square (integer). Fermat proceeded by assuming that such a triangle was possible, and obtained another of the same type, but with a *smaller* hypotenuse. Continuing this process, he obtained an infinite decreasing sequence of positive integers (the hypotenuses of the corresponding right triangles). But this is clearly impossible, which established the result. It follows as an easy corollary that $x^4 + y^4 = z^4$ has no integer solutions. See [7] and Sect. 2.5.

More important than the result was the method used to establish it, since known as the *method of infinite descent*. Its essence is this: Assume that a positive integer satisfies a given condition, and show, by an iterative process, that a smaller positive integer satisfies the same condition; then no positive integer can satisfy the condition. Logically, this is nothing but a variant of the principle of mathematical induction, but it provided Fermat (and his followers) with a powerful tool for proving many number-theoretic results. As he put it with considerable foresight, "this method will enable extraordinary developments to be made in the theory of numbers."

1.2.4 Bachet's Equation

This is the equation $x^2 + k = y^3$ (k an integer), a special case of which was considered by Bachet in his edition of the *Arithmetica*. Fermat found the (positive) solutions for $x^2 + 2 = y^3$ and $x^2 + 4 = y^3$, namely $x = 5$, $y = 3$ for the first equation, and $x = 2$, $y = 2$ and $x = 11$, $y = 5$ for the second. It is easy to verify that these are solutions of the respective equations, but it is rather difficult to show that they are the *only* (positive) solutions. Bachet's equation plays a central role in number theory to this day (see Sects. 1.9 and 3.8).

1.2.5 Pell's Equation

The Pell equation, $x^2 - dy^2 = 1$ (d a nonsquare positive integer) was noted in connection with Indian mathematics, though Fermat was likely unaware of that work. "The study of the [quadratic] form $x^2 - 2y^2$ must have convinced Fermat of the paramount importance of the equation $x^2 - Ny^2 = \pm 1$," said Weil [25, p. 92]. Fermat claimed to have shown that the equation has infinitely many integer solutions. This equation, too, has been very important in number theory, even in recent times (see Sects. 1.4 and 2.6).

1.2.6 Fermat Numbers

Having investigated when $2^n - 1$ is prime, it was natural for Fermat to consider the same question for numbers of the form $2^n + 1$. It is easy to show that for $2^n + 1$ to be prime, n must be a power of 2. Numbers of the form $2^{2^k} + 1$ are called Fermat numbers and are denoted by F_k. Fermat repeatedly asserted in correspondence that the F_k are prime for every k, although he admitted that he could not find a proof. Now, $F_1 = 5$, $F_2 = 17$, $F_3 = 257$, $F_4 = 65,537$ are all prime – they are called *Fermat primes*. However, Euler showed in 1732 that F_5 is not. The proof was not simply a matter of computation; F_5 has ten digits, and one would need a table of primes up to 100,000, unavailable to Euler, to test the primality of F_5 by "brute computational force." Euler's proof was largely theoretical: he proved that every factor of F_k must be of the form $t \times 2^{k+1} + 1$, t some integer, hence was able to show that F_5 is divisible by 641; in fact, $F_5 = 641 \times 6700417$. (In general, proving that a given very large number is composite is *much* easier than factoring it.) It was shown only in the 1870s that F_6 is composite. Various other Fermat numbers have been shown to be composite (this is a difficult problem), but none to be prime. In fact, it is thought that there are *no* Fermat primes other than the four listed above. But it is known (and easy to prove) that any two Fermat numbers are *relatively prime*. Since every integer >1 is divisible by some prime, this gives another proof (aside from Euclid's) that there are infinitely many primes. Fermat primes were shown by Gauss to be closely related to constructibility of regular polygons (see Sect. 1.6).

We conclude our account of Fermat. It is remarkable how well he chose for consideration problems that would become central in number theory. These stimulated the best mathematical minds, including those of Euler and Gauss, for the next two centuries. Without doubt, Fermat is the founder of modern number theory. See Chap. 2.

1.3 Euler

Euler was the greatest mathematician of the 18th century, and one of the most eminent of all time, "the first among mathematicians," according to Lagrange. He was also the most productive ever. Although "only" four volumes of a projected 80-volume collection of his works are on number theory, they contain priceless treasures, dealing with all existing areas in number theory and giving birth to new methods and results.

Euler was the first to take up the study of number theory in close to 100 years. His love for the subject, like Fermat's, was great. Legendre, in the preface to his 1798 book on number theory, put it thus [25, p. 325]:

> It appears ... that Euler had a special inclination towards ... [number-theoretic] investigations, and that he took them up with a kind of passionate addiction, as happens to nearly all those who concern themselves with them.

A considerable part of Euler's number-theoretic work consisted in proving Fermat's results and trying to reconstruct his and Diophantus' methods. See Chap. 2.

Euler's interest in number theory was apparently stimulated by his friend Goldbach, an amateur mathematician of Goldbach's Conjecture fame (see Sect. 1.8) with whom Euler carried on a correspondence over several decades. It began with a letter in 1729, when Euler was twenty-two, in which Goldbach asked Euler's opinion about Fermat's claim that the Fermat numbers are all prime. Euler was skeptical, but it was only two years later that he discovered the counterexample F_5 (see above). This set him on a lifelong study of Fermat's works. See [25].

Below we describe some of Euler's number-theoretic investigations, focusing on new departures in both methods and results.

1.3.1 Analytic Number Theory

The overriding reason why there was little interest in number theory among mathematicians in the 17th and 18th centuries was probably the ascendance during this period of calculus as the predominant mathematical field. A major topic was summation of series. Leibniz' result, $1 - 1/3 + 1/5 - 1/7 + \cdots = \pi/4$, fascinated mathematicians. Leibniz and the brothers Jakob and Johann Bernoulli attempted to sum the series $\sum_1^\infty 1/n^2 = 1 + 1/4 + 1/9 + 1/16 + \ldots$, without success. In 1735, Euler prevailed, showing that $1 + 1/4 + 1/9 + 1/16 + \ldots = \pi^2/6$. This was a spectacular achievement for the young Euler, with important consequences for number theory. It helped establish his growing reputation. See [25].

Euler next studied the series $\sum 1/n^{2k}$ for an arbitrary positive integer k, and proved the beautiful result that $\sum_{n=1}^\infty 1/n^{2k} = (2^{2k-1}\pi^{2k}|\mathrm{B}_{2k}|)/(2k)!$, where the B_i are the Bernoulli numbers, the coefficients in the power-series expansion $x/(e^x - 1) = \sum_1^\infty B_n x^n/n!$ ($|B_{2k}|$ denotes the absolute value of B_{2k}). The Bernoulli numbers are rational (e.g., $B_2 = 1/6$, $B_4 = -1/30$, $B_6 = 1/42$, hence $\sum 1/n^4 = \pi^4/90$ and $\sum 1/n^6 = \pi^6/945$). They turned out to be very important in number theory and elsewhere.

The series $\sum_{n=1}^\infty 1/n^{2k+1}$ was a mystery to Euler, and remained so to mathematicians of subsequent generations. Only in 1978 was it shown that $\sum_1^\infty 1/n^3$ is irrational. In November 2000, it was announced that $\sum_1^\infty 1/n^{2k+1}$ is irrational for infinitely many k; but it is still not known if $\sum_1^\infty 1/n^5$ is irrational, although one of $\sum_1^\infty 1/n^{2k+1}$, for $n = 2, 3, 4, 5$, is irrational. It is probably this lack of knowledge that persuaded Euler to study the function $\zeta(s) = \sum_1^\infty 1/n^s$ for all real $s > 1$, for which the series converges. This turned out to be a pivotal function in number theory, the *zeta function* (see Sect. 1.8). Euler soon derived what came to be called the Euler product formula, $\zeta(s) = \sum_{n=1}^\infty 1/n^s = \Pi_p 1/(1 - p^{-s})$, where the product ranges over all the primes p. This most important identity may be viewed as an analytic counterpart of the Fundamental Theorem of Arithmetic, expressing integers in terms of primes. See [2, 11].

An easy consequence of Euler's product formula is yet another proof of the infinitude of primes (if there were finitely many primes, then letting s approach 1 from the right, $\sum 1/n^s$ approaches infinity, while $\Pi 1/(1 - p^{-s})$) is finite). Another relatively easy corollary of the product formula is the divergence of $\sum 1/p$, where the sum is taken over all the primes p [2]. Since $\sum 1/n^2$ converges, this shows that the primes are "denser" than the squares in the sequence of positive integers, that is, there are, in some sense, "more" primes than squares.

Euler's introduction of analysis – the study of the continuous – into number theory – the study of the discrete – may at first appear paradoxical. However, it was a crucial development, greatly exploited in the next century. It led to the rise of a new area of study – *analytic number theory* (see Sect. 1.8). As Euler put it [25, p. 176]:

> One may see how closely and wonderfully infinitesimal analysis is related ... to the theory of numbers, however repugnant the latter may seem to that higher kind of calculus.

Another important instance in which Euler began to relate analysis to number theory was his use of "elliptic integrals" (integrals arising in finding the length of an arc of an ellipse) to study Diophantine equations of the form $y^2 = f(x)$, with $f(x)$ of degree three or four (their graphs are called elliptic curves; see Sect. 1.9 and [9,25]). More broadly, building bridges between different, seemingly unrelated, areas of mathematics is an important and powerful idea, for it brings to bear the tools of one field in the service of the other. (A very important example of bridge-building between algebra and geometry resulted in the 17th century in a new field/method, analytic geometry.) Further examples of the vital interaction between number theory and other areas will be discussed in Sects. 1.7, 1.8, and 1.9.

1.3.2 Diophantine Equations

This is a vast subfield of number theory, with diverse branches. Here are some examples of Euler's contributions.

The simplest diophantine equation is $ax + by = c$. In the early 1730s, Euler rediscovered its solution, known to Brahmagupta, Bachet, Fermat, and others. In this context he proved the important result, from which the solution of the equation follows, that if a and b are relatively prime to c, there is an x relatively prime to c such that $ax \equiv b \pmod{c}$. In connection with this work, Euler proved that for any relatively prime positive integers a and n, $a^{\varphi(n)} \equiv 1 \pmod{n}$, where $\varphi(n)$, the Euler φ-function, denotes the number of positive integers less than n and relatively prime to n. This important result generalized Fermat's Little Theorem, $a^{p-1} \equiv 1 \pmod{p}$, since for a prime p, $\varphi(p) = p - 1$.

Euler's φ-function is an example of a *multiplicative arithmetic function*, a function $f\colon N \to N$ (N the positive integers) such that $f(mn) = f(m)f(n)$ for m and n relatively prime. Other important multiplicative arithmetic functions introduced and studied by Euler were $d(n)$, the number of positive divisors of n, and $\sigma(n)$, the sum of the divisors of n. (In this notation, a number n is perfect if $\sigma(n) = 2n$.) Using the multiplicativity of the σ-function, Euler proved the converse of Euclid's theorem on perfect numbers, namely that if $2^{n-1}(2^n - 1)$ is perfect (necessarily even), then $2^n - 1$ is prime. Thus the theorems of Euclid and Euler, one proved 2,000 years after the other, characterize all *even* perfect numbers, reducing their existence to that of Mersenne primes. It is not known if there are any *odd* perfect numbers, but if there are, they are huge, larger than 10^{300} (this result dates from the 1990s).

In the latter part of his book *Elements of Algebra* (1770) Euler dealt with various Diophantine equations. An important one was $z^3 = ax^2 + by^2$, for which he developed various techniques. He specialized it to Bachet's equation, $z^3 = x^2 + 2$, which Fermat claimed to have solved. Euler's solution introduced a new, and most important, technique. He factored the right side of the equation and obtained $z^3 = x^2 + 2 = (x + \sqrt{-2})(x - \sqrt{-2})$. This was now an equation in the domain of "complex integers" of the form $Z(\sqrt{-2}) = \{a + b\sqrt{-2} : a, b \in Z\}$. These possess many of the number-theoretic properties of the ordinary integers Z (see Sect. 1.7). Euler exploited this analogy to solve Bachet's equation. Analogy, it should be noted, is a most important mathematical device (see Chap. 9). Euler was its undisputed master, using it again and again.

With this problem, Euler had taken the audacious step of introducing complex numbers into number theory, the study of the positive integers. "A momentous event had taken place," declared Weil [24, p. 242]. This foreshadowed the creation of a new field, *algebraic number theory*. While Euler had earlier wedded number theory to analysis, he now linked number theory with algebra. This bridge-building, too, would prove most fruitful in the following century (see Sect. 1.9 and Chap. 3). Here are two other examples of Euler's use of these ideas.

When his attention was drawn in the 1740s to Fermat's claim about $x^n + y^n = z^n$, he called it "a very beautiful theorem." In 1753 he wrote to Goldbach that he had proved it for $n = 3$, but he *published* a proof only in 1770, in his *Elements of Algebra*. Here he used the arithmetic (number theory) of the domain $Z(\sqrt{-3}) = \{a + b\sqrt{-3} : a, b \in Z\}$. There was, however, a considerable gap in the proof: the domain $Z(\sqrt{-3})$, unlike $Z(\sqrt{-2})$, does not have the arithmetic properties of Z. Analogy is, indeed, a powerful tool, but it must be used with great caution. Even the likes of Euler can err. Not much, however, is needed to repair the proof. Euler also applied his emerging ideas on the use of "complex integers" for studying number-theoretic problems to quadratic forms of the type $x^2 + cy^2$, by writing them as $(x + y\sqrt{-c})(x - y\sqrt{-c})$. See [9, 11, 24, 25].

1.3.3 Partitions

A partition of a positive integer n is a representation of n as a sum of positive integers. For instance, the partitions of 5 are 5, 4 + 1, 3 + 2, 3 + 1 + 1, 2 + 2 + 1, 2 + 1 + 1 + 1, and 1 + 1 + 1 + 1 + 1; the order of the summands is irrelevant. Partition theory is the study of such representations. It is a subfield of so-called *additive number theory*, which deals with the representation of integers as sums of other integers, for example, as cubes. Euler initiated the study of partitions in his great book *Introduction to the Analysis of the Infinite* (1748).

Let $p(n)$ denote the number of partitions of n. The object of partition theory is to study the properties of this and related arithmetic functions, for example $p_k(n)$, the number of partitions of n in which each summand is no greater than k. $p(n)$ is a rather complicated function, which grows enormously; for example, $p(200) = 3,972,999,029,388$. Euler's interest in partitions was stimulated by a colleague's letter that asked for the number of ways in which an integer n can be written as a sum of t distinct integers. Euler began to study $p(n)$ by introducing the important notion of its "generating function," the *formal* power series $\sum_1^\infty p(n)x^n$ (no questions of convergence involved). With its aid he proved the fundamental result that the number of partitions of an integer in which all summands are *odd* equals the number of partitions of the integer in which all summands are *distinct*. Partition theory is now an active and important area of number theory that has recently found applications in physics. A major contributor to the theory in the 20th century was the great Indian mathematician Srinivasa Ramanujan. For example, in 1918 he proved, with G. H. Hardy, that log $p(n) \sim c\sqrt{n}$, for some real number c (this means that log $p(n)/c\sqrt{n} \to 1$ as $n \to \infty$). See [9, 25].

1.3.4 The Quadratic Reciprocity Law

The quadratic reciprocity law was conjectured, but not proved, by Euler. It came to be one of the central results in number theory. It says (in the language of congruences) that there is a "reciprocity relation" between the solvability of $x^2 \equiv p \pmod{q}$ and $x^2 \equiv q \pmod{p}$ for any distinct odd primes p and q. Specifically, $x^2 \equiv p \pmod{q}$ is solvable if and only if $x^2 \equiv q \pmod{p}$ is solvable, unless $p \equiv q \equiv 3 \pmod{4}$, in which case $x^2 \equiv p \pmod{q}$ is solvable if and only if $x^2 \equiv q \pmod{p}$ is not. It is *the* fundamental law when it comes to the solvability of quadratic Diophantine equations (or quadratic congruences).

The *quadratic reciprocity law* arose from the study of representations of primes by quadratic forms $x^2 + cy^2$, in particular in connection with the representation of p by $x^2 + qy^2$ (p and q distinct odd primes). Euler had made some progress on this question in the 1740s, but gave a clear formulation of the law of quadratic reciprocity only in 1772. He attached great importance to this conjecture, proved in 1801 by Gauss, and made an important contribution by proving what came to be called the Euler Criterion: $x^2 \equiv a \pmod{p}$ is solvable (p an odd prime not dividing a) if and only if $a^{(p-1)/2} \equiv 1 \pmod{p}$. (If $x^2 \equiv a \pmod{p}$ is solvable, a is said to be a *quadratic residue* mod p; otherwise it is a quadratic nonresidue.) See [9, 11, 22, 24, 25].

These results represent only a miniscule part of Euler's contributions to number theory, which themselves are only a miniscule part of his overall contributions to mathematics; but they alone would have earned him entry to mathematics' hall of fame.

1.4 Lagrange

Euler's contemporaries took little interest in number theory, with the sole exception of Lagrange, much of whose work, especially in number theory, was directly inspired by Euler's. Lagrange became actively interested in number-theoretic problems in the 1760s, though his interest lasted less than 10 years. Among his accomplishments, three stand out: work on Pell's equation, on sums of four squares, and on binary quadratic forms.

1.4.1 Pell's Equation

Pell's equation, $x^2 - dy^2 = 1$ (d a nonsquare positive integer), is one of the most important Diophantine equations. It is a key to the solution of arbitrary quadratic diophantine equations; its solutions yield the best approximations (in some sense) to $\sqrt{d}$ (the Pell equation can be written as $(x/y)^2 = d + 1/y^2$, so that for large y, x/y is an approximation to $\sqrt{d}$); there is a 1-1 correspondence between the solutions of $x^2 - dy^2 = 1$ and the invertible elements of quadratic fields, $Q(\sqrt{d}) = \{s + t\sqrt{d} : s, t \in Q\}$; and the equation played a crucial role in the solution (in 1970) of Hilbert's Tenth Problem, the nonexistence of an algorithm for solving arbitrary Diophantine equations. (In 1900, Hilbert presented 23 problems at the International Congress of Mathematicians in Paris. These played an important role in the development of 20th-century mathematics [26].)

Pell's equation had been investigated by Fermat, Euler, and others. Lagrange gave the definitive treatment in the 1760s. This appeared as a Supplement to Euler's *Elements of Algebra*. In particular, Lagrange was the first to prove that a solution of Pell's equation always exists. In fact, he gave an explicit procedure for finding all solutions using the "continued fraction expansion" of $\sqrt{d}$ (see [1,9]). (It is one thing to prove the *existence* of solutions, quite another to *find* them.) Noteworthy was Lagrange's use of *irrational* numbers to solve equations in *integers*. He also used complex numbers to solve diophantine equations, thereby extending some of Euler's ideas. When Euler heard about these approaches, he remarked [25, p. 240]:

> I have greatly admired your method of using irrationals and even imaginary numbers in this kind of analysis which deals with nothing else than rational numbers. Already for several years I have had similar ideas.

1.4.2 Sums of Four Squares

Fermat claimed to have proved that every positive integer is a sum of four squares (of integers, some of which may be 0). Euler was captivated by this result, and tried for many years to prove it, without success. Lagrange was (of course) very pleased to

Fig. 1.1 Joseph-Louis Lagrange (1736–1813)

have succeeded where Euler had failed. He gave a proof which used Euler's identity that a product of a sum of four squares is again a sum of four squares (recall a similar identity about sums of two squares). In 1829, Jacobi used elliptic functions to give an explicit formula for the number of representations of an integer as a sum of four squares. See Sect. 2.4.

1.4.3 *Binary Quadratic Forms*

A binary quadratic form is an expression of the type $f(x, y) = ax^2 + bxy + cy^2$, with a, b, and c integers. The basic question related to such forms is: given $f(x, y)$, which integers does it "represent"? That is, for which integers n are there integers x and y such that $n = ax^2 + bxy + cy^2$? Other interesting questions deal with the number of such solutions for a given n, and an algorithm for finding them.

Fermat had considered specific cases of quadratic forms of the type $x^2 + cy^2$, and Euler studied forms of the type $ax^2 + cy^2$, but Lagrange was the first to deal with general quadratic forms. A fundamental observation was that two distinct forms can represent the same set of integers. This is the case, for example, for the forms $x^2 + y^2$ and $2x^2 + 6xy + 5y^2$; to see this, note that $2x^2 + 6xy + 5y^2 = (x + y)^2 + (x + 2y)^2$.

To deal with this phenomenon, Lagrange introduced the fundamental notion of *equivalence of forms*. Two forms $f(x, y) = ax^2 + bxy + cy^2$ and $F(X, Y) = AX^2 + BXY + CY^2$ are equivalent if there exists a transformation of the variables,

$x = sX + tY$, $y = uX + vY$, such that $sv - tu = \pm 1$, with s, t, u, v integers. It is easy to see that two equivalent forms represent the same set of integers. An important object is the *discriminant* $D = b^2 - 4ac$ of a form $f(x, y)$. It is an *invariant* of the form under equivalence; that is, if f and F are equivalent forms, they have the same discriminant; the converse fails.

Equivalence of forms is an equivalence relation, hence it divides the quadratic forms into equivalence classes. (The *terms* "equivalent" and "equivalence class" are due to Gauss.) Lagrange showed that there are only finitely many inequivalent forms for a given discriminant $D < 0$ (a form with negative discriminant is called "definite"); their number is called the *class number*, denoted by $h(D)$. He also described a procedure for finding a "simple" representative for each class, called a "reduced" form.

Lagrange applied his theory to prove results of Fermat and Euler on the representation of primes by quadratic forms, but the theory enabled him to go beyond them. For example, he showed that every prime of the form $20n + 1$ can be represented by the form $x^2 + 5y^2$. This form caused difficulty for both Fermat and Euler. Its discriminant is -20, and $h(-20) = 2$, the other inequivalent form being $2x^2 + 2xy + 3y^2$. Lagrange's comprehensive and beautiful theory of binary quadratic forms was fundamental for subsequent developments in number theory and algebra. See [1, 9, 22, 24, 25].

1.5 Legendre

Legendre was one of the most prominent mathematicians of Europe in the 19th century, although not of the stature of Euler or Lagrange. His texts were very influential. In 1798 he published his *Theory of Numbers*, the first book devoted exclusively to number theory. It underwent several editions, but was soon to be superseded by Gauss' *Disquisitiones Arithmeticae* (see Sect. 1.6).

Many of Legendre's results were found independently by Gauss, and serious priority disputes arose between them. Legendre's proofs, moreover, left much to be desired, even by mid-18th-century standards. For example, he discovered the law of quadratic reciprocity, unaware of Euler's prior discovery, and gave a proof based on what he viewed as a self-evident fact, namely the existence of infinitely many primes in any arithmetic progression $an + b$ ($n = 1, 2, 3, \ldots$, with a and b relatively prime). This was a very difficult result, proved subsequently by Dirichlet using deep methods of analysis. Legendre was chagrined when Gauss, who gave a rigorous proof, claimed the result as his own. In connection with this law, Legendre introduced the useful and celebrated *Legendre symbol* (a/p) (it does *not* denote division), with p an odd prime and a an integer not divisible by p : $(a/p) = 1$ if $x^2 \equiv a \pmod{p}$ is solvable and $(a/p) = -1$ if it is not. In terms of this symbol, the law of quadratic reciprocity can be stated succinctly as $(p/p)(q/p) = (-1)^{(p-1)(q-1)/4}$, where p and q are distinct odd primes.

An important achievement was Legendre's proof, given when he was in his 70s, of FLT for $n = 5$, and his conjecture that $\pi(x) \sim x/(A \log x + B)$, where $\pi(x)$ denotes the number of primes less than or equal to x, and $f(x) \sim g(x)$, read "$f(x)$ is asymptotic [approximately equal] to $g(x)$," means that lim $(f(x)/g(x)) = 1$ as $x \to \infty$. This conjecture was refined by Gauss and became known as the Prime Number Theorem (PNT) (see Sect. 1.8).

Finally, "one of Legendre's main claims to fame [in number theory]" (according to Weil [25, p. 327]) is the result that the equation $ax^2 + yb^2 + cz^2 = 0$ has a solution in integers not all zero if and only if $-bc$, $-ca$, $-ab$ are quadratic residues mod a, b, c, respectively, where a, b, c are integers not of the same sign, and abc is square-free.

1.6 Gauss' *Disquisitiones Arithmeticae*

1.6.1 Introduction

Gauss was the greatest mathematician of the 19th century, and number theory, the Queen of Mathematics (according to him), was his greatest mathematical love. As he put it in an 1838 letter to Dirichlet, "I place this part of mathematics [number theory] above all others (and have always done so)." His supreme masterpiece was *Disquisitiones Arithmeticae* (*Arithmetical Investigations*), published in 1801 but completed in 1798, when he was 21.

Pre-19th-century number theory consisted of many brilliant results but often lacked thematic unity and general methodology. In the *Disquisitiones* Gauss supplied both. He systematized the subject, provided it with deep and rigorous methods, solved important problems, and furnished mathematicians with new ideas to help guide their researches for much of the 19th century. The following are several of the far-reaching concepts and results in the *Disquisitiones*.

While the 18th century paid little attention to formal proof, the 19th saw the emergence of a critical spirit, in which rigor and abstraction began to play fundamental roles. Gauss was one of that spirit's early exponents. The FTA, a cornerstone of number theory, was undoubtedly known to its pioneers, but Gauss was the first to state the theorem explicitly and to give a rigorous proof. Perhaps Fermat, Euler, and others thought the result too obvious to mention, although the proof is far from trivial.

1.6.2 Quadratic Reciprocity

Gauss was also the first to define the fundamental notion of *congruence*, introducing the notation "≡" in use today. He chose it deliberately because it is similar to the notation "=" for equality. In fact, congruence has many of the same properties as equality: for example, if $a \equiv b \pmod{m}$ and $c \equiv d \pmod{m}$, then $a + c \equiv b + d \pmod{m}$. Analogy, as was mentioned, is a powerful tool, and the simple notation that Gauss introduced for congruence had a great impact on number theory. Gauss himself exploited the analogy in important ways, in particular in his theory of congruences of the second degree, $ax^2 + bx + c \equiv 0 \pmod{m}$, where a, b, c, and m are integers, with $m > 1$. This can be reduced to the congruence $ax^2 + bx + c \equiv 0 \pmod{p}$, for prime divisors p of m, and by completing the square, to the congruence $y^2 \equiv d \pmod{p}$. This was the congruence (recall) at the heart of the pivotal quadratic reciprocity law, which Gauss called the *golden theorem*. He was the first to give a rigorous proof. In fact, he gave *six* proofs, hoping that one of them might generalize to "higher" reciprocity laws, dealing with solutions of $y^n \equiv d \pmod{p}$ for $n > 2$ (see Sect. 1.7.1). Gauss considered his work on quadratic reciprocity to be one of his most important contributions to number theory.

1.6.3 Binary Quadratic Forms

By far the largest part of the *Disquisitiones* was the powerful and beautiful, but difficult, theory of binary quadratic forms. Major strides were made by Lagrange (see Sect. 1.4), but Gauss brought the theory to perfection. Here he introduced the important and deep concepts of genus and composition of forms. (Two forms are said to be in the same "genus" if there is a nonzero integer that is representable by both. If integers m and n are representable by forms f and g, respectively, then mn is representable by the "composition" of f and g.) An important criterion for representability of integers by quadratic forms is the following: if n is "properly" representable by $f(x, y) = ax^2 + bxy + cy^2$, that is, $n = ax^2 + bxy + cy^2$ for some relatively prime integers x and y, and D is the discriminant of f, then $x^2 \equiv D \pmod{4|n|}$ is solvable; and if $x^2 \equiv D \pmod{4|n|}$ is solvable, then n is properly representable by *some* form with discriminant D. This result was probably a strong motivation for Gauss' interest in quadratic residues.

According to Weil, the theory of quadratic forms "remained a stumbling-block for all readers of the *Disquisitiones* [for more than 60 years]." Dirichlet made it more accessible in his *Lectures on Number Theory*, published by his student Dedekind in 1863. This motivated a generation of mathematicians to try to come to grips with its ideas. In 1871 Dedekind reinterpreted the theory of binary quadratic forms in terms of his just-created theory of algebraic numbers; in particular he established a correspondence between quadratic forms of discriminant D and the ideals of the quadratic field $Q(\sqrt{D})$, under which the product of ideals corresponds to the composition of quadratic forms (see Sect. 1.7 and [13, p. 125]).

1.6.4 Cyclotomy

The last part of the *Disquisitiones*, a beautiful blend of algebra, geometry, and number theory, dealt with cyclotomy – the division of a circle into n equal parts. An important aspect of this work was a characterization of regular polygons constructible with straightedge (unmarked ruler) and compass: a regular polygon of n sides is so constructible if and only if $n = 2^k p_1 p_2 \ldots p_s$, where the p_i are distinct Fermat primes (see Sect. 1.2). The vertices of a regular polygon of n sides inscribed in the unit circle are the roots of the polynomial $x^n - 1 = (x-1)(x^{n-1} + x^{n-2} + \ldots + x + 1) = 0$. The polynomial $p(x) = x^{n-1} + x^{n-2} + \ldots + x + 1$, called the *cyclotomic polynomial*, was central to Gauss' characterization. It has since been an essential object in number-theoretic studies. See [5, 9, 11, 22].

1.7 Algebraic Number Theory

Algebraic number theory is the study of number-theoretic problems using the concepts and results of abstract algebra, mainly those of groups, rings, modules, fields, and ideals. In fact, some of these abstract concepts were *invented* in order to deal with problems in number theory. The initial inroads in the subject were made in the 18th century by Euler and Lagrange, in their use of "foreign" objects such as irrational and complex numbers to help solve problems about integers (Sects. 1.3 and 1.4). But the fundamental breakthroughs were achieved in the 19th century. Two basic problems provided the early stimulus for these developments: reciprocity laws and FLT.

1.7.1 Reciprocity Laws

The quadratic reciprocity law, the relationship between the solvability of $x^2 \equiv p \pmod{q}$ and $x^2 \equiv q \pmod{p}$, with p and q distinct odd primes, is (as mentioned) a central result in number theory. In the 19th century a major problem was the extension of the law to higher analogues, which would describe the relationship between the solvability of $x^n \equiv p \pmod{q}$ and $x^n \equiv q \pmod{p}$ for $n > 2$. (The cases $n = 3$ and $n = 4$ give rise, respectively, to what are called "cubic" and "biquadratic" reciprocity.) Gauss opined that such laws cannot even be conjectured within the context of the integers. As he put it: "such a theory [of higher reciprocity] demands that the domain of higher arithmetic [i.e., the domain of integers] be endlessly enlarged [11, p. 108]." This was indeed a prophetic statement.

Gauss himself began to enlarge the domain of higher arithmetic by introducing (in 1832) what came to be known as the *Gaussian integers*, $Z(i) = \{a + bi : a, b \in Z\}$. He needed these to formulate a biquadratic reciprocity law. The elements of $Z(i)$ do indeed qualify as "integers," in the sense that they obey all the crucial arithmetic properties of the "ordinary" integers Z. They can be added, subtracted, and multiplied, and, most importantly, they obey a FTA: every nonzero and noninvertible element of $Z(i)$ is a unique product of primes of $Z(i)$, called Gaussian primes. The latter are those elements of $Z(i)$ that cannot be written nontrivially as products of Gaussian integers; for example, $7+i = (2+i)(3-i)$, where $2+i$ and $3-i$ are Gaussian primes (this is not difficult to show).

A domain with a unique factorization property such as the above is called a unique factorization domain (UFD). Gauss also formulated a *cubic* reciprocity law, and to do that he introduced yet another domain of "integers," the cyclotomic integers of order 3, $C_3 = \{a + bw + cw^2 : a, b, c \in Z\}$, where $w = (-1 + \sqrt{}3\text{i})/2$ is a primitive cube root of 1 ($w^3 = 1$, $w \neq 1$). This, too, turned out to be a UFD.

1.7.2 Fermat's Last Theorem

Recall that in the 17th century Fermat proved FLT, the unsolvability in nonzero integers of $x^n + y^n = z^n$, $n > 2$, for $n = 4$. Given this result, one can readily show that it suffices to prove FLT for $n = p$, an odd prime. Over the next two centuries, the theorem was proved for only three more cases: $n = 3$ (Euler, in the 18th century), $n = 5$ (Legendre and Dirichlet, independently, in the early 19th century), and $n = 7$ (Lamé, 1837).

A general attack on FLT was made in 1847, again by Lamé. His idea was to factor the left side of $x^p + y^p = z^p$ into linear factors (as Euler had already done for $n = 3$) to obtain the equation $(x + y)(x + yw)(x + yw^2)\ldots(x + yw^{p-1}) = z^p$, where w is a primitive pth root of 1 ($w^p = 1$, $w \neq 1$). This is an equation in the domain of cyclotomic integers of order p, $C_p = -\{a_0 + a_1w + a_2w^2 + \ldots + a_{p-1} : a_i \in Z\}$. Lamé now proceeded as Euler and others had done before him: he used the arithmetic of the domain C_p and thereby "proved" FLT (the approach is analogous to Euler's solution of Bachet's equation, $x^2 + 2 = y^3$).

Well, not quite. The proof hinged on knowing that the arithmetic properties of Z do, indeed, carry over to C_p, namely, that C_p is a UFD. When Lamé presented his proof to the Paris Academy of Sciences, Liouville, who was in the audience, took the floor to point out precisely that. Lamé responded that he would reconsider his proof, but was confident that he could repair it.

Alas, this was not to be. Two months after Lamé's presentation, Liouville received a letter from Kummer informing him that while C_p is, indeed, a UFD for all $p < 23$, C_{23} is not. (It was shown in 1971 that unique factorization fails in C_p for *all* $p > 23$.) But all hope was not lost, continued Kummer in his letter [21, p. 7]:

> It is possible to rescue it [unique factorization] by introducing new kinds of complex numbers, which I have called *ideal complex numbers*. I considered long ago the application of this theory to the proof of Fermat's [Last] Theorem and I succeeded in deriving the impossibility of the equation $x^n + y^n = z^n$ [for all $n < 100$].

Kummer "rescued" unique factorization in C_p by adjoining to it "ideal numbers," and thereby established FLT for all $p < 100$. That was quite a feat, considering that during the previous two centuries FLT had been proved for only three primes. It would take another century and more for further crucial progress on FLT to be made (see Sect. 1.9 and Chap. 3). (The notion of "ideal numbers" is complicated, but the following example may give an indication of what is involved. Let D be the set of even integers. Here $100 = 2 \times 50 = 10 \times 10$, where 2, 10, 50 are primes *in* D, that is, cannot be factored in D, so that D does not possess unique factorization. If we adjoin the "ideal" number 5 to D (it does not exist in D), unique factorization will have been restored to the element 100. For then $100 = 2 \times 50 = 2 \times 2 \times 5 \times 5$ and $100 = 10 \times 10 = 2 \times 5 \times 2 \times 5$. To restore unique factorization to every element of D, infinitely many ideal numbers will have to be added.) By the way, Kummer was also very interested in higher reciprocity laws. These, too, give rise to the cyclotomic integers C_p. His introduction of ideal numbers was motivated at least as much by these considerations as by FLT.

1.7.3 Dedekind's Ideals

Kummer's work was brilliant, but it left unanswered important questions, as any good work ought to do. In particular, can one simplify his complicated theory, and, more importantly, can one extend it to other domains that arise in various number-theoretic contexts, for example, the "quadratic domains," important in the study of quadratic forms? These are the domains $Z_d = \{a + b\sqrt{d} : a, b \in Z\}$, if $d \equiv 2$ or 3 (mod 4), or $Z_d = \{a/2 + (b/2)\sqrt{d} : a$ and b are both even or both odd $\}$, if $d \equiv 1$ (mod 4). They are not, as a rule, UFDs. For instance, $Z_{-5} = \{a+b\sqrt{-5} : a, b \in Z\}$ is not, since (for example) $6 = 2 \times 3 = (1 + \sqrt{-5})(1 - \sqrt{-5})$, and 2, 3, $1 + \sqrt{-5}, 1 - \sqrt{-5}$ are primes in Z_{-5}.

The above comments give rise to two fundamental questions: (a) what are the domains for which a unique factorization theorem (UFT) is to hold? (b) what shape is such a theorem to take? It clearly cannot say that every element in such a domain as has been determined in (a) is a unique product of primes, since this would disqualify many of the quadratic domains. It took Dedekind about 20 years to answer these two questions; the first was the more difficult.

Dedekind's work, given in Supplement X to the second edition (1871) of Dirichlet's *Lectures on Number Theory*, was revolutionary in its formulation, its grand conception, its fundamental new ideas, and its modern spirit. As for our concerns here, to answer (i), namely, to determine the domains in which a UFT would obtain, Dedekind first had to define "fields of algebraic numbers"; the domains in question would be identified as distinguished subsets of these fields.

The fields of algebraic numbers needed for Dedekind's theory were sets of the form $Q(a) = \{q_0 + q_1 a + q_2 a^2 + \cdots + q_n a^n\}$, where q_i are rational numbers and a is an algebraic number (a root of a polynomial with integer coefficients). The $Q(a)$

are fields, and *all* their elements are algebraic numbers. The domains for which a UFT was sought were defined to be "the integers of $Q(a)$," those elements $I(a)$ of $Q(a)$ that are roots of "monic" polynomials with integer coefficients. (A polynomial is monic if the coefficient of the highest-degree term is 1.) Such elements are called *algebraic integers*. (For example, $\sqrt{15} + 3$ is an algebraic integer since it is a root of the polynomial $x^2 - 6x - 6$.) They behave like integers in the sense that they can be added, subtracted, and multiplied, but they do not, in general, form UFDs. They form, however, commutative rings.

The $I(a)$ are vast generalizations of the domains of integers that were considered above, namely the Gaussian integers, the cyclotomic integers, and the quadratic integers (also, of course, the ordinary integers). Having defined the $I(a)$, Dedekind's second major task was to formulate and prove a UFT in $I(a)$. It turned out to be the following: every nonzero ideal in $I(a)$ is a unique product of prime ideals. The domains $I(a)$ are examples of *Dedekind domains*. These are integral domains in which every nonzero ideal is a unique product of prime ideals. They play an important role in number theory. See [9, 13].

1.7.4 Summary

To summarize the events that we have described: After more than 2,000 years in which number theory meant the study of properties of the (positive) integers, its scope became enormously enlarged. One could no longer use the term "integer" with impunity; it had to be qualified – a "rational" (ordinary) integer, a Gaussian integer, a cyclotomic integer, a quadratic integer, or any one of an infinite species of other algebraic integers, the various $I(a)$. Moreover, powerful new algebraic tools were introduced and brought to bear on the study of these integers – fields, commutative rings, ideals, prime ideals, and Dedekind domains. A new subject emerged – algebraic number theory, of vital importance to this day. See [7, 9, 11, 13, 24] for details.

1.8 Analytic Number Theory

Algebra was not the only "foreign" subject that invaded number theory in the 19th century. Analysis was another. The bridge-building between number theory and analysis began with Euler in the 18th century, and gave rise in the 19th to a new field – analytic number theory.

1.8.1 The Distribution of Primes Among the Integers: Introduction

The broad context for the introduction of analytic methods into number theory was the problem of the distribution of primes among the integers. Euclid had shown that there are infinitely many primes, but do they follow a discernible pattern? This question baffled mathematicians for centuries.

Numerical evidence showed that the primes are spread out irregularly among the integers, in particular that they get scarcer – but not uniformly – as the integers increase in size. For example, there are 8 primes between 9,991 and 10,090 and 12 primes between 67,471 and 67,570. Furthermore, arbitrarily large gaps exist between primes. For example, it is easy to produce a sequence of (say) $10^9 - 1$ consecutive composite integers, namely $10^9! + 2,\ 10^9! + 3, \ldots, 10^9! + 10^9$. On the other hand, considerable evidence suggests that there are infinitely many pairs of primes p, q as close together as can be, namely such that $q - p = 2$; they are called *twin primes*. This apparent irregularity in the distribution of primes prompted Euler in the 18th century to observe that

> Mathematicians have tried in vain to this day to discover some order in the sequence of prime numbers, and we have reason to believe that it is a mystery which the human mind will never penetrate [8, p. 241].

1.8.2 The Prime Number Theorem

Euler's "pessimism" was in an important sense unjustified. It is true that there is no regularity in the distribution of primes considered *individually*, but there *is* regularity in their distribution when considered *collectively*. Instead of looking for a rule that will generate successive primes, one asked for a description of the number of primes in a given interval. Specifically, if $\pi(x)$ denotes the number of primes less than or equal to x, where x is any positive real number, the goal was to describe the behavior of the function $\pi(x)$. By inspecting lists of primes, Gauss conjectured that $\pi(x)$ is asymptotic to $x/\log x$, $\pi(x) \sim x/\log x$, (the log is to the base e). The irregularity in the distribution of primes would preclude an exact formula for $\pi(x)$. If we rewrite $\pi(x) \sim x/\log x$ in the form $\pi(x)/x \sim 1/\log x$, this says, roughly speaking, that the probability of picking a prime from the first x integers is approximately $1/\log x$, and that the approximation improves with the size of x. See [6].

Gauss' conjecture, made at the start of the 19th century, and now known as the *Prime Number Theorem*, was proved at the century's end. It is remarkable that such a complex distribution as exhibited by the primes would be "modeled" by such a simple formula as $x/\log x$. Davis and Hersh, in their book *The Mathematical Experience* (Birkhäuser, 1995, p. 210), refer to the PNT as "one of the finest examples of the extraction of order from chaos in the whole of mathematics."

1.8.3 *The Riemann Zeta Function*

A major step toward the proof of the PNT was taken by Riemann in the mid-19th century. His key idea was to extend the zeta function, introduced a century earlier by Euler, to complex variables, so that now $\zeta(s) = \sum_1^\infty 1/n^s$ was defined for complex numbers s. The series converges for those s for which $\text{Re}(s) > 1$ ($\text{Re}(s)$, the real part of $s = a + bi$, is a), and Euler's product formula $\zeta(s) = \sum_1^\infty 1/n^s = \prod_p 1/(1-p^{-s})$ continues to hold for these s. Most importantly, the function $\zeta(s)$ can be extended to all complex numbers by a method known as "analytic continuation." This latter function has come to be known as the *Riemann zeta function.*

Riemann showed that the study of $\zeta(s)$ leads to information about $\pi(x)$. He conjectured that all the "nontrivial" roots of $\zeta(s)$ (the trivial roots are the negative even integers) are on the line $\text{Re}(s) = 1/2$, that is, that $\zeta(s) = 0 \Rightarrow s = 1/2 + bi$, b a real number. The line $\text{Re}(s) = 1/2$ is known as the *critical line.* It was shown in 2005 that 1029.9 billion roots of $\zeta(s)$ lie on it. This conjecture, still open 150 years later, is called the *Riemann Hypothesis.* It is arguably the most celebrated unsolved problem in mathematics, and has numerous implications in all branches of number theory. In particular, Riemann showed that it implies the PNT.

But the proof of the PNT took a somewhat different route. It was given, independently, in 1896, by Hadamard and de la Vallée Poussin. They proved a much weaker result than the Riemann Hypothesis, namely that $\zeta(s) \neq 0$ for $\text{Re}(s) = 1$, and showed that this implies the PNT; in fact, it is equivalent to the PNT. Both relied on Riemann's work, as well as on advanced techniques in the theory of functions of a complex variable developed subsequently. In the 1940s, in a most unexpected development, Erdös and Selberg gave an "elementary" proof of the PNT, a proof that is far from simple but does not use complex analysis. See [2, 2a, 4,7a].

1.8.4 *Primes in Arithmetic Progression*

Euclid proved that there are infinitely many primes. This can be rephrased to say that there are infinitely many primes in the arithmetic sequence $2n + 1$ ($n = 0, 1, 2, 3, \ldots$). In 1837, Dirichlet proved a grand generalization of this result by showing that *any* arithmetic sequence $an + b$ ($n = 0, 1, 2, 3, \ldots$), with a and b relatively prime, contains infinitely many primes. To do that, he introduced far-reaching ideas from analysis – in particular, the very important L-series, $L(s, X) = \sum_{n=1}^\infty X(n)/n^s$ (s is a real number greater than 1, and the "Dirichlet character" X is a function that associates with each integer relatively prime to n an nth root of 1, and satisfies certain properties). He showed that if $L(1, \mathrm{X}) \neq 0$, where X is not the so-called "principal character," then $an + b$ has infinitely many primes (compare the proof of the PNT). He applied similar ideas from analysis to prove other results, for example, that $am^2 + bmn + cn^2$ contains infinitely many primes, where a, b, and c are fixed relatively prime integers and $m, n = 0, 1, 2, 3, \ldots$. Dirichlet was the first to introduce deep methods of analysis into number theory, providing new perspectives on the subject, and may therefore be considered the founder of analytic number theory. See [2] and (c) below.

1.8.5 More on the Distribution of Primes

Many questions about the distribution of primes remain open. For example, it is not known if $n^2 + 1$ contains infinitely many primes ($n = 0, 1, 2, 3, \ldots$), although it is easy to show that no polynomial p(n) in a single integer variable n contains *only* primes. Polynomials in two variables seem easier to handle. For example, it was shown in the 1960s that $m^2 + n^2 + 1$ yields infinitely many primes as m and n range over the positive integers. In 1996, Friedlander and Iwaniec showed that $m^2 + n^4$ contains infinitely many primes, and in 1999 Heath-Brown proved the same for $m^3 + 2n^3$. The proofs used deep ideas from analysis and other fields, and were considered remarkable achievements.

Even more remarkable are the following three results about primes proved within the last few years, using tools of analysis and other areas. The last two may provide insight into proving the *twin-prime conjecture*, the existence of infinitely many pairs of primes p, q such that $q - p = 2$. See [16] for details:

(a) There is an efficient, "polynomial time" algorithm to check for primality (Agrawal, Kayal, and Saxena, 2002).
(b) This is a technical result concerning the varying size of gaps between consecutive primes. More specifically: there are infinitely many primes for which the gap to the next prime is as small as we want compared to the average gap between consecutive primes (Goldston, Pintz, and Yildirim, 2003).
(c) The primes contain arbitrarily long arithmetic progressions (Green and Tao, 2004).

Among other unsolved problems about primes are the following: is every integer >2 a sum of two primes, as the evidence suggests? This is the celebrated *Goldbach Conjecture*, outstanding for 250 years. Are there infinitely many primes of the form $2p+1$, where p is a prime? Is there a prime between n^2 and $(n+1)^2$? Is $\pi(x+y) \leq \pi(x) + \pi(y)$ for every x and y? Are there infinitely many twin primes? Mersenne primes? Fermat primes? What is the smallest prime in the arithmetic sequence $an + b$? And the grandest question of them all: Is the Riemann Hypothesis true? When all is said and done, perhaps Euler's comments (Sect. 1.8.1) about the mysterious character of the primes are not unwarranted. See [2, 4, 6, 10, 16, 18, 20] for various aspects of this section.

1.9 Fermat's Last Theorem

The 20th century will undoubtedly be regarded as a golden age in mathematics, both for the emergence of brilliant new ideas and the solution of longstanding problems (the two are, of course, not unrelated). One of its major triumphs was Wiles' 1994 proof of FLT, outstanding for over 350 years.

1.9.1 Work Prior to That of Wiles

Various attempts to prove the theorem had been made during the previous three centuries, the most important being Kummer's in the 19th century (see Sect. 1.7 and Chap. 3). In the 1920s, Vandiver established FLT for all primes $p < 157$ (recall that about 70 years earlier Kummer had reached $p < 100$), and in 1954, he extended the result, using the SWAC calculating machine, to $p < 2{,}521$. Using modern computers and advanced theoretical mathematics, the theorem was established for $p < 125{,}000$ in 1973, and for $p < 4{,}000{,}000$ in 1993.

But computations, no matter how powerful and sophisticated, clearly cannot establish FLT for *all* exponents p. A theoretical departure was needed, and it materialized in the 1980s. While previous attempts to prove the theorem were mainly algebraic, relying on ideas of Kummer and others, this approach was geometric, having its roots (in a sense) in Diophantus' work. The crucial breakthrough came in 1985, when Frey related FLT to elliptic curves. An elliptic curve is a plane curve given by an equation of the form $y^2 = x^3 + ax^2 + bx + c$, with a, b, and c integers or rational numbers. Elliptic curves had (in effect) been studied by Diophantus and Fermat, and intensively investigated by Euler and Jacobi; the curve represented by Bachet's equation $y^2 = x^3 + k$ is an important example.

The association of elliptic curves with FLT was, according to experts, a most surprising and innovative link. Specifically, if $a^p + b^p = c^p$ holds for nonzero integers a, b, c, the associated elliptic curve, now known as the Frey Curve, is $y^2 = x(x - a^p)(x + b^p)$. Frey conjectured that if such a, b, c did exist, that is, if FLT failed, then the resulting elliptic curve would be "badly behaved:" it would be a counterexample to the Taniyama-Shimura Conjecture (TSC). Put positively, Frey conjectured that if the TSC holds, then FLT is true. The outline of a possible proof of FLT now emerged:

(a) Prove Frey's conjecture, namely that the TSC implies FLT.
(b) Prove the TSC.

The TSC was formulated by Taniyama in 1955 and refined in the 1960s by his colleague Shimura. It says that every elliptic curve is modular. The notion of modularity is technically difficult to define, but the following statement from

Harvard mathematician Barry Mazur gives a sense of the scope and depth of the TSC [23, p. 190]:

> It was a wonderful conjecture, but to begin with it was ignored because it was so ahead of its time. On the one hand you have the elliptic world, and on the other the modular world. Both these branches of mathematics had been studied intensively but separately. Then along comes the Taniyama-Shimura conjecture, which is the grand surmise that there's a bridge between the two completely different worlds. Mathematicians love to build bridges.

1.9.2 Andrew Wiles

Enter Ken Ribet of the University of California at Berkeley. In 1986 he proved Frey's conjecture that TSC implies FLT. It was a big event. Wiles was ecstatic on hearing of the proof [23, p. 205]:

> I knew that moment that the course of my life was changing because this meant that to prove Fermat's Last Theorem all I had to do was to prove the Taniyama-Shimura conjecture. It meant that my childhood dream was now a respectable thing to work on. I just knew that I could never let that go.

Work he did on it – for the next 7 years. As he relates it [14, p. 10]:

> I made progress in the first few years. I developed a coherent strategy. Basically, I restricted myself to my work and my family. I don't think I ever stopped working on it. It was on my mind all the time. Once you're really desperate to find the answer to something, you can't let go.

In 1993, Wiles was convinced that he had a proof of FLT, and he presented it in a series of three talks at a conference in Cambridge, though he did not reveal the goal of his lectures until the very end. He concluded the third lecture with the words: "And this proves FLT. I think I'll stop here." Mazur described the event [23, p. 248]: "I've never seen such a glorious lecture, full of such wonderful ideas, with such dramatic tension, and what a buildup. There was only one possible punch line." Specifically, what Wiles did was prove the TSC for an important class of elliptic curves, the "semi-stable" elliptic curves. This sufficed to prove FLT, for Ribet had earlier proved that if such curves are modular, then FLT holds. (The full TSC was proved in 1999.)

Wiles' proof was very deep and technically demanding. Ram Murty, an authority in the field, described it thus [17, p. 17]:

> By the end of the day, it was clear to experts around the world that nearly all of the noble and grand ideas that number theory had evolved over the past three and a half centuries since the time of Fermat were ingredients in the proof.

So, in a sense – without detracting from Wiles' great achievement – the proof was a grand collaborative effort of dozens of mathematicians over several centuries.

Wiles' lectures at Cambridge in June 1993 were, however, not to be the end of this 350-year odyssey. The proof was very long and complex, and required validation by experts. Many errors were found; most were easily and quickly corrected. One error, however, could not be fixed. Wiles worked for several months, without success, on repairing it, and in January 1994 sought the help of Cambridge mathematician and former student Richard Taylor. On September 19, they found the "vital fix." In October, two papers proving FLT (totaling over 120 pages) were published, one by Wiles, the other by Taylor and Wiles.

Many tributes poured in following the publication of the proof. Here are two, from eminent number-theorists Murty and Coates, respectively (the latter was Wiles' doctoral advisor at Cambridge):

> Fermat's Last Theorem deserves a special place in the history of civilization. By its simplicity it has tantalized amateurs and professionals alike, and with remarkable fecundity led to the development of many areas of mathematics such as algebraic geometry, and more recently the theory of elliptic curves and representation theory. It is truly fitting that the proof crowns an edifice composed of the greatest insights of modern mathematics (Murty [17, p. 20]).
>
> In mathematical terms, the final proof is the equivalent of splitting the atom or finding the structure of DNA. A proof of Fermat's Last Theorem is a great intellectual triumph, and one shouldn't lose sight of the fact that it has revolutionized number theory in one fell swoop (Coates [23, p. 279]).

The last word belongs to Wiles [23, p. 285]:

> I had this very rare privilege of being able to pursue in my adult life what had been my childhood dream. I know it's a rare privilege, but if you can tackle something in adult life that means that much to you, then it's more rewarding than anything imaginable. Having solved this problem, there's certainly a sense of loss, but at the same time there is this tremendous sense of freedom. I was so obsessed by this problem that for eight years I was thinking about it all the time – when I woke up in the morning to when I went to sleep at night. That's a long time to think about one thing. That particular odyssey is over. My mind is at rest.

For further details on this section see [7, 14, 17, 19, 21, 23], and Chap. 3.

References

1. W. W. Adams and L. J. Goldstein. *Introduction to Number Theory*, Prentice-Hall, 1976.
2. T. M. Apostol, *Introduction to Analytic Number Theory*, Springer-Verlag, 1976.
2a. T. M. Apostol, A centennial history of the Prime Number Theorem. In *Number Theory*, ed. by R. P. Bambah et al, Birkhäuser, 2000, pp. 1–14.
3. I. Bashmakova, *Diophantus and Diophantine Equations* (translated from the Russian by A. Shenitzer), Math. Assoc. of Amer., 1997.
4. J. Derbyshire, *Prime Obsession: Bernhard Riemann and the Greatest Unsolved Problem in Mathematics*, Joseph Henry Press, 2003.
5. L. E. Dickson, *History of the Theory of Numbers*, 3 vols., Chelsea, 1966.
6. U. Dudley, *Elementary Number Theory*, 2nd ed., W. H. Freeman, 1978.
7. H. M. Edwards, *Fermat's Last Theorem: A Genetic Introduction to Algebraic Number Theory*, Springer-Verlag, 1977.
7a. H. M. Edwards, *Riemann's Zeta Function*, Dover, 2001 (orig. 1974).

8. L. Euler, Discovery of a most extraordinary law of numbers concerning the sum of their divisors, *Opera Omnia*, Ser. 1, Vol. 2, pp. 241–253. (E175 in the Eneström index.)
9. J. R. Goldman, *The Queen of Mathematics: A Historically Motivated Guide to Number Theory*, A K Peters, 1998.
10. G. H. Hardy and E. M. Wright. *An Introduction to the Theory of Numbers*, The Clarendon Press, 1938.
11. K. Ireland and M. Rosen. *A Classical Introduction to Modern Number Theory*, Springer-Verlag, 1972.
12. V. J. Katz, *A History of Mathematics*, 3rd ed., Addison-Wesley, 2009.
13. I. Kleiner, *A History of Abstract Algebra*, Birkhäuser, 2007.
14. G. Kolata, Andrew Wiles: A math whiz battles a 350-year-old puzzle, *Math Horizons* (Winter 1993) 8–11.
15. R. Kumanduri and C. Romero, *Number Theory with Computer Applications*, Prentice-Hall, 1998.
16. D. Mackenzie and B. Cipra, *What's Happening in the Mathematical* Sciences, Vol. 6, Amer. Math. Soc., 2006.
17. R. Murty, A long-standing mathematical problem is solved: Fermat's Last Theorem, *Can. Math. Society Notes* (Sept. 1993) 16–20.
18. O. Ore, *Number Theory and its History*, McGraw-Hill, 1948.
19. P. Ribenboim, *Fermat's Last Theorem for Amateurs*, Springer, 1999.
20. P. Ribenboim, *The Book of Prime Number Records*, 2nd ed., Springer-Verlag, 1989.
21. P. Ribenboim, *13 Lectures on Fermat's Last Theorem*, Springer-Verlag, 1979.
22. W. Scharlau and H. Opolka. *From Fermat to Minkowski: Lectures on the Theory of Numbers and its Historical Development*, Springer-Verlag, 1985.
23. S. Singh, *Fermat's Enigma: The Quest to Solve the World's Greatest Mathematical Problem*, Penguin, 1997.
24. J. Stillwell, *Elements of Number Theory*, Springer, 2003.
25. A. Weil, *Number Theory: An Approach through History*, Birkhäuser, 1984.
26. B. H. Yandell, *The Honors Class: Hilbert's Problems and their Solvers*, A K Peters, 2002.
27. G. M. Ziegler, The great prime-number record races, *Notices of the Amer. Math. Soc.* 51 (2004) 414–416.

Chapter 2
Fermat: The Founder of Modern Number Theory

2.1 Introduction

Fermat, though a lawyer by profession and only an "amateur" mathematician, is regarded as the founder of modern number theory. What were some of his major results in that field? What inspired his labors? Why did he not publish his proofs? How did scholars attempt to reconstruct them? Did Fermat have a proof of Fermat's Last Theorem? What were the attitudes of seventeenth-century mathematicians to his number theory? These are among the questions we will address in this chapter.

We know that work on Fermat's Last Theorem (FLT) led to important developments in mathematics. What of his other results? How should we view them in the light of the work of subsequent centuries? These issues will form another major focus.

Number theory was Fermat's mathematical passion. His interest in the subject was aroused in the 1630s by Bachet's Latin translation of Diophantus' famous treatise *Arithmetica* (*c.* 250 AD). Bachet, a member of an informal group of scientists in Paris, produced an excellent translation, with extensive commentaries.

Unlike other fields to which he contributed, Fermat (1607–1665) had no formal publications in number theory. (Fermat's date of birth is usually given as 1601; recently it has been suggested that the correct date is 1607 [5].) His results, and very scant indications of his methods, became known through his comments in the margins of Bachet's translation and through his extensive correspondence with leading scientists of the day, mainly Carcavi, Frenicle, and Mersenne. Fermat's son Samuel published his father's marginal comments in 1670, as *Observations on Diophantus*. A fair collection of Fermat's correspondence has also survived. Both are available in his collected works [35] (see also [26]). But they reveal little of his methods and proofs. As his biographer Mahoney notes ruefully [26, pp. 284–285]:

> Fermat's secretiveness about his number theory makes the historian's task particularly difficult. In no other aspect of Fermat's career are the results so striking and the hints at

I. Kleiner, *Excursions in the History of Mathematics*,
DOI 10.1007/978-0-8176-8268-2_2, © Springer Science+Business Media, LLC 2012

> the underlying methods so meager and disappointing. It is the results – the theorems and conjectures – and not the methods that drew the attention of men such as Euler, Gauss, and Kummer.

Weil, who wrote a masterful book analyzing (among other things) Fermat's number-theoretic work, speculates about its lack of proofs [37, p. 44]:

> It is clear that he always experienced unusual difficulties in writing up his proofs for publication; this awkwardness verged on paralysis when number theory was concerned, since there were no models there, ancient or modern, for him to follow.

It must be emphasized, however, that Fermat did lay considerable stress on general methods and on proofs, as his correspondence makes clear. Weil gave plausible reconstructions of the proofs of some of Fermat's results. He did this by considering the often cryptic comments about his methods in letters to his correspondents, and, more importantly, by examining the proofs of his results in the works of Euler and Lagrange, in order to determine whether the methods used in these proofs were available to Fermat. As Weil put it in the case of one such reconstruction: "If we consult Euler ... we see that Fermat could have proceeded as follows [37, p. 64]." He cautions that "any attempt at reconstruction can be no more than a hit or miss proposition [37, p. 115]." For a *modern* interpretation of some of Fermat's number-theoretic work consult Weil [37, Chapter II, Appendices I–V].

Fermat tried to interest his mathematical colleagues, notably Huygens, Pascal, Roberval, and Wallis, in number theory by proposing challenging problems, for which he had the solutions. This was not an uncommon practice at the time. He stressed that

> Questions of this kind [i.e., number-theoretic] are not inferior to the more celebrated questions in geometry [mathematics] in respect of beauty, difficulty, or method of proof [20, p. 286].

But to no avail. Mathematicians showed little serious interest in number theory until Euler came on the scene some 100 years later. They were preoccupied with other subjects, mainly calculus. Their typical attitude during the seventeenth century was well expressed by Huygens: "There is no lack of better things for us to do [37, p. 119]." The mathematical community apparently failed to see the depth and subtlety of Fermat's propositions on numbers. And he provided little help in that respect.

2.2 Fermat's Intellectual Debts

What number-theoretic knowledge was available to Fermat when he started his investigations? Mainly what was in Euclid's *Elements* and Diophantus' *Arithmetica* [20, 21]. There is no evidence (as far as we can ascertain) that Fermat knew of the considerable Indian, Chinese, or Moslem contributions to number theory – on, for example, linear diophantine equations, the Chinese remainder theorem, and Pell's equation [34].

In books VII–IX of the *Elements* Euclid introduced some of the main concepts of the subject, such as divisibility, prime and composite integer, greatest common divisor, and least common multiple. He also established some of its major results, among them the Euclidean algorithm, the infinitude of primes, results on perfect numbers, and what some historians consider to be a version of the Fundamental Theorem of Arithmetic [2].

Diophantus' *Arithmetica* differs radically in style and content from Euclid's *Elements*. It contains no axioms or formal propositions and proofs. It has, instead, about 200 problems, each giving rise to one or more indeterminate equations – now called *Diophantine equations*, many of degree two or three. These are (in modern terms) equations in two or more variables, with integer coefficients, for which the solutions sought are integers or rational numbers. Diophantus sought rational solutions; nowadays we are usually interested in integer solutions.

In fact, our interest in integer solutions follows that of Fermat, who, contrasting his work with that of Diophantus, noted that "arithmetic has, so to speak, a special domain of its own, the theory of integral numbers [13, p. 25]." (Of course, Euclid, as well as Indian and Chinese mathematicians, dealt with *integers* in studying number-theoretic problems.) It should be stressed, however, that the study of *rational* solutions of Diophantine equations has become important in the last 100 years or so, with the penetration into number theory of the methods of algebraic geometry. Another of Fermat's legacies is his quest for *all* solutions of a given Diophantine equation; Diophantus was usually satisfied with a single solution.

We now come to discuss some of Fermat's major results, commenting on their sources and on developments arising from them.

2.3 Fermat's Little Theorem and Factorization

Fermat's little theorem (Flt) states that $a^p - a$ is divisible by p for any integer a and prime p, or, equivalently, that $a^{p-1} - 1$ is divisible by p provided that a is not divisible by p. In post-1,800 terms, following Gauss' introduction of the congruence notation, we can write the above as $a^{p-1} \equiv 1 \pmod{p}$, provided that $a \not\equiv 0 \pmod{p}$. Fermat stated several versions of this result, one of which he sent to Frenicle in 1640 [37, p. 56]:

> Given any prime p, and any geometric progression $1, a, a^2$, etc., p must divide some number $a^n - 1$ for which n divides $p - 1$; if then N is any multiple of the smallest n for which this is so, p divides a^N.

Fermat is thought to have arrived at Flt by studying perfect numbers [13, p. 119, 37, pp. 54, 189]. Euclid showed that if $2^n - 1$ is prime then $2^{n-1}(2^n - 1)$ is perfect (Proposition IX.36). This result presumably prompted Fermat to ask about the divisors of $2^n - 1$, which led him to the special case $a = 2$ of Flt, that is, that $2^{p-1} - 1$ is divisible by p, and thence to the general case.

Fletcher [14, 15] examines the correspondence between Frenicle and Fermat in 1640, and concludes that it was Frenicle's challenge to Fermat (delivered via

Fig. 2.1 Pierre de Fermat (1607–1665)

Mersenne, who often acted as intermediary) concerning a specific perfect number that was responsible for Flt. Frenicle asked: "And if he (Fermat) finds that it is not much effort for him to send you a perfect number having 20 digits, or the next following it [15, p. 150]." Fermat responded that there is no such number, basing his answer on Flt. He wrote to Mersenne that "he would send [the proof] to Frenicle if he did not fear [it] being too long [37, p. 56]." In his book, Weil speculates how Fermat's proof might have gone, sketching two versions [37, pp. 56–57].

The dual problems of primality testing and factorization of large numbers are vital nowadays. The oldest method of testing if an integer n is prime, or finding a factor if n is composite, is by trial: test if there are divisors of n up to $\sqrt{n}$. The Sieve of Eratosthenes, devised c. 230 BC for finding all primes up to a given integer, is based on this idea.

Fermat, too, was concerned with such problems. Note, for example, his interest in determining the primality of the Mersenne numbers, $2^n - 1$, and of what we now call Fermat numbers, $2^{2^n} + 1$. In 1643, in a letter probably addressed to Mersenne, he proposed the following problem [26, p. 326]:

> Let a number, for example, 2,027,651,281, be given me and let it be asked whether it is prime or composite, and, in the latter case, of what numbers it is composed.

In the same letter Fermat answered his own query by outlining what came to be known as *Fermat's factorization method*. It was inspired by his interest in the problem of representing integers as differences of two squares.

The factorization method is based on the observation that an odd number $n > 3$ can be factored if and only if it is a difference of two squares: If $n = ab$, with $a \geq b > 1$, let $x = (a + b)/2$, $y = (a - b)/2$, then $n = x^2 - y^2$. Since n is odd, so are a and b, hence x and y are integers. The converse is obvious.

The algorithm works as follows: Given an integer n to be factored (we can assume without loss of generality that it is odd), we begin the search for possible x and y satisfying $n = x^2 - y^2$, or $x^2 - n = y^2$, by finding the smallest x such that $x \geq \sqrt{n}$. We then consider successively $x^2 - n$, $(x + 1)^2 - n$, $(x + 2)^2 - n, \ldots$ until we find an $m \geq \sqrt{n}$ such that $m^2 - n$ is a square. The process must terminate in such a value, at worst with $m = [(n + 1)/2]$, yielding the trivial factorization $n \times 1$ (which comes from $[(n + 1)/2]^2 - n = [(n - 1)/2]^2$), in which case n is prime.

Fermat's factorization algorithm is efficient when the integer to be factored is a product of two integers which are close to one another.

2.3.1 *A Look Ahead*

As we mentioned, Fermat did not publish any proofs of his number-theoretic results, save one (see below). Most, including Flt, were proved by Euler in the next century. In 1801, Gauss gave an essentially group-theoretic proof of Flt, without using group-theoretic terminology. For a proof of the theorem using dynamical systems, see the recent article by Iga [22].

Fermat's little theorem turned out to be one of his most important results. It is used throughout number theory (an entire chapter of Hardy and Wright [19] discusses consequences of the theorem), so it is anything but a "little theorem," although the term has historical roots. For example, it can be used to prove that if -1 is a quadratic residue mod p, p an odd prime, that is, if $x^2 \equiv -1 \pmod{p}$ is solvable, then $p \equiv 1 \pmod{4}$; and it can be used to show that a given number p is composite, without finding its factors, by finding a "small" a not divisible by p that does not satisfy Flt, though this is, in general, computationally not very efficient [31]. Moreover, Flt

> contain[s] the key idea behind two of today's most powerful algorithms for factoring numbers with large prime factors, the Quadratic Sieve and the Continued Fraction Algorithms [10, p. 58].

The converse of Flt is false, so the theorem cannot be used as a test of primality. But refinements and extensions of the theorem are at the basis of several primality tests. Here is one: The positive integer n is prime if and only if there is an a such that $a^{n-1} \equiv 1 \pmod{n}$ and $a^{(n-1)/q} \not\equiv 1 \pmod{n}$ for all primes q dividing $n - 1$ [3, p. 267]. A generalization of Flt to integers of cyclotomic fields was used by Adleman, Pomerance, and Rumely to yield a "deterministic algorithm [9, p. 547]" for testing for primality (1983), and the extension of the theorem to polynomials was the starting point for the recent (2002) spectacular achievement of Agrawal, Kayal, and Saxena in devising a test of primality in *polynomial time* [25, p. 52]. The test is rather slow, and of little practical value, but the result is of great theoretical interest [9]. The books by Bach and Shalit [3], Bressoud [10], and Riesel [31] deal with issues of primality and factorization.

2.4 Sums of Squares

In Problem III.19 of the *Arithmetica*, which asks "to find four numbers such that the square of their sum *plus* or *minus* any one singly gives a square," Diophantus remarked that since 5 and 13 are sums of two squares, and $65 = 5 \times 13$, 65 is also a sum of two squares [20, p. 167]. He most likely had the identity $(a^2+b^2)(c^2+d^2) = (ac + bd)^2 + (ad - bc)^2$ in mind. (This was proved by Viète in the late sixteenth century using his newly created algebraic notation.) In Problem VI.14, "To find a right-angled triangle such that its area *minus* the hypotenuse or *minus* one of the perpendiculars gives a square," Diophantus noted in passing that "This equation we cannot solve because 15 is not the sum of two [rational] squares" [20, p. 237]. His remarks in these problems appear to have prompted Bachet to ask which integers are sums of two squares, namely, for which integers n is the Diophantine equation $n = x^2 + y^2$ solvable.

Fermat took up the challenge. He reduced the question to asking which *primes* are sums of two squares, and claimed to have shown (recall that he gave no proofs) that every prime of the form $4k + 1$ is a sum of two squares, in fact, a unique such sum. He also stated results on the number of representations (if any) of an arbitrary integer as a sum of two squares [37, p. 70].

In a letter to Huygens in 1659, Fermat gave a slight indication of how he had proved the proposition about representing primes as sums of two squares, a result he had announced about 20 years earlier. He used, he said, his "method of infinite descent" (discussed in the next section), showing that if the proposition were not true for some prime, it would also not be true for a smaller prime, "and so on until you reach 5" [37, p. 67]. Weil observes (charitably to Fermat, we think) that "this may not have seemed quite enlightening to Huygens," adding that

> We are in a better position, because Euler, in the years between 1742 and 1747, constructed a proof precisely of that kind; it is such that we may with some verisimilitude attribute its substance to Fermat [37, p. 67].

Weil proceeds to sketch Euler's proof.

The problem about sums of two squares is one of the first topics Fermat studied, and it led him to other important results, for example, that

(a) Every prime of the form $8n + 1$ or $8n + 3$ can be written as $x^2 + 2y^2$.
(b) Every prime of the form $3n + 1$ can be written as $x^2 + 3y^2$.
(c) Every integer is a sum of four squares.

Other related questions he considered are cited by Weil [37, pp. 59–61, 69–75, 80–92].

2.4.1 A Look Ahead

The above results were extended in various directions in subsequent centuries:

1. Sums of kth powers

Fermat was proud to have shown that every integer is a sum of four squares, noting Descartes' failure to do so [26, p. 346]. The proposition was probably already known to Diophantus and was formally conjectured by Bachet. Euler was captivated by this result and tried for many years to prove it, without success. It was left to Lagrange to give a proof (in 1770).

A natural question suggested itself: Is every integer a sum of kth powers? Waring stated (in 1782) that every integer is a sum of nine cubes, nineteen 4th powers, "and so on" [19, p. 297]. The following came to be known as Waring's Problem: Given a positive integer k, does the equation $n = x_1^k + x_2^k + \cdots + x_s^k$ hold for every integer n, where s depends on k but not on n? If so, what is the smallest value of s for a given k? (This is usually denoted by $g(k)$.)

Waring's Problem was solved only in 1909, by Hilbert, who proved the *existence* of s for each k without determining the *value* of $g(k)$ for various k. Before that time the value of $g(k)$ was known only for about half a dozen values of k. In particular, it was known that $g(3) = 9$ and $g(4) = 19$, so Waring's statement turned out to have been correct [12]. It is now known that $g(k) = 2^k + [(3/2)^k] - 2$, provided that $2^k\{(3/2)^k\} + [(3/2)^k] \leq 2^k$, where for any real number x, $[x]$ denotes the greatest integer not exceeding x, and $\{x\} = x - [x]$. A similar result holds when the above inequality fails [36, p. 301]. However, this is not the end of the story as far as Waring's problem is concerned. A recent survey article by Vaughan & Wooley includes a bibliography of 162 items [36]. Hardy and Wright [19] has an entire chapter devoted to the classical theory.

Much work has also been done since Fermat's time on the representation of integers as sums of *squares*. For example, which integers are sums of *three* squares?

Can the above results on sums of squares be extended to algebraic integers? Some of this work is very subtle and related to Artin and Schreier's work in the 1920s on formally real fields. (A field is *formally real* if -1 cannot be represented as a sum of squares of elements in the field.) Artin used the theory of formally real fields to settle Hilbert's 17th Problem, posed at the International Congress of Mathematicians in Paris in 1900, which asks if every positive definite rational function in n variables over the reals is a sum of squares of rational functions. A recent book by Yandell is devoted to Hilbert's Problems [39].

2. Primes of the form $x^2 + ny^2$

Euler proved Fermat's results about the representation of primes in the form x^2+ny^2 for $n = 1, 2$, and 3, but he had difficulty with the case $n = 5$, essentially because the class number of the quadratic forms x^2+y^2, x^2+2y^2, and x^2+3y^2 is 1, while that of $x^2 + 5y^2$ is 2 [11, 13, p. 18]. (Fermat, too, realized that the case $n = 5$ was different from those for which $n = 1, 2$, and 3 [13, p. 18].) However, studying problems about the representation of primes in the form $x^2 + ny^2$ led Euler to conjecture the *quadratic reciprocity law*, the relationship between the solvability of $x^2 \equiv p \pmod q$ and $x^2 \equiv q \pmod p$, p and q odd primes [1]. This was because of the following result: $p|x^2 + ny^2$ and $(x, y) = 1$ if and only if $z^2 \equiv -n \pmod p$ has a solution; that is, $-n$ is a quadratic residue (mod p) [11, p. 13].

The problem of representing primes in the form $x^2 + ny^2$ for arbitrary n is very difficult, and was solved only in the twentieth century using high-powered tools of class field theory. It is the subject of an entire book by Cox [11].

3. Binary quadratic forms

A binary quadratic form is an expression of the type $ax^2 + bxy + yc^2$, with a, b, and c integers. The question of the representation of integers by binary quadratic forms, namely, given a fixed form $ax^2 + bxy + cy^2$, determining the integers n such that $n = ax^2 + bxy + cy^2$ for some integers x and y, became one of the central topics in number theory, studied intensively by Lagrange and treated masterfully by Gauss in his *Disquisitiones Arithmeticae*. This was an outgrowth of the investigations of Fermat and Euler as outlined above; see [1, 17, 37], and Chap. 1.

2.5 Fermat's Last Theorem

> It is impossible for a cube to be written as a sum of two cubes or a fourth power to be written as a sum of two fourth powers or, in general, for any number which is a power greater than the second to be written as a sum of two like powers. I have a truly marvellous demonstration of this proposition, which this margin is too narrow to contain [13, p. 2].

This is Fermat's famous note, written (perhaps in the 1630s) in the margin of Bachet's translation of Diophantus' *Arithmetica* alongside his Problem II.8, which asks "to divide a given square into two squares" [20, p. 144]. Symbolically, it says that $z^n = x^n + y^n$ has no positive integer solutions if $n > 2$. This came to be known as *Fermat's Last Theorem*. (As we mentioned, Fermat made many assertions in number theory without proof; all but one were later proved by Euler, Lagrange, and others. The exception – the last unproved "result" – was presumably the reason for the name "Fermat's Last Theorem." Of course, we now have a proof of that too.)

Fermat never published his "marvellous demonstration," and some very prominent mathematicians, among them Weil and Wiles, believe that he was probably mistaken in thinking he had a proof, and that perhaps he later realized this [30, pp. 74–75, 37, p. 104]. For it was only in the margin of Diophantus' *Arithmetica* that Fermat claimed to have proved FLT for arbitrary n. In later correspondence on this problem, he referred only to his having proofs of the theorem for $n = 3$ and $n = 4$ (see [16]). As Weil put it [37, p. 104]:

> For a brief moment perhaps, and perhaps in his younger days, he must have deluded himself into thinking that he had the principle of a general proof; what he had in mind on that day can never be known.

Fermat's only published proof in number theory was of a proposition whose immediate corollary is a proof of FLT for $n = 4$. The proposition in question states that the area of a right-angled triangle with integer sides cannot be a square (of an integer), that is, if $x^2 + y^2 = z^2$ for nonzero integers x, y, z, there is no integer u such that $(1/2)xy = u^2$. This problem was inspired by those in Diophantus' *Arithmetica*, Book VI, each of whose 26 problems asks for a right-angled triangle satisfying given conditions. Fermat's proof was found by his son, Samuel, in the margin of Fermat's copy of the *Arithmetica*, and was included in his *Observations on Diophantus* (Observation 45), posthumously published by Samuel. The proof is ambiguous in places, but Fermat noted that "The margin is too small to enable me to give the proof completely and with all detail" (!) [13, p. 12].

In the proof just mentioned, Fermat introduced the *method of infinite descent*. That is, he showed that if there exists some positive integer u satisfying the above conditions, then there is a positive integer $v < u$ satisfying the same conditions. Repeating this process ad infinitum clearly leads to a contradiction.

Fermat was very proud of his method of infinite descent, using it (he said) in the proofs of many of his number-theoretic propositions. He predicted that "this method will enable extraordinary developments to be made in the theory of numbers" [20, p. 293]. In an account of his number-theoretic work sent to Huygens in 1659 he gave more details [37, p. 75]:

> As ordinary methods, such as found in the books, are inadequate to proving such difficult propositions, I discovered at last a most singular method ... which I called *infinite descent*. At first I used it only to prove negative assertions, such as ... "there is no right-angled triangle of numbers whose area is a square." ... To apply it to affirmative questions is much harder, so that, when I had to prove that "Every prime of the form $4n + 1$ is a sum of two squares," I found myself in a sorry plight. But at last such questions proved amenable to my method ....

2.5.1 *A Look Ahead*

Fermat's method of infinite descent is logically only a variant of the Principle of Mathematical Induction, but it provided Fermat, and indeed his successors, with a powerful tool for proving number-theoretic results. The method of infinite descent may be likened, conceptually, to Dirichlet's pigeonhole principle: both are mathematically trivial observations with far-reaching ramifications.

In the eighteenth century, FLT was proved for only one exponent, $n = 3$, by Euler, using the method of infinite descent (there was, however, a gap in his proof). In fact, the method of infinite descent was used in all subsequent proofs of FLT, for various values of the exponent n. In the nineteenth century, attempts to prove FLT motivated the introduction of ideal numbers by Kummer, and later of ideals by Dedekind, giving rise also to such fundamental algebraic concepts as ring, field, prime ideal, unique factorization domain, and Dedekind domain. These developments led, in the hands of Dedekind and Kronecker, to the founding in the 1870s of *algebraic number theory*, the marriage of number theory and abstract algebra. In the twentieth century, FLT entered the mainstream of mathematics by becoming linked with a profound mathematical problem, the Shimura–Taniyama Conjecture, which says that every elliptic curve is modular. This, in turn, led to Wiles' 1994 proof of FLT, using deep ideas from various branches of mathematics (see Chap. 3 and [24]).

2.6 The Bachet and Pell Equations

The two equations are, respectively, $x^2 + k = y^3$, k any integer, and $x^2 - dy^2 = 1$, d a nonsquare positive integer. These equations, along with the Pythagorean equation $x^2 + y^2 = z^2$ and the Fermat equation $x^n + y^n = z^n$, $n > 2$, are perhaps the most important Diophantine equations. Fermat studied all of the above.

2.6.1 *Bachet's Equation*

A special case of the Bachet equation, $x^2 + 2 = y^3$, appears in Diophantus' *Arithmetica* (Problem VI.17). He wants "To find a right-angled triangle such that the area added to the hypotenuse gives a square, while the perimeter is a cube." In the course of solving it, he reduces the problem, saying that "Therefore we must find some square which, when 2 is added to it, becomes a cube" [20, p. 241]. The equation $x^2 + k = y^3$ was considered by Bachet, who raised the question of its solvability.

Fermat gave the solution $x = 5$, $y = 3$ for $x^2 + 2 = y^3$ and the solutions $x = 2$, $y = 2$, and $x = 11$, $y = 5$ for $x^2 + 4 = y^3$. In both cases he used infinite descent, he claims. Of course it is easy to see that these are solutions of the respective equations, but it is rather difficult to show that they are the *only* (positive) solutions, which is what Fermat had in mind. He challenged his colleagues to confirm these results: "I don't know," he wrote, "what the English will say of these negative propositions or if they will find them too daring. I await their solution and that of M. Frenicle ..." [26, p. 343].

Frenicle "could hardly believe" Fermat's claims, which he "found too daring and too general" [26, p. 343]. As for the English, Wallis responded (via Digby, to whom Fermat had sent his letter) as follows [26, p. 345]:

> I say ... [about] his recent negative propositions ... [that] I am not particularly worried whether they are true or not, since I do not see what great consequence can depend on their being so. Hence, I will not apply myself to investigating them. In any case, I do not see why he displays them as something of a surprising boldness that should stupefy either M. Frenicle or the English; for such negative conditions are very common and very familiar to us.

Mahoney has the following take on this [26, p. 345]:

> Wallis' overwhelming sense that number theory consisted essentially of wearying computations closed his mind to the promises Fermat was making about the new arithmetic.

2.6.2 A Look Ahead

Mordell noted that "[The Bachet equation $x^2 + k = y^3$] has played a fundamental role in the development of number theory [29, p. 238]." It has been studied for the past 300 years. Special cases were solved by various mathematicians throughout the eighteenth and nineteenth centuries. Euler introduced a fundamental new idea to solve $x^2 + 2 = y^3$ by factoring its left-hand side, which yielded the equation $(x + \sqrt{2}i)(x - \sqrt{2}i) = y^3$. The result was an equation in a domain D of "complex integers," where $D = \{a + b\sqrt{2}\mathrm{i} : a, b \in Z\}$. This was the first use of complex numbers – "foreign objects" – in number theory. The ideas involved in the solution of the equation entailed consideration of whether D is a unique factorization domain, and were part of the development which gave rise in the nineteenth century to *algebraic number theory*. See [1, 13, 22], and Chap. 3 for details.

In the 1920s, Mordell showed that $x^2 + k = y^3$ has finitely many (integer) solutions for each k (it may have none, for example, $x^2 - 45 = y^3$ [29, p. 239]), and in the 1960s Baker and Stark gave explicit bounds for x and y in terms of k, so that in theory all solutions for a given k can be found by computation. Moreover, Baker notes that

> techniques have been devised which, for a wide range of numerical examples, render the problem of determining the complete list of solutions in question accessible to machine computation [4, p. 45].

Bachet's equation is an important example of an elliptic curve. (An *elliptic curve* is a plane curve represented by the equation $y^2 = ax^3 + bx^2 + cx + d$, where a, b, c, d are integers or rational numbers, and the cubic polynomial on the right side of the equation has distinct roots.) In fact,

> [the Bachet equation], special as it may seem, is a central player in the Diophantine drama and in a certain sense 'stands for' the arithmetic theory of elliptic curves. One of the objects of this article is to give hints about why the [Bachet] equation plays this central role (Mazur [28, p. 196]).

Fermat dealt with many Diophantine equations, all, except for the Fermat equation $x^n + y^n = z^n$, of genus 0 or 1 [37, p. 104]. (For a sufficiently smooth curve given by a polynomial equation of degree n, the *genus* is $(n-1)(n-2)/2$; see also [7, p. 13].) Most of these define elliptic curves – algebraic curves of genus 1. The study of elliptic curves has involved the use of powerful methods, including those of algebraic geometry [7,23,28,29]:

> The theory of elliptic curves, and its generalization to curves of higher genus and to abelian varieties, has been one of the main topics in modern number theory. Fermat's name, and his method of infinite descent, are indissolubly bound with it; they promise to remain so in the future (Weil [37, p. 124]).

2.6.3 *Pell's Equation*

Pell's equation, $x^2 - dy^2 = 1$, was known in part of the ancient world [12]. (The equation was inappropriately named by Euler after the British mathematician John Pell.) Special cases were considered by the Greeks, and the Indians of the Middle Ages had a procedure for solving the general case, as did British mathematicians of the seventeenth century [37].

Weil asserts that "the study of the [quadratic] form $x^2 - 2y^2$ must have convinced Fermat of the paramount importance of the equation $x^2 - Ny^2 = \pm 1$ [37, p. 92]." (The equation $x^2 - dy^2 = -1$ is also sometimes known as Pell's equation.) Edwards counters that "it is impossible to reconstruct the way in which Fermat was led to this problem [13, p. 27]."

Fermat challenged mathematicians to show that Pell's equation has infinitely many solutions for each d. This is how he phrased it [20, p. 286]:

> Given any number whatever that is not a square, there are also given an infinite number of squares such that, if the square is multiplied into the given number and unity is added to the product, the result is a square.

He was aware of Brouncker's and Wallis' solutions of Pell's equation, but found them wanting, lacking a "general demonstration [26, p. 328]." What he had in mind is a proof that the equation always has a solution, in fact, infinitely many solutions, and that the known methods of finding solutions yield all of them. Fermat declared that he had such a demonstration, though he did not divulge it, other than to indicate that it involved his method of infinite descent [26, p. 350]. He also employed a "method of ascent" to obtain new solutions from given ones [37, pp. 105, 112].

Fermat challenged Frenicle to solve the equation $x^2 - 61y^2 \doteq 1$. "He must have known, of course, that the smallest solution [of this equation is] (1766319049, 226153980)," says Weil [37, p. 97]. There is no discernible pattern to the sizes of the minimal solutions of Pell's equation. For example, the minimal solution of $x^2 - 75y^2 = 1$ is (26, 3). (The *minimal solution* of Pell's equation, the so-called "fundamental solution," is one in terms of which all others can be expressed [1].)

2.6.4 A Look Ahead

The definitive treatment of Pell's equation was given by Lagrange in the latter part of the eighteenth century. He was the first to prove that it has a solution for every nonsquare positive integer d, and to give a procedure for finding all solutions for a given d by means of the continued fraction expansion of $\sqrt{d}$ – another use of "foreign objects" in number theory. There are, indeed, infinitely many solutions for each d [6, 17].

Pell's equation has continued to play an important role in number theory. For example:

1. It is a key to the solution of arbitrary quadratic Diophantine equations, as well as other Diophantine equations [29].
2. Its solutions yield the best approximation (in some sense) to $\sqrt{d}$: Pell's equation $x^2 - dy^2 = 1$ can be written as $(x/y)^2 = d + 1/y^2$, so that for large y, x/y is an approximation to $\sqrt{d}$. This may already have been realized by the Greeks [6, 12, 33].
3. There is a 1–1 correspondence between the solutions of $x^2 - dy^2 = 1$ and the invertible elements of the domain of integers of the quadratic field $Q(\sqrt{d}) = \{s + t\sqrt{d} : s, t \text{ rational}\}$ [1, 38].
4. The equation played a crucial role in the solution (in 1970) of Hilbert's 10th Problem, the nonexistence of an algorithm for solving arbitrary Diophantine equations [39].

For these and other reasons, the Pell equation has been studied extensively, but much remains to be done [38, p. 428]:

> The current state of the art in solving the Pell equation [computationally] is far from satisfactory. In spite of the enormous progress that has been made on this problem in the last few decades, we are still without answers to many fundamental questions. However, we are, it seems, beginning to understand what the questions should be.

2.7 Conclusion

We have considered only some of Fermat's contributions to number theory. These comprise results, methods, and concepts considered only casually, if at all, before Fermat. Moreover, they turned out to have applications in various number-theoretic contexts and became harbingers of significant departures in number theory in succeeding centuries. Without doubt, these accomplishments entitle Fermat to be known as the founder of modern number theory.

In 1659, Fermat wrote a four-page letter to Carcavi, intended for Huygens, which he titled "An account of new discoveries in the science of numbers," and in which he meant to give a brief summary of some of his accomplishments in number theory. We conclude with his reflections, taken from the last paragraph [26, p. 351]:

> *Perhaps posterity will thank me for having shown it that the ancients did not know everything*, and this account will pass into the mind of those who come after me as a "passing of the torch to the next generation".

References

1. W. W. Adams and L. J. Goldstein, *Introduction to Number Theory*, Prentice-Hall, 1976.
2. A. G. Agargün and E. M. Özkan, A historical survey of the Fundamental Theorem of Arithmetic, *Hist. Math.* 28 (2001) 207-214.
3. E. Bach and J. Shallit, *Algorithmic Number Theory*, Vol. 1, MIT Press, 1996.
4. A. Baker, *Transcendental Number Theory*, Cambridge Univ. Press, 1990.
5. K. Barner, How old did Fermat become?, *NTM, Intern. Jour. Hist. and Ethics of Natur. Sc., Techn. and Med.* 8 (4) (October 2001).
6. E. J. Barbeau, *Pell's Equation*, Springer, 2003.
7. I. G. Bashmakova, *Diophantus and Diophantine Equations*, Math. Assoc. of Amer., 1997. (Translated from the Russian by A. Shenitzer.)
8. E. T. Bell, *Men of Mathematics*, Simon and Schuster, 1937.
9. F. Bornemann, PRIMES is in P: A breakthrough for 'everyman', *Notices of the Amer. Math. Soc.* 50 (2003) 545-552.
10. D. M. Bressoud, *Factorization and Primality Testing*, Springer, 1989.
11. D. A. Cox, *Primes of the Form $x^2 + ny^2$: Fermat, Class Field Theory, and Complex Multiplication*, Wiley, 1989.
12. L. E. Dickson, *History of the Theory of Numbers*, 3 vols., Chelsea, 1966.
13. H. M. Edwards, *Fermat's Last Theorem: A Genetic Introduction to Algebraic Number Theory*, Springer, 1977.
14. C. R. Fletcher, A reconstruction of the Frenicle-Fermat correspondence, *Hist. Math.* 18 (1991) 344–351.
15. C. R. Fletcher, Fermat's theorem, *Hist. Math.* 16 (1989) 149–153.
16. K. Fogarty and C. O'Sullivan, Arithmetic progressions with three parts in prescribed ratio and a challenge of Fermat, *Math. Mag.* 77 (2004) 283–292.
17. J. R. Goldman, *The Queen of Mathematics: A Historically Motivated Guide to Number Theory*, A K Peters, 1998.
18. E. Grosswald, *Representation of Integers as Sums of Squares*, Springer, 1985.
19. G. H. Hardy and E. M. Wright, *An Introduction to the Theory of Numbers*, 4th ed., Oxford Univ. Press, 1959.

20. T. L. Heath, *Diophantus of Alexandria: A Study in the History of Greek Algebra*, 2nd ed., Dover, 1964. (Contains a translation into English of Diophantus' *Arithmetica*, a 130-page Introduction to Diophantus' and related work, and a 60-page Supplement on Fermat's number-theoretic work.)
21. T. L. Heath (ed.), *The Thirteen Books of Euclid's Elements*, 3 vols., 2nd ed., Dover, 1956.
22. K. Iga, A dynamical systems proof of Fermat's little theorem, *Math. Mag.* 76 (2003) 48-51.
23. K. Ireland and M. Rosen, *A Classical Introduction to Modern Number Theory*, Springer, 1982.
24. I. Kleiner, *A History of Abstract Algebra*, Birkhäuser, 2007.
25. D. Mackenzie and B. Cipra, *What's Happening in the Mathematical Sciences*, Amer. Math. Soc., 2006.
26. M. S. Mahoney, *The Mathematical Career of Pierre de Fermat*, 2nd ed., Princeton Univ. Press, 1994.
27. B. Mazur, Mathematical perspectives, *Bull. Amer. Math. Soc.* 43 (2006) 309–401.
28. B. Mazur, Questions about powers of numbers, *Notices Amer. Math. Soc.* 47 (2000) 195–202.
29. L. J. Mordell, *Diophantine Equations*, Academic Press, 1969.
30. C. J. Mozzochi, *The Fermat Diary*, Amer. Math. Soc., 2000.
31. H. Riesel, *Prime Numbers and Computer Methods for Factorization*, 2nd ed., Birkhäuser, 1994.
32. W. Scharlau and H. Opolka, *From Fermat to Minkowski: Lectures on the Theory of Numbers and its Historical Development*, Springer, 1985.
33. J. Stillwell, *Elements of Number Theory*, Springer, 2003.
34. J. Stillwell, *Mathematics and its History*, 2nd ed., Springer, 2002.
35. P. Tannery and Ch. Henry (eds.), *Oeuvres de Fermat*, 4 vols., Gauthier-Villars, 1891-1912, and a *Supplément*, ed. by C. de Waard, 1922.
36. C. Vaughan and T. D. Wooley, Waring's problem: a survey. In *Number Theory for the Millennium III*, ed. by M. A. Bennett et al, A K Peters, 2002, pp. 301-340.
37. A. Weil, *Number Theory: An Approach through History, from Hammurapi to Legendre*, Birkhäuser, 1984.
38. H. C. Williams, Solving the Pell equation. In *Number Theory for the Millennium III*, ed. by M. A. Bennett et al, A K Peters, 2002, pp. 397-435.
39. B. H. Yandell, *The Honors Class: Hilbert's Problems and their Solvers*, A K Peters, 2002.

Chapter 3
Fermat's Last Theorem: From Fermat to Wiles

3.1 Introduction

When historians come to judge the mathematics of the twentieth century, I am confident that they will regard it as a golden age, for both the emergence of brilliant new ideas and the solution of longstanding problems (the two are, of course, not unrelated). In the latter category, Fermat's Last Theorem (FLT) is neither the most ancient nor the latest example. In the late 1990s, Thomas Hales solved Kepler's Sphere-Packing Problem, posed in 1611, and Grigori Perelman proved the Poincaré Conjecture, proposed in 1904. Of course, the Riemann Hypothesis, the Goldbach Conjecture, and other outstanding problems are still unresolved.

Here are quotations from the two main protagonists in the drama associated with FLT:

> It is impossible to separate a cube into two cubes or a fourth power into two fourth powers or, in general, any power greater than the second into powers of like degree. I have discovered a truly marvelous demonstration, which this margin is too narrow to contain ([29], pp. 145–146).
>
> One morning in late May, Nada was out with the children and I was sitting at my desk thinking about the remaining family of elliptic equations. I was casually looking at a paper of Barry Mazur's, and there was one sentence there that just caught my attention. It mentioned a nineteenth-century construction, and I suddenly realized that I should be able to use that to make the Kolyvagin-Flach method work on the final family of elliptic equations. I went on into the afternoon and I forgot to go down for lunch, and by about three or four o'clock I was really convinced that this would solve the last remaining problem. It got to about teatime and I went downstairs and Nada was very surprised that I'd arrived so late. Then I told her—I'd solved Fermat's Last theorem [32, p. 243].

Both statements, by Fermat and Wiles, respectively – about 360 years apart – purport to have proved FLT. Wiles, as we know, published a proof, although his initial proof contained a major error which took 18 months to set right. But I'm getting ahead of myself.

I. Kleiner, *Excursions in the History of Mathematics*,
DOI 10.1007/978-0-8176-8268-2_3, © Springer Science+Business Media, LLC 2012

The aim of this chapter is to relate something of what happened in the three-and-a-half centuries between these two pronouncements, and, in particular, to describe some of the drama and the ideas connected with Wiles' proof.

3.2 The First Two Centuries

We begin at the beginning, with Fermat. His famous claim was that the equation $x^n + y^n = z^n$ has no (nonzero) integer solutions if $n > 2$. In the nineteenth century this pronouncement came to be known as FLT. (Fermat made many assertions in number theory without proof; all but one were later proved by Euler, Lagrange, and others. The exception – the last unproved "result" – was presumably the reason for the name "FLT.") Fermat made the claim in the 1630s, in the margin of Diophantus' book *Arithmetica* (c. 250 AD), alongside his Problem 8, Book II, which said: Given a number which is a square, write it as a sum of two squares. As for Fermat's "truly marvelous demonstration," it was, of course, never published (see Sect. 2.5).

Fermat did publish a proof for $n = 4$, the simplest exponent to deal with. He accomplished this by introducing the *method of infinite descent*, which has turned out to be important in the proofs of many number-theoretic results. The idea is to assume that the equation $x^4 + y^4 = z^4$ does have a solution for some positive integers a, b, c, and to show that it then has a solution for positive integers u, v, w, with $w < c$. Repeating this process ad infinitum leads to a contradiction, since it introduces an infinite descending sequence of positive integers; see [15] for details.

The fact that FLT holds for $n = 4$ implies that it also holds for $n = 4k$, k any positive integer. For, if $x^{4k} + y^{4k} = z^{4k}$ for some integers x, y, z, then $(x^k)^4 + (y^k)^4 = (z^k)^4$ for the integers x^k, y^k, z^k. The same type of argument shows that if FLT holds for $n = p$, then it holds for $n = pk$. Since any integer >2 is either a multiple of 4 or a multiple of an odd prime, Fermat's proof for $n = 4$ implies that it suffices to prove FLT for odd primes p.

A proof of FLT for $n = 3$ was given by Euler about 1760, over 100 years after Fermat's proof for $n = 4$. Euler's argument, however, contained a significant gap, not noticed by anyone at that time [15].

At the end of the eighteenth century the Paris Academy offered a prize for a proof of FLT. In 1816, Olbers, Gauss' astronomer friend, suggested that he compete for the prize. This was 15 years after Gauss' publication of the *Disquisitiones Arithmeticae*, which established him as one of the foremost mathematicians of his time. Gauss responded as follows ([28], p. 3):

> I am very much obliged for your news concerning the Paris prize. But I confess that Fermat's theorem as an isolated proposition has very little interest for me, because I could easily lay down a multitude of such propositions, which one could neither prove nor dispose of.

It appears that Gauss did not consider FLT a fruitful problem. (But proofs of the theorem for $n = 3$ and 5 were found among his unpublished notes; see [6, pp. 90–91]. This raises an interesting question: What *is* a good mathematical problem? Of course, individual mathematicians choose a problem to work on because it interests them; but how are they going to get their colleagues interested in it?

The following three major criteria for what makes a good problem are likely not in dispute:

(a) The solution of the problem has important consequences. The Riemann Hypothesis more than qualifies under this criterion.
(b) New ideas are introduced in attempts to solve the problem. This, as it turned out, was undoubtedly true of FLT, but of course one knows that only in retrospect. In this sense, Gauss seems to have misjudged the problem, as we shall see.
(c) The problem is connected with some other important problem or issue. This turned out to be the case for FLT, but it was not apparent in the nineteenth century.

The next strides in the proof of FLT were made by Legendre and Dirichlet, who, around 1825, independently established the theorem for $n = 5$. In 1839, Lamé proved it for $n = 7$. Incidentally, in 1832 Dirichlet showed that FLT holds for $n = 14$ but could not prove it for $n = 7$; the latter result, as we noted, implies the former.

3.3 Sophie Germain

The first important breakthrough on FLT was made in 1823 by the French mathematician Sophie Germain. She proved the following useful result, using relatively elementary methods: If p and $2p + 1$ are both prime, then $x^p + y^p = z^p$ has no solutions for which xyz is not divisible by p. Largely as a result of this theorem, it has been customary to divide the proofs of FLT for various values of p into two cases, the so-called *Case I*, in which none of x, y, z is divisible by p, and *Case II*, in which at least one of x, y, z is divisible by p. For example, it follows from Germain's result that Case I of FLT is true for the primes 5 and 11. Case II is usually regarded as much harder than Case I [15, 28].

Legendre extended Germain's theorem to the following: Case I of FLT holds for the prime exponent p provided that one of $4p + 1$, $8p + 1$, $10p + 1$, $14p + 1$, or $16p + 1$ is also prime. Germain and Legendre were now able to establish the first case of FLT for all primes $p < 100$. In 1977, Terjanian showed that the first case holds for all even exponents $2p$ ([28], p. 20).

An interesting problem is whether there are infinitely many "Sophie Germain primes," primes p for which $2p + 1$ is also prime. "This question is of the same order of difficulty as the well-known 'twin-prime' problem" [28, p. 56].

3.4 Lamé

In over 200 years, FLT was proved for only four exponents – 3, 4, 5, and 7(!) On March 1, 1847, a dramatic event occurred at a meeting in the Paris Academy of Sciences. Lamé announced that he had proved FLT for *all* exponents, and presented a brief outline of the proof. Before describing its gist, let us consider the following simpler problem, whose solution embodies the essential elements of Lamé's proof.

3.4.1 Pythagorean Triples

The problem of finding all primitive pythagorean triples, namely all integer solutions of $x^2 + y^2 = z^2$ with x, y, z relatively prime, was mentioned briefly in Sects. 1.1 and 2.5. While there are elementary solutions of this problem, the following method is instructive for our purposes. We factor the left side of $x^2 + y^2 = z^2$ to obtain $(x + yi)(x - yi) = z^2$. This is now an equation in the domain of "complex integers" of the form $G = \{a + bi : a, b \in Z\}$, the so-called *Gaussian integers*. It turns out that we can do number theory in G just as in Z. In particular, a "Fundamental Theorem of Arithmetic" holds in G, namely, every nonzero, noninvertible element of G is a unique product of primes. It follows that if a product of two relatively prime elements in G is a square, then each element is a square. The same result holds with the exponent 2 replaced by any exponent > 2 (see [2, 20]).

Since x, y, z are relatively prime in Z, it can be shown that $x + yi$ and $x - yi$ are relatively prime in G. Because their product is a square, each must be a square. In particular, $x + yi = (a + bi)^2$, where $a, b \in Z$. Thus $x + yi = (a^2 - b^2) + 2abi$, and comparing real and imaginary parts we get $x = a^2 - b^2$, $y = 2ab$. Since $z^2 = x^2 + y^2$, it follows that $z = a^2 + b^2$. So the solutions of $x^2 + y^2 = z^2$ are $x = a^2 - b^2$, $y = 2ab$, $z = a^2 + b^2$, $a, b \in Z$. Conversely, it can be shown that these are solutions for *every* choice of integers a and b. For x, y, z relatively prime, a and b must be relatively prime and of opposite parity. The resulting formula yields all primitive pythagorean triples (see [20]).

Two important ideas are implicit in this solution:

(a) Embedding a problem about integers in a domain of "complex integers." The notion of embedding a problem formulated in a given domain in a larger domain is a common and important mathematical technique. Hadamard's dictum that the shortest path between two truths in the real domain passes through the complex domain is, indeed, illuminating.

(b) Transforming an additive problem into a multiplicative one; in this case, $x^2 + y^2 = z^2$ into $(x+yi)(x-yi) = z^2$. This, too, is an important and not uncommon device. Multiplicative problems in number theory are, in general, much easier to deal with than additive ones, especially in the presence of a Fundamental Theorem of Arithmetic.

3.4.2 *Lamé's Proof*

Now to a sketch of Lamé's proof of FLT.

Assume that the equation $x^p + y^p = z^p$ *has* integer solutions (p an odd prime). Factor its left side to obtain $(x + y)(x + yw)(x + yw^2) \ldots (x + yw^{p-1}) = z^p$ (**), where w is a primitive p-th root of 1 (that is, w is a root of the equation $x^p = 1$, $w \neq 1$).

This is now an equation in the domain $D_p = \{a_0 + a_1 w + \cdots + a_{p-1} w^{p-1} : a_i \in Z\}$ of so-called *cyclotomic integers*. Lamé claimed that if the factors on the left-hand side of (**) are pairwise relatively prime in D_p, then, since their product is a p-th power, each must be a p-th power. From this a contradiction can be derived using Fermat's method of infinite descent by finding integers u, v, w such that $u^p + v^p = w^p$, with $w < z$. If the factors are not relatively prime, then by a suitable division by some element a, one obtains the relatively prime factors $(x + y)/a$, $(x + yw)/a$, $(x + yw^2)/a, \ldots, (x + yw^{p-1})/a$, and the proof proceeds analogously [5, 15].

After Lamé's presentation, Liouville, who was in the audience, took the floor and noted what seemed to him to be a gap in the proof, namely Lamé's contention that if a product of relatively prime factors is a p-th power, each must be a p-th power. The result is, indeed, true for the integers, Liouville observed, but it remains to be shown that it is also true for the cyclotomic integers. Lamé agreed that further consideration was needed, but was convinced that he had the right approach to the proof.

What was required for the proof was a Fundamental Theorem of Arithmetic in D_p [15, 28]. This is the subject of the next section.

3.5 Kummer

About two months after Lamé's presentation to the Academy, Liouville received a letter from Kummer confirming the grounds for his skepticism about Lamé's proof [28, p. 7]:

> Encouraged by my friend M. Lejeune Dirichlet, I take the liberty of sending you a few copies of a dissertation which I wrote three years ago.... In these memoirs, which I beg you to accept as a sign of my deep esteem, you will find developments concerning certain points in the theory of complex numbers composed of roots of unity, that is, roots of the equation $r^n = 1$, which have been recently the subject of some discussions at your illustrious Academy, at the occasion of an attempt by M. Lamé to prove the last theorem of Fermat.
>
> Concerning the elementary proposition for these complex numbers, that a *composite complex number may be decomposed into prime factors in only one way,* which you regret so justly in this proof, which is also lacking in some other points, I may assure you that *it does not hold in general* for complex numbers of the form $a_0 + a_1 r + a_2 r^2 + \ldots + a_{n-1} r^{n-1}$, but it is possible to rescue it, by introducing new kinds of complex numbers, which I have called *ideal complex numbers*.

I considered already long ago the applications of this theory to the proof of Fermat's theorem and I succeeded in deriving the impossibility of the equation $x^n + y^n = z^n$ [for all $n < 100$].

Kummer says three fundamental things here:

(a) Unique factorization fails, in general, in D_p. He showed that it already fails for $p = 23$. It was shown in 1971 by Uchida that unique factorization fails in D_p for *all* $p \geq 23$.
(b) Unique factorization can be "rescued" in D_p by introducing "ideal numbers."
(c) Using unique factorization in the extended cyclotomic domains containing the ideal numbers, one can prove FLT for all primes $p < 100$. Kummer proved more. Specifically, he showed that FLT holds for all "regular" primes, a prime being *regular* if it does not divide the *class number* of D_p; equivalently, if it does not divide the numerators of the *Bernoulli numbers* $B_2, B_4, \ldots, B_{p-3}$ (for definitions of class number and Bernoulli numbers see [26,28]). He then showed that all but three of the primes <100 are regular; the irregular primes were handled separately. Incidentally, it was shown in 1915 that there are infinitely many irregular primes; it is not known if there are infinitely many regular primes [15,26,28].

The following examples of nonunique factorization into primes in various domains and its restoration by the addition of "ideal" elements illustrate some of Kummer's ideas in more elementary contexts.

(a) $D = 2Z$, the even integers. Here $100 = 2 \times 50 = 10 \times 10$, where 2, 10, 50 are primes in D (they cannot be factored in D).
(b) $D =$ all polynomials over the reals (say) of degree > 1. Here $x^6 = x^2 \cdot x^2 \cdot x^2 = x^3 \cdot x^3$, with x^2 and x^3 prime in D.
(c) $D = \{a + b\sqrt{5}i : a, b \in Z\}$. Here $6 = 2 \times 3 = (1 + \sqrt{5}i)(1 - \sqrt{5}i)$, and it can be readily shown that 2, 3, $1 \pm \sqrt{5}i$ are prime in D. This example was given by Dedekind in the 1870s. In the first two examples D is not an integral domain, but its multiplicative structure illustrates well nonunique factorization [2].

As for rescuing unique factorization:
In (a) adjoin the "ideal number" 5.
In (b) adjoin the "ideal polynomial" x.
In (c) adjoin the "ideal numbers" $\sqrt{2}$, $(1 + \sqrt{5}i)/\sqrt{2}$ and $(1 - \sqrt{5}i)/\sqrt{2}$. We then have: $6 = 2 \times 3 = \sqrt{2} \times \sqrt{2}[(1 + \sqrt{5}i)/\sqrt{2}]$ and $6 = (1 + \sqrt{5}i)(1 - \sqrt{5}i) = \sqrt{2} \times [(1 + \sqrt{5}i)/\sqrt{2}] \times \sqrt{2} \times [(1 - \sqrt{5}i)/\sqrt{2}]$. Unique factorization has been restored (to the element 6) in D [20].

Kummer's work saw the emergence of a new subject – *algebraic number theory*, foreshadowed in earlier works of Gauss, Eisenstein, and Jacobi in connection with higher reciprocity laws [15, 19]. Moreover, Kummer's work on ideal numbers was vastly extended by Dedekind through his introduction of ideals, "one of the most decisive advances of modern algebra" [6, p. 91]. Dedekind, along with Kronecker, brought algebraic number theory to maturity [6,20,26]. Thus FLT acted as an incentive to the introduction of important mathematical concepts and results. More generally, "elementary" number theory has inspired the construction of deep theories that have illuminated mathematics well beyond the problems which gave them birth.

3.6 Early Decades of the Twentieth Century

Many technical results about FLT were obtained in the period 1850–1950, but there were no major breakthroughs. Here is a very small sample of such results [28]:

(a) Case I of FLT holds for infinitely many pairwise relatively prime exponents (Maillet, 1897).
(b) If p is a prime such that $2^{p-1} \not\equiv 1 (\bmod\ p^2)$, then Case I of FLT holds for p (Wieferich, 1909).
(c) If the so-called "second factor" of the class number of D_p ([28], p. 27) is not divisible by p, and if none of the Bernoulli numbers B_{2np} $(n = 1, 2, \ldots, (p-3)/2)$ is divisible by p^3, then Case II of FLT holds for p (Vandiver, 1929).
(d) If $p \equiv 1 (\bmod\ 4)$ and p does not divide the Bernoulli numbers B_{2s} for all odd s with $2 \le 2s \le p-3$, then FLT holds for p (Vandiver, 1929).

Using some of these and other results, Vandiver was able to establish by the end of the 1920s that FLT holds for all primes $p < 157$ (recall that about 70 years earlier Kummer had reached $p < 100$). Using the SWAC calculating machine, Vandiver in 1954 extended the result to $p < 2{,}521$ [28, p. 202].

At the turn of the twentieth century, Hilbert was asked why he never attempted to prove FLT. Here is his response [33, p. 69]:

> Before beginning I should have to put in three years of intensive study, and I haven't that much time to squander on a probable failure.

Contrast this with Gauss' statement about why he did not compete for the Paris prize offered for a proof of FLT: Gauss claimed the problem did not interest him, Hilbert that it was too difficult.

In 1908, the mathematician Paul Wolfskehl bequeathed a prize for a proof of FLT, valued at 100,000 marks (the equivalent of $1,000,000 by today's standards). This came to be known as the *Wolfskehl Prize*. His stipulation was that if the prize were not awarded by September 13, 2007, no subsequent claim would be accepted. It seems, certainly in retrospect, that Wolfskehl had a good sense of the difficulty of the problem, giving mathematicians another 100 years to come up with a proof [3].

In a lighter vein, mathematicians at the mid-twentieth century would likely have empathized with the following sentiments [14]:

> M. Fermat—what have you done?
> Your simple conjecture has everyone
> Churning out proofs,

Which are nothing but goofs!
Could it be that your statement's an erudite spoof?
A marginal hoax
That you've played on us folks?
But then you're really not known for your practical jokes.
Or is it true
That you knew what to do
When n was greater than two?
Oh then why can't we find
That same proof … are we blind?
You must be reproved, for I'm losing my mind.

3.7 Several Results Related to FLT, 1973–1993

We briefly list here a number of results about FLT – apart from those leading directly to Wiles' proof – obtained in the second half of the twentieth century.

(a) The age of the computer

In 1973 Wagstaff proved that FLT holds for all exponents $p < 125{,}000$, and 20 years later Buhler, Crandall, Ernvall, and Metsänkylä pushed the result to $p < 4{,}000{,}000$. These proofs did not use merely the "brute force" of the computer, but were a mix of sophisticated theoretical mathematics combined with sophisticated use of computations. Specifically, methods were developed to determine the *irregular* primes up to the indicated limits, and subsequently FLT was shown to hold for these primes (recall that Kummer had established FLT for all *regular* primes) [7,28,36].

(b) The Mordell Conjecture

In 1922, Mordell conjectured that there are only finitely many points with rational coordinates on an algebraic curve of genus greater than one (for a definition of genus see [4, 11]). Gerd Faltings proved the conjecture in 1983 using high-powered methods of algebraic geometry, developed only in the second half of the twentieth century. This was a major feat, for which Faltings was awarded the Fields Medal – the mathematical counterpart of the Nobel Prize. Now, the equation $x^n + y^n = z^n$ has genus 0 for $n = 2$ and genus greater than 1 for $n > 2$, so an immediate corollary of Mordell's Conjecture – now a theorem – is that *for each* $n > 2$ FLT has at most finitely many solutions [11, 29].

(c) Miyoka

Building on ideas of Faltings, and making connections between number theory and differential geometry, the Japanese mathematician Yoichi Miyoka announced in 1988 that he had proved FLT. Don Zagier, who was in the audience at the Max Planck Institute where Miyoka presented an outline of his proof, observed that "Miyoka's proof is very exciting, and some people feel that there is a very good chance that it is going to work. It's still not definite, but it looks fine so far"

[32, p. 232]. Two months later Faltings found an error in the proof. Many ideas in Miyoka's purported proof, however, remain important [11, 32].

3.8 Some Major Ideas Leading to Wiles' Proof of FLT

3.8.1 Elliptic Curves

We are about to enter the promised land. A key breakthrough, which less than ten years later would lead to a proof of FLT, came in 1985, when Gerhard Frey related FLT to elliptic curves – "a most surprising and innovative link" [13, p. 3]. Specifically, if $a^p + b^p = c^p$ holds for nonzero integers a, b, c, the associated elliptic curve – now known as the *Frey Curve* – is $y^2 = x(x - a^p)(x + b^p)$.

Number theory and geometry, in particular Diophantine equations and geometry, have been associated for about two millennia. In fact, it has been argued that the methods of Diophantus (c. 250 AD) for the solution of Diophantine equations could be viewed as geometric; they came to be known as the "tangent and secant methods" [4].

An *elliptic curve* is a plane curve given by an equation of the form $y^2 = x^3 + ax^2+bx+c$, where a, b, c are integers or rational numbers, and the cubic polynomial on the right side of the equation has distinct roots. (The coefficients may also be taken to be real or complex numbers, in fact elements in any field, although this is not of interest in our study.) A famous result of Siegel says that this equation has finitely many *integer* solutions, although it may have infinitely many *rational* solutions.

Elliptic curves had been studied by Diophantus and Fermat, and intensively investigated by Euler and Jacobi [4]. The name "elliptic curves" reflects their connection with elliptic functions, studied deeply in the nineteenth century. See [4], [31, p. 25], on [34, p. 228] for an explanation of this connection, and [4, p. 77], on [29, p. 148] for reasons why elliptic curves are important in number theory; "practical" uses of elliptic curves in the factoring of large integers into primes were found in the last few decades. The significant connection of elliptic curves with FLT came to light only in the 1980s.

3.8.2 Number Theory and Geometry

Before pursuing this connection, we want to present an elementary example of the use of geometry – the secant method – to solve a Diophantine equation, namely to find all solutions in integers of $x^2 + y^2 = z^2$ (cf. the *algebraic* solution of this equation in Sect. 3.4.1).

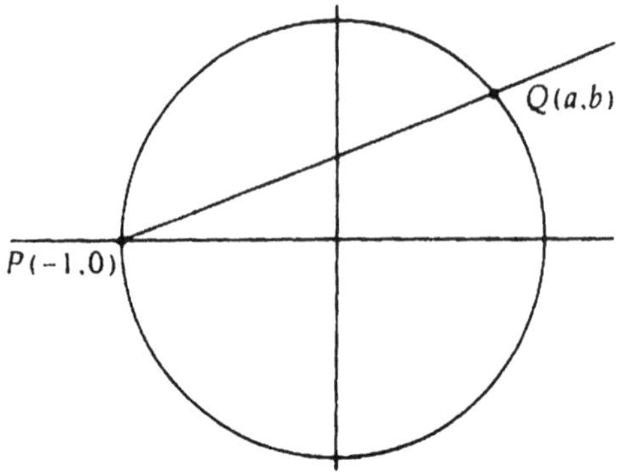

Fig. 3.1 Solving $x^2 + y^2 = z^2$ by geometry

Divide both sides of the equation by z^2 to get $(x/z)^2 + (y/z)^2 = 1$. Solving $x^2 + y^2 = z^2$ in Z is equivalent to solving $u^2 + v^2 = 1$ in Q. Geometrically, the latter requires finding all points with rational coordinates on the unit circle; they are called *rational points* [4, 30].

Suppose we are given a fixed rational point on the unit circle, say $P(-1, 0)$. If $Q(a, b)$ is any other rational point on the circle, the line determined by P and Q has rational slope, $b/(a + 1)$. Conversely, any line through $(-1, 0)$ with rational slope t intersects the circle in another rational point (a, b). So, to find all rational points on the unit circle is to find the points of intersection of the unit circle with all lines PQ having rational slope t.

We thus solve $u^2 + v^2 = 1$ and $v = t(u + 1)$ simultaneously for u and v and get $u = (1 - t^2)/(1 + t^2)$, $v = 2t/(1 + t^2)$.

Letting $t = m/n$, we find the *integer* solutions of $x^2 + y^2 = z^2$ to be $x = n^2 - m^2$, $y = 2nm$, $z = n^2 + m^2$.

The same method can be used to find all *rational* points on any quadratic curve, provided that one can find *one* rational point on the curve, and to find (at least in theory) all rational points on a *cubic* curve, provided one can find *two* rational points on the curve. The former problem is elementary, the latter is part of a rich theory; see [4, 30, 31].

3.8.3 *The Shimura-Taniyama Conjecture*

Back to Frey's key idea, namely the association of the elliptic curve $y^2 = x(x - a^p)(x + b^P)$ with the equation $a^p + b^p = c^p$. Frey conjectured that if there are indeed integers a, b, c such that $a^p + b^p = c^p$, then the resulting elliptic curve $y^2 = x(x - a^p)(x + b^P)$ is "badly behaved." It is a counterexample to the so-called *Shimura-Taniyama Conjecture* (STC). Frey's conjecture, reformulated by Serre, is known as the *Epsilon Conjecture* (EC). Put positively, the EC says that if the STC holds, then FLT is true.

The outline of a possible proof of FLT now emerged:

(a) Prove the EC, namely that the STC $\Rightarrow$ FLT.
(b) Prove the STC.

The STC was formulated by Taniyama in 1955 and refined in the 1960s by his friend and colleague Shimura (and by Weil). It says that *every elliptic curve is modular.* The notion of modularity is technically difficult to define, "but essentially it means that there is a formula for the number of solutions of the curve's cubic equation in each finite number system" [10, p. 4]; see also [11, 18, 24, 29]. As for the STC, it "represents a deep connection between algebra and analysis" [29, p. 152]. The following two statements by Barry Mazur give a very good sense of its scope and depth:

> [The conjecture] plays a structural and deeply influential role in much of our thinking and our expectations in Arithmetic.... Although it is undeniably a conjecture 'about arithmetic,' it can be phrased variously, so that in one of its guises, one thinks of it as being also deeply 'about' integral transforms in the theory of one complex variable; in another as being 'about' geometry [24, pp. 594, 596].
>
> It was a wonderful conjecture..., but to begin with it was ignored because it was so ahead of its time. When it was first proposed it was not taken up because it was so astounding. On the one hand you have the elliptic world, and on the other you have the modular world. Both these branches of mathematics had been studied intensively but separately.... Then along comes the Taniyama-Shimura conjecture, which is the grand surmise that there's a bridge between the two completely different worlds. Mathematicians love to build bridges [32, p. 190].

There are, of course, innumerable examples in mathematics of bridge-building, among the best known and most important being that between algebra and geometry, viz. analytic geometry. In this paper we built bridges between number theory and algebra, and number theory and geometry.

The STC was not only most surprising, it was also very important – in the sense that if true, it had innumerable and very significant consequences [24]. Thus a counterexample to the conjecture would have devastating consequences – much more severe than a counterexample to FLT! (Recall that the EC says that a counterexample to FLT would imply a counterexample to the STC.)

Enter Ken Ribet of the University of California at Berkeley. In 1986, he proved the EC. This was, of course, a big event. As Ribet relates it [32, p. 201]:

> It was the crucial ingredient that I had been missing and it had been staring me in the face.... I was completely enthralled... I sort of casually mentioned to a few people [at the 1986 International Congress of Mathematicians in Berkeley] that I'd proved that the Taniyama-Shimura conjecture implies Fermat's Last Theorem. It spread like wildfire and soon large groups of people knew; they were running up to me asking, *Is it really true you've proved that Frey's elliptic equation is not modular?*

For a sketch of the ideas involved in Ribet's proof of the EC see [11]; see also [18, 32].

3.9 Andrew Wiles

For most of its 350-year history, FLT was not part of mainstream mathematics – in the sense that it had no *direct* link with important parts of mathematics. Ribet's proof of the EC changed all that. "What Ribet [did]," Wiles noted, "was to link FLT with a problem in mathematics [the STC] that would never go away" [9, p. 1133]. On hearing of Ribet's proof, Wiles was ecstatic [32, p. 205]:

> It was one evening at the end of the summer of 1986 when I was sipping tea at the house of a friend. Casually in the middle of a conversation he told me that Ken Ribet had proved the link between Taniyama-Shimura and Fermat's Last Theorem. I was electrified. I knew that moment that the course of my life was changing because this meant that to prove Fermat's Last Theorem all I had to do was to prove the Taniyama-Shimura conjecture. It meant that my childhood dream was now a respectable thing to work on. I just knew that I could never let that go. I just knew that I would go home and work on the Taniyama-Shimura conjecture.

Work on it he did – for the next 7 years – in secret, which is most unusual in mathematics, though perhaps understandable under the circumstances. As Princeton colleague and then Chair of the Department Simon Kochen put it [21, p. 10]:

> If he [Wiles] said he was working on Fermat's last theorem, people would look askance. And if you start telling people who are experts, you end up collaborating with them. He wanted to do it on his own.

Here is some of what happened in the next 7 years, as told by Wiles [21, p. 10]:

> I made progress in the first few years. I developed a coherent strategy.... Basically, I restricted myself to my work and my family. I don't think I ever stopped working on it. It was on my mind all the time. Once you're really desperate to find the answer to something, you can't let go.

Only in the seventh year did he bring into his confidence his Princeton colleague Nicholas Katz, "who agreed to serve as a sort of sounding board for Dr Wiles" [21, p. 10]. At the end of 7 years, Eureka! [32, p. 244]:

> By May 1993 I was convinced that I had the whole of Fermat's Last Theorem in my hands. I still wanted to check the proof some more, but there was a conference which was coming up at the end of June in Cambridge, and I thought that would be a wonderful place to announce the proof—it's my old hometown, and I'd been a graduate student there.

The conference – on number theory – was organized by John Coates, Wiles' thesis advisor. It brought together some of the world's top experts in the subject. Wiles asked Coates to arrange for him to give a series of three lectures, one on each of the three-day conference. The title of his proposed talks was "Elliptic curves, modular forms, and Galois representations" – no mention of FLT. Only during the third talk did it become apparent – to the experts in the audience – that a proof of FLT was the likely outcome of the talks. Ribet describes the historic event [32, p. 248]:

> I came relatively early and I sat in the front row with Barry Mazur. I had my camera with me just to record the event. There was a very charged atmosphere and people were very excited. We certainly had the sense that we were participating in a historic moment.... The tension

Fig. 3.2 Andrew Wiles (1953–)

had built up over the course of several days. There was this marvelous moment when we were coming close to a proof of Fermat's Last Theorem.

And from Harvard colleague Barry Mazur [32, p. 248]:

> I've never seen such a glorious lecture, full of such wonderful ideas, with such dramatic tension, and what a buildup. There was only one possible punch line.

Indeed, Wiles concluded his third lecture with the sentence: "And this proves FLT; I think I'll stop here" [32, p. 249]. Specifically, what Wiles did was prove the STC for an important class of elliptic curves, the so-called *semi-stable* elliptic curves. (Roughly speaking, an elliptic curve is semi-stable if whenever a prime p divides the discriminant of the cubic defining the curve, exactly two of its roots are congruent modulo p; see [11, 29]). That is, he showed that every semi-stable elliptic curve is modular. Ribet had earlier proved a strong form of the EC, namely that if every semi-stable elliptic curve is modular, FLT is true. (The *full* STC was proved in 1999 [12].) For a sketch of the ideas involved in Wiles' proof see [11, 18, 22, 23].

Wiles' work is very deep and technically very demanding. "The finished proof is still rough going even for the experts" [9, p. 1134]. The following two statements by R. Murty give a sense of its profundity:

> By the end of the day, it was clear to experts around the world that nearly all of the noble and grand ideas that number theory had evolved over the past three and a half centuries since the time of Fermat were ingredients in the proof [25, p. 17].

So, in a sense, Wiles' proof was a grand collaborative effort of dozens of mathematicians over several centuries!

> The work is extremely deep, involving the latest ideas from a score of different fields, including the theories of group schemes, crystalline cohomology, Galois representations, deformation theory, Gorenstein rings, (geometric) Euler systems and many others [25, p. 16].

Behold the simplicity of the question and the complexity of the answer! The problem belongs to number theory – a question about positive integers. But what area does the proof come from? It is unlikely one could give a satisfactory answer, for the proof brings together many important areas – a characteristic of recent mathematics.

Wiles' lectures at Cambridge in June 1993 were, however, not to be the end of this 350-year odyssey. Here is some of what happened next.

The news of Wiles' proof of FLT electrified the mathematical world. E-mail messages started circulating incessantly. The news also made a great splash in the media – a rare event when it comes to mathematical news. Wiles' proof made the front pages of the *New York Times*. It was also featured in *Newsweek* and *Time*, and it made the *NBC Nightly News* that evening. *People* magazine listed Wiles among "the 25 most important people of the year."

After the celebrations were over, the business of checking the proof began. Wiles submitted a 200-page paper proving FLT to *Inventiones Mathematicae*. *Six* mathematicians were assigned to referee it – most unusual (normally there are 1–3 referees), but warranted under the circumstances. Many errors were found; most were easily and quickly corrected. One error, however, found by Katz, could not be fixed. But it was not divulged to the mathematical community – much was at stake! After some months, when no proof or announcement of an impending proof was forthcoming, rumors began to circulate. Was Wiles' proof destined to the same fate as Fermat's? Lamé's? Miyoka's?

On December 4, 1993, five months after his extraordinary announcement at Cambridge that he had proved FLT, Wiles issued the following e-mail note on a mathematical bulletin board [32, p. 264]:

> In view of the speculation on the status of my work on the Taniyama-Shimura conjecture and Fermat's Last Theorem, I will give a brief account of the situation. During the review process a number of problems emerged, most of which have been resolved, but one in particular I have not settled. ... I believe that I will be able to finish this in the near future using the ideas explained in my Cambridge lectures.

In January 1994, on the advice of Princeton colleague Peter Sarnak, Wiles sought the help of Cambridge mathematician Richard Taylor, his former PhD student. The preceding and ensuing months must have been most trying for Wiles, as we can surmise, and as Simon Singh confirms [32, pp. 275, 265, 273]:

> The last fourteen months [July 1993-August 1994] had been the most painful, humiliating period of [Wiles'] mathematical career.... The pleasure, passion, and hope that carried him through the years of secret calculations were replaced with embarrassment and despair.... After eight years of unbroken effort and a lifetime's obsession, Wiles was prepared to admit defeat. He told Taylor that he could see no point in continuing with their attempts to fix the proof.... Taylor... suggested they persevere for one more month.

Nine months after Wiles and Taylor started to work on repairing the proof they found "the vital fix" [33, p. 73]. Wiles recalls the clinching insight [33, p. 73]:

> It was so incredibly beautiful; it was so simple and so elegant. The first night I went back home and slept on it. I checked through it again the next morning, and I went down and

> told my wife, 'I've got it. I think I've found it'. And it was so unexpected she thought I was talking about a children's toy or something, and she said, 'Got what?' I said, 'I've fixed my proof. I've got it.'

On October 25, 1994, two papers proving FLT were released for publication, one by Wiles, the other by Taylor and Wiles, as follows:

(a) Wiles: Modular elliptic curves and FLT, *Annals of Mathematics* 142 (1995) 443–551.
(b) Taylor and Wiles: Ring-theoretic properties of certain Hecke algebras, *Annals of Mathematics* 142 (1955) 553–572.

3.10 Tributes to Wiles

The 1994 International Congress of Mathematicians (ICM) was held in Zürich in August. Had Wiles filled the gap in his proof of FLT prior to the start of the Congress, he would undoubtedly have received the Fields Medal at the Congress. At the next ICM in Berlin, in 1998, he was not eligible for the medal, being over 40. He was, however, awarded a one-time Special Tribute – the "International Mathematical Union Silver Plaque." On June 27, 1997, he collected the Wolfskehl prize – ten years before its expiry and now worth $50,000 (recall that in 1907 it was valued at $1,000,000; see Sect. 3.6).

The following appreciations of Wiles' work come from some of the foremost experts in the subject:

> To complete his [proof] Wiles needed to draw on and further develop many modern ideas in mathematics. In particular, he had to tackle the Shimura-Taniyama conjecture, an important 20th-century insight into both algebraic geometry and complex analysis. In doing so, Wiles forged a link between these major branches of mathematics. Henceforth insights from either field are certain to inspire new results in the other. Moreover, now that this bridge has been built, other connections between distant mathematical realms may emerge (Singh and Ribet [33, p. 68]).
>
> In mathematical terms, the final proof is the equivalent of splitting the atom or finding the structure of DNA. A proof of Fermat is a great intellectual triumph, and one shouldn't lose sight of the fact that it has revolutionized number theory in one fell swoop. For me, the charm and beauty of Andrew's work has been that it has been a tremendous step for number theory (Coates [32, p. 279]).
>
> Fermat's Last Theorem deserves a special place in the history of civilization. By its simplicity it has tantalized amateurs and professionals alike, and with remarkable fecundity led to the development of many areas of mathematics such as algebraic geometry, and more recently the theory of elliptic curves and representation theory. It is truly fitting that the proof crowns an edifice composed of the greatest insights of modern mathematics (Murty [25, p. 20]).

This statement surely belies Gauss' claim that FLT was not an interesting problem to work on! Even the greatest among mathematicians can misjudge.

The last word belongs to Wiles [32, p. 285]:

> I had this very rare privilege of being able to pursue in my adult life what had been my childhood dream. I know it's a rare privilege, but if you can tackle something in adult life that means that much to you, then it's more rewarding than anything imaginable. Having solved this problem, there's certainly a sense of loss, but at the same time there is this tremendous sense of freedom. I was so obsessed by this problem that for eight years I was thinking about it all the time—when I woke up in the morning to when I went to sleep at night. That's a long time to think about one thing. That particular odyssey is now over. My mind is at rest.

3.11 Is There Life After FLT?

I want to mention here two major ideas related to FLT.

The first is the so-called *ABC Conjecture*, formulated by Masser and Oesterlé in 1985 [27, p. 364]:

> Let A and B be relatively prime integers with $C = A + B$. Let $\mathrm{R}(ABC)$ be the product of the distinct prime factors of ABC. Then, for any $\varepsilon > 0$ there exists $\mathrm{k}(\varepsilon) > 0$ such that $C < \mathrm{k}(\varepsilon)\mathrm{R}(ABC)^{1+\varepsilon}$.

This innocent-looking statement is a most important conjecture; in particular, it implies FLT [4, p. 38]. More importantly, Dorian Goldfeld, one of the experts in the field, notes that

> The ABC Conjecture is the most important unsolved problem in Diophantine analysis. ... [It] promises to provide a new way of expressing Diophantine problems, one that translates an infinite number of Diophantine equations into a single mathematical statement [17, pp. 38, 39].

The second development related to FLT is the *Langlands Program* (LP), a series of deep and far-reaching conjectures, formulated by Langlands in the 1960s, relating various areas of mathematics, in particular number theory, algebra, and analysis [8, 16]. I will not describe the LP since, in the words of Stephen Gelbart,

> To merely state the conjectures [of the LP] correctly requires much of the machinery of class field theory, the structure theory of algebraic groups, the representation theory of real and p-adic groups, and (at least) the language of algebraic geometry. In other words, though the promised rewards are great, the initiation process is forbidding [16, p. 178].

A very special case of the LP is the STC (now a theorem), relating the elliptic and modular worlds, thus, in particular, number theory and analysis. Since the STC implies FLT, FLT is also a very special case of the LP.

References

1. A. D. Aczel, *Fermat's Last Theorem: Unlocking the Secret of an Ancient Mathematical Problem*, Four Walls Eight Windows, 1996.
2. W. W. Adams and L. J. Goldstein, *Introduction to Number Theory*, Prentice-Hall, 1976.
3. K. Barner, Paul Wolfskehl and the Wolfskehl Prize, *Notices Amer. Math. Soc.* 44 (1997) 1294–1303.
4. I. G. Bashmakova, *Diophantus and Diophantine Equations* (translated from the Russian by A. Shenitzer), Math. Assoc. of America, 1997.
5. Z. I. Borevich and I. R. Shafarevich, *Number Theory*, Academic Press, 1966.
6. N. Bourbaki, *Elements of the History of Mathematics*, Springer-Verlag, 1994.
7. J. Buhler, R. Crandell, R. Ernvall, and T. Metsänkylä, Irregular primes and cyclotomic invariants to four million, *Math. Comp.* 61 (1993) 151–153.
8. B. Cipra, New heights for number theory. In *What's Happening in the Mathematical Sciences*, Amer. Math. Soc., Vol. 5, 2000, pp. 3–11.
9. B. Cipra, Princeton mathematician looks back on Fermat proof, *Science* 268 (26 May 1995) 1133–1134.
10. B. Cipra, A truly remarkable proof. In *What is Happening in the Mathematical Sciences*, Amer. Math. Soc., Vol. 4, 1994, pp. 3–7.
11. D. A. Cox, Introduction to Fermat's Last Theorem, *Amer. Math. Monthly* 101 (1994) 3–14.
12. H. Darmon, A proof of the full Taniyama-Shimura-Weil Conjecture is announced, *Notices Amer. Math. Soc.* 46 (1999) 1397–1401.
13. K. Devlin, F. Gouvêa, and A Granville, Fermat's Last Theorem, a theorem at last, *MAA Focus* 13 (August 1993) 3–4.
14. J. P. Dowling, Fermat's Last Theorem, *Math. Mag.* 59 (1986) 76.
15. H. M. Edwards, *Fermat's Last Theorem: A Genetic Introduction to Algebraic Number Theory*, Springer-Verlag, 1977.
16. S. Gelbart, An elementary introduction to the Langlands Program, *Bulletin Amer. Math. Soc.* 10 (1984) 177–219.
17. D. Goldfeld, Beyond the Last Theorem, *The Sciences* 36 (March/April 1996) 34–40.
18. F. Gouvêa, A marvellous proof, *Amer. Math. Monthly* 101 (1994) 203–222.
19. K. Ireland and M. Rosen, *A Classical Introduction to Modern Number Theory*, Springer-Verlag, 1982.
20. I. Kleiner, The roots of commutative algebra in algebraic number theory, *Math. Mag.* 68 (1995) 3–15.
21. G. Kolata, Andrew Wiles: A math whiz battles 350-year-old puzzle, *Math. Horizons* (Winter 1993) 8–11.
22. J. Kramer, Über den Beweis der Fermat-Vermutung II, *El. Math.* 53 (1998) 45–60.
23. J. Kramer, Über die Fermat-Vermutung, *El. Math.* 50 (1995) 12–25.
24. B. Mazur, Number theory as gadfly, *Amer. Math. Monthly* 98 (1991) 593–610.
25. R. Murty, A long standing mathematical problem is solved: Fermat's Last Theorem, *Can. Math. Soc. Notes* 25 (Sept. 1993) 16–20.
26. H. Pollard and H. G. Diamond, *The Theory of Algebraic Numbers*, 2nd ed., Math. Assoc. of Amer., 1975.
27. P. Ribenboim, *Fermat's Last Theorem for Amateurs*, Springer-Verlag, 1999.
28. P. Ribenboim, *13 Lectures on Fermat's Last Theorem*, Springer-Verlag, 1979.
29. K. A. Ribet and B. Hayes, Fermat's Last Theorem and modern arithmetic, *American Scientist* 82 (March-April 1994) 144–156.
30. J. H. Silverman, *A Friendly Introduction to Number Theory*, Prentice Hall, 1997.
31. J. H. Silverman and J. Tate, *Rational Points on Elliptic Curves*, Springer-Verlag, 1992.

32. S. Singh, *Fermat's Enigma: The Quest to Solve the World's Greatest Mathematical Problem*, Penguin, 1997.
33. S. Singh and K. Ribet, Fermat's last stand, *Scientific Amer.* 277 (November 1997) 68–73.
34. J. Stillwell, *Mathematics and Its History*, 2nd ed., Springer-Verlag, 2002.
35. A. van der Poorten, *Notes on Fermat's Last Theorem*, Wiley, 1996.
36. S. M. Wagstaff, The irregular primes to 125000, *Math. Comp.* 32 (1978) 583–591.

Part B
Calculus/Analysis

Chapter 4
History of the Infinitely Small and the Infinitely Large in Calculus, with Remarks for the Teacher

4.1 Introduction

The infinitely small and the infinitely large – in one form or another – are essential in calculus. In fact, they are among the distinguishing features of calculus compared to some other branches of mathematics, for example algebra. They have appeared throughout the history of calculus in various guises: infinitesimals, indivisibles, differentials, evanescent quantities, moments, infinitely large and infinitely small magnitudes, infinite sums, power series, limits, and hyperreal numbers. And they have been fundamental at both the technical and conceptual levels – as underlying tools of the subject and as its foundational underpinnings. We will consider examples of these aspects of the infinitely small and large as they unfolded in the history of calculus from the seventeenth through the twentieth centuries. This will, in fact, entail discussing central issues in the development of calculus.

We will also present brief "didactic observations" at relevant places in the historical account. For elaboration you may consult some of the many works dealing with the interface between the history and the teaching of mathematics. The following address, at least in part, aspects of calculus/analysis: [10, 19, 21, 23, 29a, 36, 59, 65, 66]. More general works on the role and uses of the history of mathematics in its teaching are [26, 53, 57, 58, 69] and Chaps. 11–14 of this book.

The invention (discovery?) of calculus is one of the great intellectual achievements of civilization. Calculus has served for three centuries as the principal quantitative tool for the investigation of scientific problems. It has given precise (mathematical) expression to such fundamental concepts as motion, continuity, variability, and the infinite (in some of its aspects) – notions that have formed the basis for much scientific and philosophical speculation since ancient times. Physics and modern technology would be impossible without calculus. The most important equations of mechanics, astronomy, and the physical sciences in general are differential and integral equations – outgrowths of the calculus of the seventeenth century. Other major branches of mathematics (in addition to differential and integral equations) derived from calculus in succeeding centuries are real analysis,

I. Kleiner, *Excursions in the History of Mathematics*,
DOI 10.1007/978-0-8176-8268-2_4,

complex analysis, differential geometry, and calculus of variations. Calculus is also fundamental in probability, topology, Lie group theory, and aspects of algebra, geometry, and number theory. In fact, mathematics as we know it today would be inconceivable without the ideas of calculus. Moreover, the fundamental contribution of mathematics in general, and of calculus in particular, to a new, mechanistic interpretation of nature, begun in the early seventeenth century by Galileo and Descartes, and greatly furthered by Newton's prodigious *Principia* of 1687 and by the monumental *Mécanique Analytique* of Lagrange (1788) and *Mécanique Céleste* of Laplace (5 vols, 1799–1825), led to philosophies of mechanism, determinism, and materialism whose influence, in one form or another, is still with us today.

Newton and Leibniz independently invented calculus during the last third of the seventeenth century. However, practically all of the prominent mathematicians of Europe around 1650 could solve many of the problems in which elementary calculus is now used. At the same time, it took another two centuries following the invention of calculus to provide it with rigorous foundations.

But what *is* calculus? No definition is likely to capture the rich and multi-faceted nature of the subject. What is important to note is that calculus includes three major elements: a set of *rules* or algorithms: a "calculus," a *theory* to explain why the rules work, and *applications* (of the theory and the rules) to fundamental problems in science. Here we can only touch on *some* aspects of the evolution of this great subject, especially those relating to the manifestations of the infinite.

4.2 Seventeenth-Century Predecessors of Newton and Leibniz

4.2.1 Introduction

The Renaissance (c. 1400–1600) saw a flowering and vigorous development of the arts, literature, music, architecture, the sciences, and – not least – mathematics. It witnessed the decisive triumph of positional decimal arithmetic, the introduction of algebraic symbolism, the solution by radicals of the cubic and quartic, the free use (if not full understanding) of irrational numbers, the introduction of complex numbers, the rebirth of trigonometry, the establishment of a relationship between mathematics and the arts through perspective drawing, and a revolution in astronomy, later to prove of great significance for mathematics. A number of these developments were necessary prerequisites for the rise of calculus. So was the discovery of analytic geometry by Descartes and Fermat in the early decades of the seventeenth century.

The Renaissance also saw the full recovery and serious study of the mathematical works of the Greeks, especially Archimedes' masterpieces. His calculations of areas, volumes, and centers of gravity were an inspiration to many mathematicians of that period. Some went beyond Archimedes in attempting systematic calculations of the centers of gravity of solids. But they used the classical "method of exhaustion" of the Greeks, which was conducive neither to the discovery of results nor to the development of algorithms. The temper of the times, however, was such that most mathematicians were far more interested in results than in proofs. (Rigor, noted Cavalieri in the 1630s, "is the concern of philosophy and not of geometry" [45, p. 383].) And to obtain results mathematicians devised new methods for the solution of calculus-type problems. These were based on geometric, algebraic, and arithmetic ideas, often in interplay. We give two examples.

4.2.2 Cavalieri

A major tool for the investigation of calculus problems was the notion of an *indivisible*. The idea of an indivisible – in the form, for example, of an area being composed of a sum of infinitely many lines – was embodied in Greek atomistic conceptions and was also part of medieval scientific thought. Mathematicians of the seventeenth century fashioned indivisibles into a powerful tool for the investigation of area and volume problems.

Indivisibles were used in calculus by Galileo and others in the early seventeenth century, but it was Cavalieri who, in his influential *Geometry of Indivisibles* of 1635, shaped the vague concept of indivisibles into a useful technique for the determination of areas and volumes. His method entails considering a geometric figure as composed of an infinite number of indivisibles *of lower dimension*. Thus a surface consists of an infinite number of equally spaced parallel lines, and a solid of an infinite number of equally spaced parallel planes. The procedure for finding the area (or volume) of a figure is to compare it to a second figure of equal height (or width), whose area (or volume) is known, by setting up a one-to-one correspondence between the indivisible elements of the two figures and using *Cavalieri's Principle*: If the corresponding indivisible elements are always in a given ratio, then the areas (or volumes) of the two figures are in the same ratio.

For example, it is easy to show that the ordinates of the ellipse $x^2/a^2+y^2/b^2 = 1$ are to the corresponding ordinates of the circle $x^2 + y^2 = a^2$ in the ratio $b : a$ (see Fig. 4.1), hence the area of the ellipse $=$ $(b/a)\times$ the area of the circle $= \pi ab$.

4.2.3 Fermat

Fermat was the first to tackle systematically the problem of tangents. In the 1630s he devised a method for finding tangents to any polynomial curve. The following example illustrates his approach.

Suppose we wish to find the tangent to the parabola $y = x^2$ at some point (x, x^2). Let $x+e$ be another point on the x-axis and let s denote the *subtangent* to the curve at the point (x, x^2) (see Fig. 4.2). Similarity of triangles yields $x^2/s = k/(s+e)$. Fermat notes that k is "adequal" to $(x+e)^2$ (presumably meaning "as nearly equal

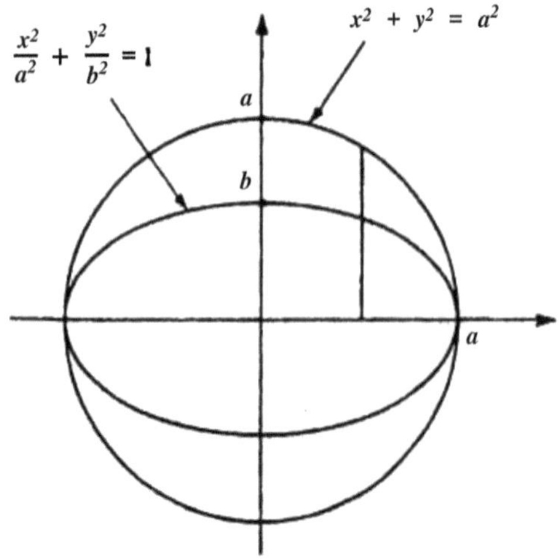

Fig. 4.1 Use of Cavalieri's Principle to find the area of an ellipse

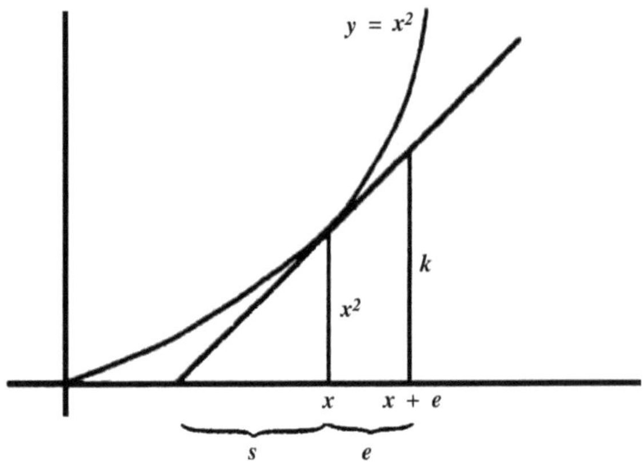

Fig. 4.2 Fermat's method of finding the tangent to $y = x^2$ at an arbitrary point on the curve

as possible," although Fermat does not say so). Writing this as $k \cong (x + e)^2$, we get $x^2/s \cong (x + e)^2/(s + e)$. Solving for s we have $s \cong ex^2/[(x + e)^2 - x^2] = ex^2/e(2x + e) = x^2/(2x + e)$. It follows that $x^2/s \cong 2x + e$. Note that x^2/s is the slope of the tangent to the parabola at (x, x^2). Fermat now "deletes" the e and claims that the slope of the tangent is $2x$.

Fermat's method was severely criticized by some of his contemporaries, notably Descartes. They objected to his introduction and subsequent suppression of the "mysterious e." Dividing by e meant regarding it as not zero. Discarding e implied treating it as zero. This is inadmissible, they rightly claimed. But Fermat's mysterious e embodied a crucial idea – the giving of a "small" increment to a variable. And it cried out for the limit concept, which was introduced formally only 200 years later. Fermat, however, considered his method to be purely *algebraic*.

The above examples give us a glimpse of the near-century of vigorous investigations in calculus prior to the work of Newton and Leibniz. Mathematicians plunged boldly into almost virgin territory – the mathematical infinite – where a more critical age might have feared to tread. They produced a multitude of powerful, if nonrigorous, infinitesimal techniques for the solution of area, volume, and tangent problems. What, then, was left for Leibniz and Newton to do?

4.3 Newton and Leibniz: The Inventors of Calculus

4.3.1 Introduction

The study of calculus (aside from applications) addresses two major themes – the algorithmic and the theoretical – which answer the questions *how* and *why*, respectively. Thus, calculus contains a well-developed technical machinery for the solution of important problems, both pure and applied, as well as a body of theoretical results underlying the techniques. It was primarily to the former of these two aspects of calculus that Newton and Leibniz contributed. More specifically, they

(a) Invented the general concepts of derivative ("fluxion," "differential") and integral. It is one thing to compute areas of curvilinear figures and volumes of solids using ad hoc methods but quite another to recognize that such problems can be subsumed under a single concept, namely the integral. The same applies to the distinction between the finding of tangents, maxima and minima, and instantaneous velocities on the one hand, and the concept of derivative on the other.
(b) Recognized differentiation and integration as inverse operations. Although several mathematicians before Newton and Leibniz – Fermat, Roberval, Torricelli, Gregory, and especially Barrow – noted the relation between tangent and area problems, mainly in specific cases, the clear and explicit recognition, in its complete generality, of what we now call the Fundamental Theorem of Calculus belongs to Newton and Leibniz.
(c) Devised a notation and developed algorithms to make calculus the powerful computational instrument it is.
(d) Extended the range of applicability of the methods of calculus. While in the past the techniques of calculus were applied mainly to polynomials, often only of low degree, they were now applicable to "all" functions, algebraic and transcendental.

It may be appropriate at this point to say a few words about anticipations and discoveries in mathematics. Hardly ever, if at all, does a mathematical theory, even a concept or result, arise full-grown in the mind of a single mathematician. Mathematical ideas *evolve* over time, although it is often not a smooth or continuous evolution – there are false starts, trials and errors, and failures as well as successes. But there is a great difference between the discovery of *instances* of a given concept and awareness of the concept in its full generality. Such awareness is usually accompanied by a recognition of the *significance* of the concept and its *exploitation and application*.

In the case of calculus, it was Newton and Leibniz who distilled the basic concepts of derivative and integral from their numerous instances in the works of their predecessors, recognized the significance of these concepts by embedding them in an algebraic-algorithmic apparatus, and applied them in many new situations. At the same time, it was a propitious period for the great synthesis. In the words of the noted historian Dirk Struik [64, p. 106]:

> A general method of differentiation and integration, derived in the full understanding that one process is the inverse of the other, could only be discovered by men who mastered the geometric methods of the Greeks and of Cavalieri, as well as the algebraic methods of Descartes and Wallis. Such men could have appeared only after 1660, and they actually did appear in Newton and Leibniz.

4.3.2 Didactic Observation

"Mathematical discoveries, like the springtime violets in the woods, have their season, which no human effort can retard or hasten" [3, p. 263]. So said Farkas Bolyai to his son Janos, one of the discoverers of non-Euclidean geometry. And just as there is a right time for mathematical synthesis in history, so there should be one in pedagogy. The predecessors of Newton and Leibniz did not synthesize mainly because they lacked enough examples which would have warranted a synthesis. It is a commonplace, but it bears repeating, that we should give students examples – many examples, in different contexts – before we define, generalize, or prove.

And now, to some examples of the calculus of Newton and Leibniz. We observe first that basic to their work in the subject was the notion of an *infinitesimal*. This was not formally defined, but was understood to be an "infinitely small" quantity, less than any finite quantity but not zero.

4.3.3 Newton

Newton developed three different versions of his calculus, apparently searching for the best approach to the subject; or perhaps, as has also been suggested, each version was to serve a different purpose – to derive results effectively, to supply useful algorithms, or to give convincing proofs. Thus Newton used infinitesimals – largely a geometric approach, "fluxions" – a kinematic approach, and finally "prime and ultimate ratios" – his most rigorous, "algebraic" approach. The three methods were not always kept apart when applied to the solution of various problems. See [68].

It is important to note that the calculus of Newton (and of Leibniz) is a calculus of *variables* and equations relating these variables; it is *not* a calculus of *functions*.

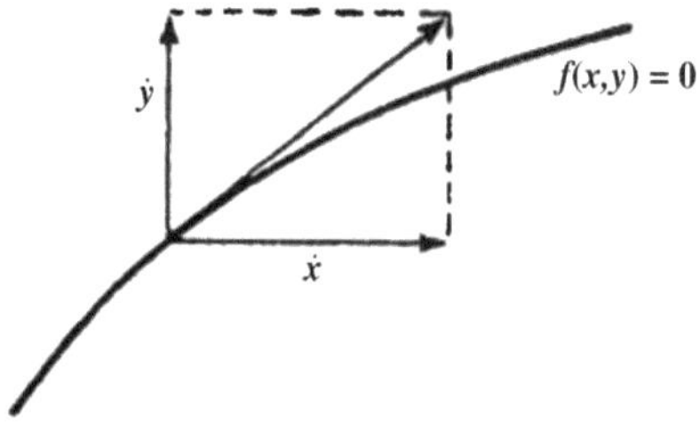

Fig. 4.3 Newton's view of the motion of a point on a curve, decomposed into horizontal and vertical motions

In fact, the notion of function as an explicit mathematical concept arose only in the early eighteenth century. Newton calls his variables "fluents" – the image is geometric and kinematic, of a quantity undergoing continuous change, for example a point "flowing" continuously along a curve. The variables are implicitly considered as functions of time.

Newton's basic concept is that of a "fluxion," denoted by $\dot{x}$; it is the instantaneous rate of change (instantaneous velocity) of the fluent x, in our notation dx/dt. The instantaneous velocity is not defined, but is taken as intuitively understood. Newton aims rather to show how to *compute* $\dot{x}$.

Since Newton regards the motion of a point on a curve, with equation $f(x, y) = 0$ say, as the composition of horizontal and vertical motions with velocities $\dot{x}$ and $\dot{y}$, respectively (see Fig. 4.3), and since the direction of motion of a point on the curve is along the tangent to the curve, it follows that the slope of the tangent to the curve $f(x, y) = 0$ at a point (x, y) on the curve is $\dot{y}/\dot{x}$. But $\dot{y} = dy/dt$, $\dot{x} = dx/dt$, hence $\dot{y}/\dot{x} = dy/dx$ (our notation). That is, the slope of the tangent – the derivative – is a quotient of fluxions.

The following is an example of the computation of the tangent to a curve with equation $x^3 - ax^2 + axy - y^3 = 0$ at an arbitrary point (x, y) on the curve. Newton lets o be an infinitesimal period of time. Then $\dot{x}$o and $\dot{y}$o are infinitesimal increments in x and y, respectively. (For, we have distance = velocity × time = $\dot{x}$o or $\dot{y}$o, assuming with Newton that the instantaneous velocities $\dot{x}$ and $\dot{y}$ of the point (x, y) moving along the curve remain constant throughout the infinitely small time interval o.) Newton calls $\dot{x}$o and $\dot{y}$o *moments*, a "moment" of a fluent being the amount by which it increases in an infinitesimal time period.

Thus $(x + \dot{x}\mathrm{o},\ y + \dot{y}\mathrm{o})$ is a point on the curve infinitesimally close to (x, y). In Newton's words: "so that if the described lines [coordinates] be x and y in one moment, they will be $x + \dot{x}\mathrm{o}$ and $y + \dot{y}\mathrm{o}$ in the next." Substituting $(x + \dot{x}\mathrm{o},\ y + \dot{y}\mathrm{o})$ into the original equation and simplifying by deleting $x^3 - ax^2 + axy - y^3$ (which equals zero) and dividing by o, we get:

$$3x^2\dot{x} - 2ax\dot{x} + ay\dot{x} + ax\dot{y} - 3y^2\dot{y} + 3x\dot{x}^2\mathrm{o} - a\dot{x}^2\mathrm{o} + a\dot{x}\dot{y}\mathrm{o} - 3\dot{y}y^2\mathrm{o} + \dot{x}^3\mathrm{o}^2 - \dot{y}^3\mathrm{o}^2 = 0.$$

Newton now discards the terms involving o, noting that they are "infinitely lesse" than the remaining terms. This yields an equation relating x and y, namely $3x^2\dot{x} - 2ax\dot{x} + ay\dot{x} + ax\dot{y} - 3y^2\dot{y} = 0$. From this relationship, we can get the slope of the

Fig. 4.4 Isaac Newton (1642–1727)

tangent to the given curve at any point $(x, y) : \dot{y}/\dot{x} = (3x^2 - 2ax + ay)/(3y^2 - ax)$. This procedure is quite general, Newton notes, and it enables him to obtain the slope of the tangent to *any* algebraic curve.

The problem of what to make of the symbol "o" remained: is it zero? a finite quantity? infinitely small? Newton's dilemma was not unlike Fermat's a half-century earlier. He attempted to clarify matters with his theory of ultimate ratios, to be discussed below.

Power series were a fundamental tool in Newton's calculus. He thought of them as the infinite decimals of analysis and claimed that "the operations of computing in numbers and with variables are closely similar" [38, p. 545]. Power series were to him but infinite polynomials on which one could operate as on ordinary polynomials. Central to Newton's use of power series in calculus was the binomial theorem, which he extended to fractional and negative exponents.

Newton applied power-series methods to problems of integration of "badly behaved" functions – both algebraic and transcendental – where it did not seem possible to evaluate their integrals directly. For example, to integrate $\sqrt{1-x^3}$ – shown in the nineteenth century not to be integrable in finite terms – Newton would expand $(1-x^3)^{1/2}$ in a power series using the binomial theorem and integrate the resulting series term by term.

Newton was the first in the West to derive power-series expansions of the trigonometric functions, known to Indian mathematicians 300 years earlier, and he used them to find the areas under the cycloid and the quadratrix. He also expanded the exponential and logarithmic functions in power series. For example, since $1/(1 + x) = 1 - x + x^2 - x^3 + x^4 - \ldots$, we can integrate both sides to get $\log(1+x) = x - x^2/2 + x^3/3 - x^4/4 + \ldots$. Newton never questioned the legitimacy of term-by-term integration of the infinite sum on the right side, although he appears to have been aware of the issue of convergence [38, p. 550].

4.3.4 *Leibniz*

Leibniz' ideas on calculus evolved gradually, and like Newton, he wrote several versions, giving expression to his ripening thoughts. Central to all of them was the concept of "differential," although that notion too had different meanings for him at different times.

Leibniz views a *curve* as *a polygon with infinitely many sides, each of infinitesimal length.* (Recall the Greek conception of the circle as a polygon of infinitely many sides.) With such a curve is associated an infinite (discrete) sequence of abscissas $x_1, x_2, x_3, \ldots$, and an infinite sequence of ordinates $y_1, y_2, y_3, \ldots$, where (x_i, y_i) are the coordinates of the points of the curve.

The difference between two successive values of x is called the *differential* of x and is denoted by $\mathrm{d}x$; similarly for $\mathrm{d}y$. The differential $\mathrm{d}x$ is a fixed non-zero quantity, infinitesimally small in comparison with x – in effect, an infinitesimal. There is a *sequence* of differentials associated with the curve, namely the sequence of differences $x_i - x_{i-1}$ associated with the abscissas $x_1, x_2, x_3, \ldots$ of the curve [23, pp. 258, 261].

The sides of the polygon constituting the curve are denoted by $\mathrm{d}s$ (again, there are infinitely many such infinitesimal $\mathrm{d}s$'s). This gives rise to Leibniz' famous *characteristic triangle* with infinitesimal sides $\mathrm{d}x$, $\mathrm{d}y$, $\mathrm{d}s$ satisfying the relation $(\mathrm{d}s)^2 = (\mathrm{d}x)^2 + (\mathrm{d}y)^2$ (Fig. 4.5). The side $\mathrm{d}s$ of the curve (polygon) is taken as coincident with the tangent to the curve (at the point x). As Leibniz puts it [42, pp. 234–235]:

> We have only to keep in mind that to find a *tangent* means to draw a line that connects two points of the curve at an infinitely small distance, or the continued side of a polygon with an infinite number of angles, which for us takes the place of the *curve*. This infinitely small distance can always be expressed by a known differential like $\mathrm{d}s$.

The slope of the tangent to the curve at the point (x, y) is thus $\mathrm{d}y/\mathrm{d}x$ – an actual *quotient* of differentials, which Leibniz calls the *differential quotient.*

Leibniz' integral is an infinite sum of infinitesimal rectangles with base $\mathrm{d}x$ and height y (Fig. 4.6). The "left-over" triangles, Leibniz notes, "are infinitely small compared with the said rectangles, [and] may be omitted without risk" [23, p. 257]. These "left-over" triangles are Leibniz' characteristic triangles, which may thus be viewed as a link between differentiation and integration. His very suggestive notation for the integral (a result of several less successful attempts) is $\int y\mathrm{d}x$ ($\int$ is an elongated S, denoting a "sum"). Like Newton, Leibniz *computes* his integrals by antidifferentiation.

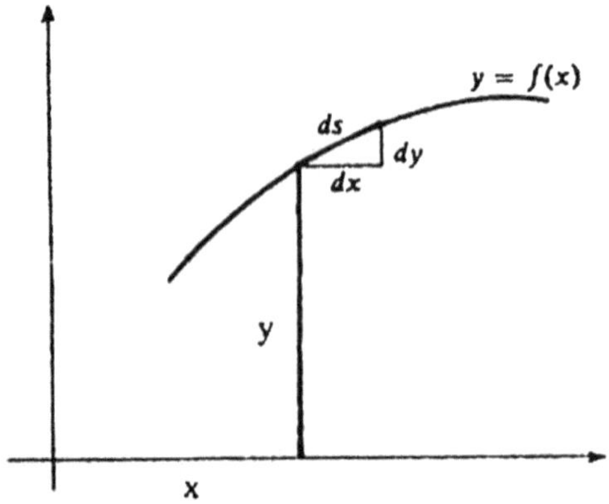

Fig. 4.5 Leibniz' characteristic triangle with infinitesimal sides $\mathrm{d}x$, $\mathrm{d}y$, $\mathrm{d}s$

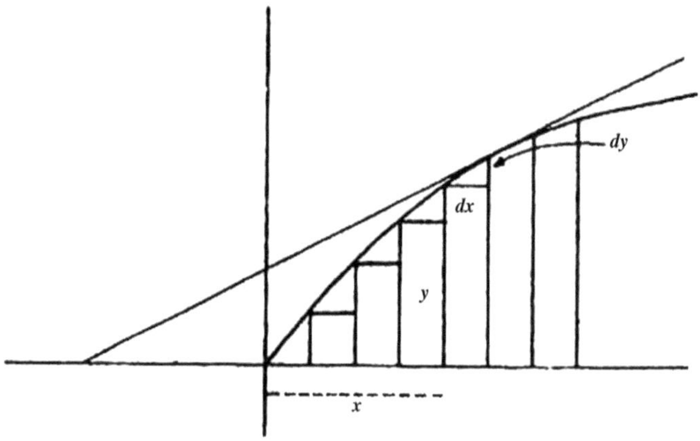

Fig. 4.6 Leibniz' integral as an infinite sum of infinitesimal rectangles

Leibniz searched for some time to find the right rules for differentiating products and quotients. When he found them, the "proofs" were easy. Thus $\mathrm{d}(xy) = (x + \mathrm{d}x)(y + \mathrm{d}y) - xy = xy + x\mathrm{d}y + y\mathrm{d}x + (\mathrm{d}x)(\mathrm{d}y) - xy = x\mathrm{d}y + y\mathrm{d}x$. Leibniz omits $(\mathrm{d}x)(\mathrm{d}y)$, noting that it is "infinitely small in comparison with the rest" [23, p. 255].

As a second example of Leibniz' calculus, let us find the tangent at a point (x, y) to the conic $x^2 + 2xy = 5$. Replacing x and y by $x + \mathrm{d}x$ and $y + \mathrm{d}y$, respectively, and noting that $(x + \mathrm{d}x, y + \mathrm{d}y)$ is a point on the conic "infinitely close" to (x, y), we get $(x + \mathrm{d}x)^2 + 2(x + \mathrm{d}x)(y + \mathrm{d}y) = 5 = x^2 + 2xy$. Simplifying, and discarding $(\mathrm{d}x)(\mathrm{d}y)$ and $(\mathrm{d}x)^2$, which are negligible in comparison with $\mathrm{d}x$ and $\mathrm{d}y$, yields $2x\mathrm{d}x + 2x\mathrm{d}y + 2y\mathrm{d}x = 0$. Dividing by $\mathrm{d}x$ and solving for $\mathrm{d}y/\mathrm{d}x$ gives $\mathrm{d}y/\mathrm{d}x = (-x-y)/x$. This is of course what we would get by writing $x^2+2xy = 5$ as $y = (5-x^2)/2x$ and differentiating this *functional* relation. (Recall that Leibniz' calculus predates the emergence of the function concept.)

We see in the above two examples how Leibniz's choice of a felicitous notation enabled him to arrive very quickly at reasonable convictions, if not rigorous proofs, of important results. But his symbolic notation served not only to *prove* results; it also greatly facilitated their *discovery*. For example, it is far from clear how one is to obtain the rule for differentiating the composite function $f(g(x))$ – the chain rule. But, setting $y = g(x)$ and $z = f(y)$, Leibniz' differential notation and the consideration of the derivative as a quotient immediately yield $\mathrm{d}z/\mathrm{d}x$ (the derivative of $f(g(x))) = (\mathrm{d}z/\mathrm{d}y)(\mathrm{d}y/\mathrm{d}x)$. This form also suggests a modern proof, namely to replace $\mathrm{d}x$, $\mathrm{d}y$, $\mathrm{d}z$ by Δx, Δy, Δz and take limits – remaining wary of pitfalls, of course.

Results derived (with a bad conscience?) in a first-year calculus course through the use of infinitesimals (differentials) are a legacy of Leibniz. For example, from Leibniz' product rule $\mathrm{d}(xy) = x\mathrm{d}y + y\mathrm{d}x$ we immediately get, by integrating, the formula for integration by parts. Thus $\int \mathrm{d}(xy) = \int x\mathrm{d}y + \int y\mathrm{d}x$, hence $xy = \int x\mathrm{d}y + \int y\mathrm{d}x$ or $\int x\mathrm{d}y = xy - \int y\mathrm{d}x$.

Leibniz' striving for an efficient notation for his calculus was part and parcel of his endeavor to find a "universal characteristic" – a symbolic language capable of reducing all rational discourse to routine calculation. As the above examples suggest, he succeeded brilliantly as far as calculus is concerned. C. H. Edwards puts it thus [23, p. 232]:

> [Leibniz'] infinitesimal calculus is the supreme example in all of science and mathematics, of a system of notation and terminology so perfectly mated with its subject as to faithfully mirror the basic logical operations and processes of the subject.

4.3.5 *Didactic Observation*

We take symbolism for granted. Mathematics without a well-developed notation would be inconceivable to us. We should note, however, that mathematics evolved for at least three millennia with hardly any symbols! In fact, as the historian K. Pederson observed [33, p. 47]:

> An important reason why mathematicians [of the early seventeenth century] failed to see the general perspectives inherent in their various methods [for solving calculus problems] was probably the fact that to a great extent they expressed themselves in ordinary language without any special notation and so found it difficult to formulate the connections between the problems they dealt with.

So, as we have said, a good notation helps not only in the proof of results but also in their *discovery*. Leibniz' calculus prevailed over Newton's largely because of his well-chosen notation, which, he said, "offers truths ... without any effort of the imagination." The pedagogical benefits for calculus are strikingly expressed by C. H. Edwards [23, p. 232]:

> It is hardly an exaggeration to say that the calculus of Leibniz brings within the range of an ordinary student problems that once required the ingenuity of an Archimedes or a Newton.

4.4 The Eighteenth Century: Euler

4.4.1 Introduction

Brilliant as the accomplishments of Newton and Leibniz were, their respective versions of calculus consisted largely of loosely connected methods and problems, and were not easily accessible to the mathematical public (such as it was). The first systematic introduction to the Leibnizian (differential) calculus was given in 1696 by L'Hospital in his text *The Analysis of the Infinitely Small, for the Understanding of Curved Lines.* Calculus was further developed during the early decades of the eighteenth century, especially by the Bernoulli brothers Jakob and Johann. Several books appeared during this period, but the subject lacked a focus. The main contemporary concern of calculus was with the geometry of curves – tangents, areas, volumes, lengths of arcs (cf. the title of L'Hospital's text). Of course Newton and Leibniz introduced an algebraic apparatus, but its motivation and the problems to which it was applied were geometric or physical, having to do with curves. In particular, this was (as we already noted) a calculus of variables related by equations, rather than a calculus of functions.

A fundamental conceptual breakthrough, still with us today, was achieved by Euler around the mid-eighteenth century. It was to make the concept of *function* the centerpiece of calculus. Thus calculus is not about curves, asserted Euler, but about functions. The derivative and the integral are not merely abstractions of the notions of tangent or instantaneous velocity on the one hand and of area or volume on the other – they are the basic concepts of calculus, to be investigated in their own right.

Euler was not the first to introduce the notion of function, but he was the first to make it central by regarding calculus as the branch of mathematics that deals with functions (see Chap. 5). But a "decree," even by an Euler, could not change mathematical practice overnight. Mathematicians of the eighteenth century did not readily embrace functions as central to their subject, especially since variables seemed to serve them well.

4.4.2 Didactic Observation

Calculus without functions! That may appear as a heresy. We certainly *can* teach calculus without function, as Newton, Leibniz, and their immediate successors have shown. Should we? If the circumstances warrant, it may make good pedagogical sense: geometry and kinematics as motivation, and variables and equations as machinery, are a potent combination. Of course if students are familiar with functions, there are considerable conceptual and technical gains in bringing them to the foreground, as Euler began to do. For one, functions, unlike equations, make clear the important distinction between independent and dependent variables, hence also that between domain and range. And, of course, functions are indispensable in more advanced courses.

4.4.3 The Algebraization of Calculus

Euler promoted his views on functionality in an influential textbook in 1748, *Introductio in Analysin Infinitorum* [24]. The entire approach is algebraic; there is not a single diagram in the first volume of the two-volume treatise. Power series play a fundamental role – they provide an "algebraic" apparatus for the subsequent treatment of calculus. This algebraization of calculus lasted for close to a century, until the work of Cauchy in the 1820s.

The following examples of Euler's work will give us a good sense of his artistry – nay, wizardry – in the use of infinitesimal methods.

1. Behold his uncanny derivation of the power-series expansion of $\sin x$:

   Use the binomial theorem to expand the left-hand side of the identity $(\cos z + \mathrm{i}\ \sin z)^n = \cos(nz) + \mathrm{i}\ \sin(nz)$. Equate the imaginary part to $\sin(nz)$ to obtain

$$\begin{aligned}\sin(nz) &= n(\cos z)^{n-1}(\sin z) - [n(n-1)(n-2)/3!](\cos z)^{n-3}(\sin z)^3\\ &\quad + [n(n-1)(n-2)(n-3)(n-4)/5!](\cos z)^{n-5}(sin\,z)^5 - \ldots \end{aligned} \tag{4.1}$$

   Now let n be an infinitely large integer and z an infinitely small number (Euler sees no need to explain what these are). Then $\cos z = 1,\ \sin z = z, n(n-1)(n-2) = n^3, n(n-1)(n-2)n(n-3)(n-4) = n^5 \ldots$ (again no explanation from Euler, although of course we can surmise what he had in mind).

   Equation (4.1) now becomes

$$\sin(nz) = nz - (n^3z^3)/3! + (n^5z^5)/5! - \ldots.$$

   Let now $nz = x$. Euler claims that x is finite since n is infinitely large and z infinitely small. This finally yields the power-series expansion of the sine function: $\sin x = x - x^3/3! + x^5/5! - \ldots$. It takes one's breath away!
2. Euler derived power-series expansions of the exponential and logarithmic functions similarly. Here is how he used the latter expansion to find the differential (derivative) of $\log x$:

   $\mathrm{d}(\log x) = \log(x + \mathrm{d}x) - \log x = \log(1 + \mathrm{d}x/x) = \mathrm{d}x/x - (\mathrm{d}x)^2/2x^2 + (\mathrm{d}x)^3/3x^3 - \ldots = \mathrm{d}x/x$, since, Euler argued, $(\mathrm{d}x)^2, (\mathrm{d}x)^3, \ldots$ are incomparably small in comparison with $\mathrm{d}x$, hence can be deleted.
3. We now present Euler's brilliant discovery (derivation) of the famous formula $1 + 1/2^2 + 1/3^2 + 1/4^2 + \ldots = \pi^2/6$ – a result which eluded the likes of Leibniz and Jakob Bernoulli:

The roots of $\sin x$ are $0, \pm\pi, \pm 2\pi, \pm 3\pi, \ldots$. These are also the roots of the "infinite polynomial" $x - x^3/3! + x^5/5! - \ldots$, which is the power-series expansion of $\sin x$.

Dividing by x, hence eliminating the root $x = 0$, implies that the roots of

$$1 - x^2/3! + x^4/5! - \ldots \quad \text{are } \pm\pi, \pm 2\pi, \pm 3\pi, \ldots.$$

Now, the infinite polynomial obtained by expansion of the infinite product $[1 - x^2/\pi^2][1 - x^2/(2\pi)^2][1 - x^2/(3\pi)^2]\ldots$ has precisely the same roots and the same constant term as $1 - x^2/3! + x^4/5! - \ldots$, hence the two infinite polynomials are identical (cf. the case of "ordinary" polynomials):

$$1 - x^2/3! + x^4/5! - \ldots = [1 - x^2/\pi^2][1 - x^2/(2\pi)^2][1 - x^2/(3\pi)^2]\ldots$$

Comparing coefficients of x^2 on both sides yields $-1/3! = -[1/\pi^2 + 1/(2\pi)^2 + 1/(3\pi)^2 + \ldots]$. (To see how the coefficient of x^2 on the right is obtained, imagine that you had finitely many terms.) Rearranging terms, we finally get

$$1 + 1/2^2 + 1/3^2 + \ldots = \pi^2/6.$$

This formal, algebraic style of analysis, used so brilliantly by Euler and practiced by most eighteenth-century mathematicians, is breathtaking. It accepted as articles of faith that what is true for convergent series is true for divergent series, what is true for finite quantities is true for infinitely large and infinitely small quantities, and what is true for polynomials is true for power series.

What made mathematicians put their trust in the power of symbols, and in such a broad "principle of continuity" – the belief that what held in a given context will continue to hold in what appear to be similar contexts? (see Chap. 9) First and foremost, the use of such formal methods led to important results. A strong intuition by the leading mathematicians of the time kept errors to a minimum. Moreover, the methods were often applied to problems, the reasonableness of whose solutions "guaranteed" the correctness of the results and, by implication, the correctness of the methods. There was also a belief, shared by Newton, that mathematicians were simply uncovering God's grand mathematical design of nature. (This belief, by the way, had at least to some extent been abandoned by the end of the eighteenth century: When Laplace gave Napoleon a copy of his *Mécanique Céleste*, Napoleon is said to have remarked: "M. Laplace, they tell me you have written this large book on the system of the universe and have never even mentioned its Creator," whereupon Laplace replied: "Sire, I have no need of this hypothesis" [45, p. 621].

4.4.4 *Didactic Observation: Discovery and Proof*

It was not uncommon for mathematicians of the seventeenth and eighteenth centuries to resort to mathematical techniques which were at best questionable, often inconsistent. They usually also recognized that their methods were unsatisfactory, but were willing to tolerate them because they yielded correct results. Justification of otherwise inexplicable notions on the grounds that they yield useful results has occurred frequently in the evolution of mathematics Of course, out of confusion emerged in time clarity and understanding (see Chaps. 7–10).

Textbooks usually present the end product of mathematical activity, but of course before one can *prove* one has to *discover*. And the method of discovery of a given result may differ radically from its method of demonstration. The examples we have presented from (for example) the works of Fermat, Leibniz, and Euler give us a glimpse of mathematical discovery by great masters. Is there a moral for pedagogy in all this? See the remarks in Sect. 4.6.4.

4.5 Foundational Issues in the Seventeenth and Eighteenth Centuries

4.5.1 Introduction

The issue of rigorous foundations for calculus began with gropings in the early seventeenth century and concluded with a "final" resolution in the 1870s. This rather slow evolution toward a logical grounding is not atypical in the history of mathematics. Rigor, formalism, and the logical development of a concept, result, or theory usually come at the *end* of a process of mathematical evolution. In the case of calculus, mathematicians achieved very impressive results during the seventeenth and eighteenth centuries by intuitive, heuristic reasoning, and therefore had no compelling reasons to put their subject on firm foundations. This does not mean that there was no concern during these two centuries for the logic behind the algorithms of calculus; and there were attempts, albeit unsuccessful, to supply it.

Mathematicians of the seventeenth and eighteenth centuries realized that the subject they were creating was not on firm ground. They were well aware, for example, that infinitesimals do not obey the Archimedean axiom and hence must be viewed with suspicion – the axiom being basic to the Greek theory of proportion which, in turn, was fundamental to seventeenth-century algebra and geometry. (The Archimedean axiom says that given two positive real numbers a and b, there exists a positive integer n such that $na > b$. But if a is an infinitesimal and $b = 1$, then $na < 1$ for every positive integer n.) Newton especially was concerned about this point.

When discussing issues of rigor in their work on calculus, mathematicians would often claim that it could all be set right, if they wanted to bother, by the rigorous Greek method of exhaustion; but the method was complex, hence impractical. Cavalieri (recall) left rigor to the philosophers, but he once (in the manner of a philosopher) likened line indivisibles of a plane surface to parallel threads of a woven fabric, and surface indivisibles of a solid to parallel pages of a book. This of course did not enhance the respectability of his methods. Fermat believed that he had a simple *algebraic* process, with a clear geometric interpretation, for finding tangents. The typical contemporary attitude toward the foundations of calculus was well expressed by Huygens at mid-seventeenth century [23, pp. 98–99]:

> In order to achieve the confidence of the experts it is not of great interest whether we give an absolute demonstration or such a foundation of it that after having seen it they do not doubt that a perfect demonstration can be given. I am willing to concede that it should appear in a clear, elegant, and ingenious form, as in all works of Archimedes. But the first and most important thing is the mode of discovery itself, which men of learning delight in knowing. Hence it seems that we must above all follow that method by which this can be understood and presented most concisely and clearly.

4.5.2 Newton and Leibniz

Both Newton and Leibniz attempted to give rigorous explanations of their methods, although they recognized that the strength of their work in calculus was *not* its ability to give logical backing to their algorithmic procedures. Newton affirmed of his fluxions that they were "rather briefly explained than narrowly demonstrated" [23, p. 201]. Leibniz said of his differentials that "it will be sufficient simply to make use of them as a tool that has advantages for the purpose of calculation, just as the algebraists retain imaginary roots with great profit" [23, p. 265]. (During this period complex numbers had no greater logical legitimacy than infinitesimals.)

Aside from the very definition of "fluxion," the major foundational weakness of Newton's calculus was the procedure for the computation of fluxions using the "indefinitely small quantity o." What is the status of this o (it was asked)? Is it zero? If so, how can one divide by it? If it is not zero, what right does one have to eventually disregard it, treating it as if it *were* zero? In the *Principia*, Newton tried to resolve this difficulty by means of his theory of "prime and ultimate ratios" – a device for dealing with limits of ratios of geometric quantities couched in the language of synthetic geometry. Lemma I of the *Principia* announces the important new concept of "ultimate equality," which is Newton's attempt to define a limit [7, p. 197]:

> Quantities and the ratios of quantities, which in any finite time converge continually to equality, and before the end of that time approach nearer to one another by any given difference, become ultimately equal.

Among the applications Newton gives of this notion in the following: Given a chord of the arc AB of a curve and a corresponding segment AD of the tangent to the curve at A (Fig. 4.7), Newton asserts that if the points A and B approach one another and meet, "the ultimate ratio of the arc, chord, and tangent, any one to any other, is the ratio of equality" [7, p. 197]. He then goes on to discuss "ultimate ratios of evanescent quantities" – in our terminology the limit of the ratio of quantities approaching zero, namely the derivative [23, p. 225]:

> By the ultimate ratio of evanescent quantities is to be understood the ratio of the quantities not before they vanish, nor afterwards, but with which they vanish.... Those ultimate ratios

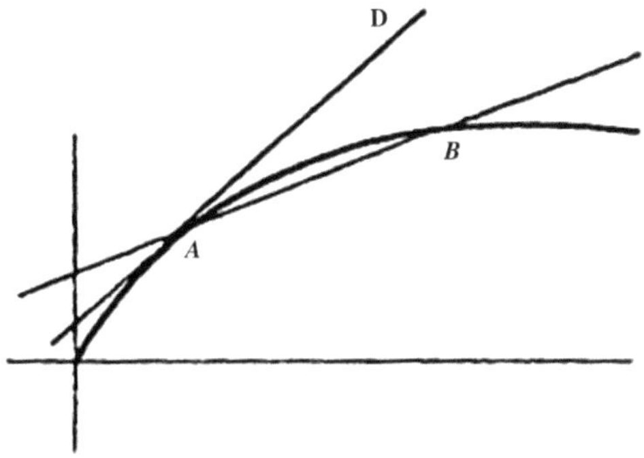

Fig. 4.7 Newton' ultimate ratio of curve, arc, and tangent, any one to any other

> with which quantities vanish are not truly the ratios of ultimate quantities, but limits toward which the ratios of quantities decreasing without limit do always converge; and to which they approach nearer than by any given difference, but never go beyond, nor in effect attain to, till the quantities are diminished *in infinitum*.

For example, to find the tangent to the curve $y = x^2 + 3x + 2$ using this (mature) version of his calculus, Newton would proceed as follows:

Given a "small" increment o in the variable x, the corresponding increment in the variable y is $[(x + \mathrm{o})^2 + 3(x + \mathrm{o}) + 2] - [x^2 + 3x + 2]$, hence the ratio of the increments is $\{[(x + \mathrm{o})^2 + 3(x + \mathrm{o}) + 2] - [x^2 + 3x + 2]\} : \mathrm{o}$. When simplified this gives $(2x + 3 + \mathrm{o})\mathrm{o} : \mathrm{o}$, which yields, on canceling the o, $(2x + 3 + \mathrm{o}) : 1$. Letting o vanish one obtains the "ultimate ratio of evanescent quantities" – that is, quantities which (Newton says) are "approaching zero" – to be $(2x + 3) : 1$. In our notation, $\mathrm{d}y/\mathrm{d}x = 2x + 3$.

Evidently Newton comes close here to our concept of limit, although his definitions are rather vague, to say the least. For example, what does "ultimately equal" mean? Does "never go beyond" suggest that a variable cannot oscillate about its limit? What does "nor in effect attain to, till the quantities are diminished *in infinitum*" imply about whether, and (if so) "when," the limit is reached? Newton did not provide answers to these questions, nor did he develop these ideas sufficiently to justify his algorithmic procedures.

Leibniz had several distinct approaches to resolving the problem of differentials. At times he admitted their actual existence, viewing them as infinitely small nonzero quantities, smaller than any real quantity. (He once remarked that $\mathrm{d}x$ may be supposed to stand to x in the proportion of a grain of sand to the earth.) At other times he viewed differentials as "fictions useful to abbreviate and to speak universally" [23, p. 264], where the abbreviations could be fleshed out using Eudoxus' method of exhaustion. At still other times he assumed that differentials are finite quantities which may cause errors, but that these errors can be made as small as one pleases. He says [7, p. 215]:

> If one preferred to reject infinitely small quantities, it was possible instead to assume them to be as small as one judges necessary in order that ... the error produced should be of no consequence, or less than any given magnitude.

Leibniz did not pursue any of these approaches in detail. In any case, he viewed the question of the existence of differentials as entirely distinct from the question of their utility in solving important problems, and of the latter he had no doubt. See [51].

4.5.3 *Berkeley and d'Alembert*

The uncertainties about the logical foundations of calculus persisted throughout the eighteenth century but did not set back the subject's rapid development. At the same time, most contemporary mathematicians did attempt to deal with foundational issues. These became somewhat more pressing after the forceful and incisive criticism of the calculus of Newton, and to some extent also that of Leibniz, by Bishop George Berkeley in a 1734 essay entitled *The Analyst*, subtitled *A Discourse Addressed to an Infidel Mathematician*. Berkeley resented, even feared, the support which Newtonian science gave to materialism, and proceeded to try to discredit calculus as the chief component of that science, thus hoping to rebut the negative views expressed by scientists on matters of religion. "He who can digest a second or third fluxion," Berkeley asserted, "or a second or third difference, need not, methinks, be squeamish about any point in divinity" [23, p. 293]. Berkeley's main – and correct – criticism centered on the use made of infinitesimals in calculus [23, p. 294] and [45, p. 428]:

> And what are these same evanescent Increments? They are neither finite Quantities, nor Quantities infinitely small, nor yet nothing. May we not call them the Ghosts of departed Quantities?... By virtue of a two-fold mistake you arrive, though not at a science, yet at the truth.

A noteworthy response to Berkeley's criticism was given by d'Alembert in 1754 in an article entitled "Différentiel" in the famous French *Encyclopédie*. D'Alembert replaced Newton's conception of the derivative as an ultimate ratio by an explicit definition of the derivative as the limit of a quotient of increments: "The differentiation of equations consists merely in finding the limit of the ratio of the finite differences of the two quantities contained in the equation" [23, p. 295]. To d'Alembert,

> One quantity is the limit of another if the second can approach the first nearer than by a given quantity, so that the difference between them is absolutely inassignable [7, p. 247].

Note that d'Alembert speaks of the limit of a *quantity*, not of a function, and that he does not permit the quantity to oscillate about its limit. D'Alembert, moreover, did not work out the consequences of his ideas on limits, although he observed prophetically that "the theory of limits is the true metaphysics of the calculus" [45, p. 433]. His contemporaries paid little attention. D'Alembert, too, realized that his venture did not suffice to put calculus on firm foundations, advising his students to "persist and faith will come to you" [45, p. 433]. His contribution was useful, however, in bringing limits to the attention of other mathematicians on the (European) continent, where infinitesimals and differentials reigned supreme.

4.5.4 Euler

Euler was well aware of the inconsistencies in calculus of the infinitely small, and in his classic *Institutiones Calculi Differentialis* of 1755 devoted large parts of the Preface and of Chap. 2 to a discussion of these problems. He claimed that infinitely small quantities are all equal to zero, but that two quantities, both equal to zero, can have a well-determined finite ratio. In conformity with his formalistic view of mathematics, Euler stipulated that there are different orders of zero, and that the subject matter of the (differential) calculus is to determine the (finite) values of the ratios 0/0. He put it thus [64, p. 125]:

> Therefore there exist infinite orders of infinitely small quantities, which, though they all $= 0$, still have to be well distinguished among themselves, if we look at their mutual relation, which is explained by a geometric ratio.

Euler's approach was essentially a heuristic procedure for finding the ratio 0/0 rather than a serious attempt at dealing with foundations. Nor was he greatly concerned with the latter.

4.5.5 Lagrange

The Berlin Academy offered a prize in 1784, hoping that "it can be explained how so many true theorems have been deduced from a contradictory supposition [that is, the existence of infinitesimals]" [32, p. 41]. The most elaborate response to this challenge came from Lagrange, who formulated his ideas on the subject in two books: *Théorie des fonctions analytique* (1797) and *Leçons sur le calcul des fonctions* (1801).

Lagrange attempted to give a rigorous foundation to calculus by reducing it to algebra, eliminating from it all references to infinitesimals or limits. To him *this idea* represented the true principles of calculus. His books, he said, were to contain

> the principal theorems of the differential calculus without the use of the infinitely small or vanishing quantities or limits and fluxions, and reduced to the art of algebraic analysis of finite quantities [45, p. 430].

The lack of rigor in the use of infinitesimals was well recognized (cf. Berkeley's critique, with which Lagrange was familiar). As for limits, Lagrange (and others) had difficulty understanding what happened to the ratio $\Delta y/\Delta x$ "as it reaches its limit." This concern was clearly expressed by Lazare Carnot, who in 1797 wrote an essay on the foundations of calculus entitled "Reflections on the metaphysics of the infinitesimal calculus" [64, p. 134]:

> That method [of limits] has the great inconvenience of considering quantities in the state in which they cease, so to speak, to be quantities; for though we can always well conceive the ratio of two quantities, as long as they remain finite, that ratio offers to the mind no clear and precise idea, as soon as its terms become, the one and the other, nothing at the same time.

Lagrange's starting point was to "prove" that any function $f(x)$ can be represented by a power series in h (except possibly at a finite number of values of x), as follows: $f(x+h) = f(x) + p(x)h + q(x)h^2 + r(x)h^3 + \ldots$. This he intended to show with the aid of a *purely algebraic* process. Taylor (and others) had derived this so-called Taylor series early in the eighteenth century using finite differences and a limit process. Lagrange purported to derive it algebraically without the use of limits (cf. Fermat's work on tangents, Sect. 4.2.3).

Lagrange calls the coefficient $p(x)$ of h in the above expansion of $f(x+h)$ the "first derived function" of $f(x)$ and denotes it by $f'(x)$. He then remarks, and later shows using questionable procedures, that only a little knowledge of the differential calculus is needed to recognize that $f'(x)$ is the derivative (differential quotient) of $f(x)$. Thus he *defines* the derivative of $f(x)$ to be the coefficient of h in the power-series expansion of $f(x+h)$, which he claims to be able to obtain algebraically for "any" function.

Lagrange next shows that in the above series expansion of $f(x+h)$, $q(x)$ can be derived from $p(x)$, $r(x)$ from $q(x)$, etc. by the same process (except for multiplication by a constant) by which $p(x)$ was derived from $f(x)$ – that is, by expanding algebraically $f'(x+h)$ in a power series in h, etc.. Denoting the first derived function of $f'(x)$ by $f''(x)$, the first derived function of $f''(x)$ by $f'''(x)$, and so on, this means that $q(x) = c_2 f''(x), r(x) = c_3 f'''(x), \ldots$. Lagrange shows that $c_n = 1/n!$ and thus claims to have obtained the Taylor series by purely algebraic means.

From our perspective Lagrange's scheme has fundamental drawbacks. One *does* need infinite processes to derive the Taylor expansion of a function, and many functions are not so expandable. Moreover, as Cauchy showed two decades later, even if a function *has* a Taylor-series representation, the series may not represent the function for all values of the variable within the domain of definition of the function. The example Cauchy gave was $f(x) = e^{-1/x^2}$, if x is nonzero, and $f(x) = 0$, if $x = 0$. Here $f(x)$ is represented by its Taylor series *only* at $x = 0$.

But Lagrange's ideas must be viewed in the context of his time. Algebraic analysis, in which power series – viewed as infinite polynomials – played a central role, was predominant in the eighteenth century. Every function encountered in practice was expandable in a power series, and if there were exceptions, this was not considered significant! In the contemporary setting there was, indeed, a coherence to Lagrange's program. (This coherence is seen today in the context of complex analysis, where, for example, the "defect" noted above concerning a function which is not the sum of its Taylor series does not occur.)

For us Lagrange's major contribution to the clarification of the foundations of calculus was his focus on the functional notation for derivatives, as contrasted with the fluxional and differential notations. This implied a clear and explicit recognition, perhaps for the first time, that the derivative of a function is yet another function. Calculus now became a calculus of functions and their derivatives rather than a calculus of fluxions and differentials (cf. Sects. 4.3 and 4.4).

4.6 Calculus Becomes Rigorous: Cauchy, Dedekind, and Weierstrass

4.6.1 Introduction

We now come to the decisive period in the evolution of a rigorous foundation for calculus, embodied in the works of Cauchy, Bolzano, Dedekind, and Weierstrass. Recall that seventeenth-century calculus was largely geometric and that of the eighteenth century was grounded in algebra. The period under discussion, which began in 1821, may be considered as based on arithmetic. Its distinguishing foundational features were:

1. The emergence of the notion of limit as the underlying concept of calculus.
2. The recognition of the important role played by inequalities in definitions and proofs.
3. The acknowledgement that the validity of results in calculus must take into account questions of the domain of definition of a function. (In the eighteenth century a theorem of calculus was usually regarded as universally true by virtue of the *formal* correctness of the underlying algebra.), and
4. The realization that for a logical foundation of calculus one must have a clear understanding of the nature of the real number system, and that this understanding should be based on an arithmetic rather than a geometric conception of the continuum of real numbers.

4.6.2 Cauchy

Cauchy's seminal work in the rigorization of calculus was begun in his famed *Cours d'Analyse* of 1821 (see [15]) and continued in two texts of (respectively) 1822 and 1829. He selected a few fundamental concepts, namely limit, continuity, convergence, derivative, and integral, established the limit concept as the one on which to base all the others, and derived by fairly modern and rigorous means the major results of calculus. That this sounds commonplace to us today is in large part a tribute to Cauchy's program – a grand design, brilliantly executed. In fact, most of the concepts just mentioned were either not recognized (as *we* understand them) or not clearly formulated before Cauchy's time.

What impelled Cauchy to make such a fundamental departure from established practice? Several reasons can be advanced:

1. Cauchy was well aware of Lagrange's foundational work in calculus. Although he used some of Lagrange's technical advances, he was strongly opposed to the latter's grand conception of basing the foundations of calculus on its reduction to algebra. In fact, Cauchy's aim was to *eliminate* algebra as a basis for calculus [42, pp. 247–248]:

> As for my methods, I have sought to give them all the rigor which is demanded in [Euclidean] geometry, in such a way as never to run back to reasons drawn from what is usually given in algebra. Reasons of this latter type, however commonly they are accepted, above all in passing from convergent to divergent series and from real to imaginary quantities, can only be considered, it seems to me, as inductions, apt enough sometimes to set forth the truth, but ill according with the exactitude of which the mathematical sciences boast. We must even note that they suggest that algebraic formulas have an unlimited generality, whereas in fact the majority of these formulas are valid only under certain conditions and for certain values of the quantities they contain.

2. Two very important "practical" problems – the vibrating-string problem and the heat-conduction problem (of the eighteenth and early nineteenth centuries, respectively) – raised questions about central issues in calculus (see Chap. 5). In connection with the latter problem, Fourier startled the mathematical community of the early nineteenth century with his work on what came to be known as *Fourier series*. Fourier claimed that *any* function f defined over $(-l,\ l)$ is representable over this interval by a series of sines and cosines:

$$f(x) = a_0/2 + \sum_{n=1}^{\infty} \{a_n \cos[(n\pi x)/l] + b_n \sin[(n\pi x)/l]\},$$

where a_n, b_n are given by

$$a_n = 1/l \int_{-l}^{l} f(t) \cos[(n\pi x)/l] \mathrm{d}t, b_n = 1/l \int_{-l}^{l} f(t) \sin[(n\pi x)/l] \mathrm{d}t.$$

Euler and Lagrange knew that *some* functions have such representations. The "principle of continuity" of eighteenth- and early-nineteenth-century mathematics (see Chap. 9) suggested that the above cannot be true for *all* functions: since sin and cos are continuous and periodic, the same must be true of a sum of such terms (recall that finite and infinite sums were viewed analogously). Fourier's result was, indeed, only partially correct, and it set off vigorous attempts to find conditions under which it held. To this end the concepts of convergence, continuity, and integral had to be clarified, and this Cauchy proceeded to do.

3. Near the end of the eighteenth century a major *social* change occurred within the community of mathematicians. While in the past they were often attached to royal courts, most mathematicians after the French Revolution earned their livelihood by teaching. Cauchy was a teacher at the influential École Polytechnique in Paris, founded in 1795. It was customary at that institution for an instructor

who dealt with material not in standard texts to write up notes for students on the subject of his lectures. The result, in Cauchy's case, was his *Cours d'Analyse* and two subsequent treatises. Since mathematicians presumably think through the fundamental concepts of the subject they are teaching much more carefully when writing for students than when writing for colleagues, this too might have been a contributing factor in Cauchy's careful analysis of the basic concepts underlying calculus.

4. The above reasons aside, it seems "natural," at least from an historical perspective, that an exploratory period be followed by reflection and consolidation. Geometry in ancient Greece is a case in point. As for calculus, after close to 200 years of vigorous growth with little thought given to foundations, the subject was ripe for careful logical scrutiny. Moreover, taking rigor seriously was "in the air" in the early part of the nineteenth century. Bolzano and Abel, in addition to Cauchy, in analysis, Peacock and De Morgan in algebra, and Gauss in all branches of mathematics, were early proponents of the new critical attitude.

We now give a brief sketch of Cauchy's contributions to the resolution of foundational questions, focusing on the concepts of limit, continuity, derivative, integral, and convergence.

4.6.2.1 Limit

Cauchy's definition of the limit concept is as follows [42, p. 247]:

> When the successive values attributed to a variable approach indefinitely a fixed value, eventually differing from it by as little as one wishes, that fixed value is called the *limit* of all the others.

We note that unlike Newton and d'Alembert, Cauchy does not refer to what happens when the variable reaches its limit, nor does he say that it cannot oscillate about its limit. And although he speaks of the limit of a *variable* rather than of a function, he had in mind the limit of the dependent variable $f(x)$ of the function f. His definition of limit is, of course, not the modern ε–δ definition, but he does use ε–δ arguments in proofs of various results involving limits.

4.6.2.2 Continuity

Cauchy (along with Bolzano) was the first to give an essentially modern definition of continuity [23, pp. 310–311]:

> The function $f(x)$ will be, between the two limits assigned to the variable x, a *continuous* function of this variable if, for each value of x between these limits, the numerical value of the difference $f(x+a) - f(x)$ decreases indefinitely with α. In other words, *the function f(x) will remain continuous with respect to x between the given limits if, between these limits, an infinitely small increment of the variable always produces an infinitely small increment of the function itself.*

Note that Cauchy uses infinitesimals in his formulation of continuity (and elsewhere). To him, however, an infinitesimal is a *variable* whose limit is zero rather than a *constant*, as in seventeenth-and eighteenth-century common usage [42, p. 247]:

> When the successive absolute values of a variable decrease indefinitely in such a way as to become less than any given quantity, that variable becomes what is called an *infinitesimal*. Such a variable has zero for its limit.

Thus Cauchy's use of infinitesimals can be viewed as shorthand for a more complicated statement involving limits, and his definition of continuity can be rephrased to say that f is continuous if $|f(x+a) - f(x)|$ tends to zero as a tends to zero.

4.6.2.3 Derivative

The evolution of the concept of derivative reflects the evolution of calculus as a whole – close to 300 years of progressive maturation, but not without fumblings and errors, beginning around 1600 and culminating in the 1870s with calculus essentially in its present form. Judith Grabiner nicely summarizes this process [31, p. 195]:

> The derivative was first *used*; it was then *discovered*; it was then *explored and developed*; and it was finally *defined*.

Indeed, the derivative (as tangent) was used by Fermat and others in the first half of the seventeenth century, was discovered by Newton and Leibniz (as fluxion and differential, respectively) in the latter part of that century, was vigorously explored and developed in the eighteenth century, and was defined in the nineteenth. The definition of the derivative, too, was given in stages – by Lagrange in the 1790s, in algebraic language, by Cauchy in the 1820s, in terms of limits and infinitesimals, and finally by Weierstrass in the 1870s, in terms of epsilons and deltas. Here is Cauchy's definition [23, p. 313]:

> When a function $y = f(x)$ remains continuous between two given limits of the variable x, and when one assigns to such a variable a value enclosed between the two limits at issue, then an infinitely small increment assigned to the variable produces an infinitely small increment in the function itself. Consequently, if one puts $\Delta x = i$, the two terms *of the ratio of differences* $\Delta x/\Delta y = [f(x+i)-f(x)]/i$ will be infinitely small quantities. But though these two terms will approach the limit zero indefinitely and simultaneously, the ratio itself can converge towards another limit, be it positive or be it negative. This limit, when it exists, has a definite value for each particular value of x; but it varies with x.... The form of the new function which serves as the limit of the ratio $[f(x+i) - f(x)]/i$ will depend on the form of the proposed function $y = f(x)$. In order to indicate this dependence, one gives the new function the name *derived function*, and designates it with the aid of an accent by the notation y' or $f'(x)$.

Although rather verbose, the definition is sufficiently precise to meet the high standards of rigor which Cauchy set for himself.

4.6.2.4 Integral

During the eighteenth century the integral was viewed as an area, or as an antiderivative evaluated at upper and lower limits. Although the idea of the integral as some sort of limit of a sum was familiar, it was usually used only in approximating integrals when the antiderivative could not be easily found. In the early nineteenth century, the work of Fourier on the representation of functions as trigonometric series whose coefficients were given as integrals forced a careful analysis of the notion of integral.

Cauchy was the first to provide a clear definition of the integral of a continuous function essentially as we give it today – as a limit of sums. He then proved that the integral of such a function exists. This enabled him to give a proof of the fundamental theorem of calculus without relying on the notion of the integral as an area. (The notion of area was considered self-evident in the eighteenth and preceding centuries.) Cauchy's work on integration saw a fundamental shift of focus from the indefinite integral (as an antiderivative) to the definite integral (as a limit of sums).

4.6.2.5 Convergence

Series of numbers and functions were used freely and frequently in the seventeenth and eighteenth centuries, with little concern for their convergence. The objective was to get results. Euler, for example, was aware that he was using divergent series, but was unperturbed if they produced interesting results. With the work on Fourier series, the results themselves began to be questioned. "Divergent series," claimed Abel in the early nineteenth century, "are the invention of the devil. By using them, one may draw any conclusion he pleases, and that is why these series have produced so many fallacies and so many paradoxes" [45, p. 973]. Cauchy banned divergent series from analysis. In his *Cours d'Analyse* of 1821 he presented the first systematic study of the convergence of infinite series. (In 1816 Gauss had given a careful treatment of the convergence of the hypergeometric series.) Cauchy gave the definition of convergence of an infinite series in terms of the existence of the limit of the sequence of partial sums, and derived some of the standard convergence tests, for example the ratio and root tests.

4.6.3 *Dedekind and Weierstrass*

Cauchy's new proposals for the rigorization of calculus generated their own problems and enticed a new generation of mathematicians to tackle them. The two major foundational difficulties with his approach were:

1. His verbal definitions of limit and continuity and his frequent use of the language of infinitesimals. Cauchy's definitions of limit, continuity, and infinitesimal

suggest continuous motion – an intuitive idea. Moreover, his formulations blur the crucial distinction between, and the placement of, the universal and existential quantifiers that precede x, ε, and δ in a modern definition of limit and continuity. These shortcomings were likely the sources of two major errors: Cauchy failed to distinguish between pointwise and uniform continuity of a function and between pointwise and uniform convergence of an infinite series of functions. In some of his "proofs" only the former is assumed in each case while the latter is needed.

2. His intuitive appeals to geometry in proving the existence of various limits. Since Cauchy's definitions of the fundamental concepts of calculus were given in terms of limits, proofs of the existence of limits of various sequences and functions were of crucial importance. The existence of many of these limits followed from the "completeness property" of the real numbers, which states, in one of its embodiments, that any bounded increasing sequence of real numbers has a limit. Cauchy took this fundamental result for granted on intuitive geometric grounds. He used it in proofs of such basic results as the existence of the integral of a continuous function, the convergence of (so-called) Cauchy sequences, and the Intermediate Value Theorem.

Dedekind and Weierstrass (among others) determined to remedy this unsatisfactory mixture of arithmetic-algebraic formulations and intuitive geometric justifications. Dedekind's expression of the prevailing state of affairs is revealing [18, pp. 1–2]:

> As professor in the Polytechnic School in Zürich I found myself for the first time obliged to lecture upon the elements of the differential calculus and felt more keenly than ever before the lack of a really scientific foundation for arithmetic. In discussing the notion of the approach of a variable magnitude to a fixed limiting value, and especially in proving the theorem that every magnitude which grows continually, but not beyond all limits, must certainly approach a limiting value, I had recourse to geometric evidence. Even now such resort to geometric intuition in a first presentation of the differential calculus I regard as exceedingly useful, from the didactic standpoint, and indeed indispensable if one does not wish to lose too much time. But that this form of introduction to the differential calculus can make no claim to being scientific, no one will deny. For myself this feeling of dissatisfaction was so overpowering that I made the fixed resolve to keep meditating on the question till I should find a purely arithmetic and perfectly rigorous foundation for the principles of infinitesimal analysis. The statement is so frequently made that the differential calculus deals with continuous magnitude, and yet an explanation of this continuity is nowhere given. Even the most rigorous expositions of the differential calculus do not base their proofs upon continuity but, with more or less consciousness of the fact, they either appeal to geometric notions or those suggested by geometry, or depend upon theorems which are never established in a purely arithmetic manner. Among these, for example, belongs the above-mentioned theorem, and a more careful investigation convinced me that this theorem, or any one equivalent to it, can be regarded in some way as a sufficient basis for infinitesimal analysis. It then only remained to discover its true origin in the elements of arithmetic and thus at the same time to secure a real definition of the essence of continuity.

Establishing theorems in a "purely arithmetic" manner implied what came to be known as the "arithmetization of analysis" (the *term* is due to Felix Klein). Since the inception of calculus, and even in Cauchy's time, the real numbers were viewed geometrically, without explicit formulation of their properties. Since the real numbers are in the foreground or background of much of analysis, proofs of many theorems were of necessity geometric and intuitive. Dedekind's and Weierstrass' astute insight recognized that a rigorous, arithmetic definition of the real numbers would resolve the major obstacle in supplying a rigorous foundation for calculus. Indeed, such a definition was given by Dedekind and Weierstrass, as well as by Cantor and others, around the early 1870s.

The other remaining foundational task was to give a precise "algebraic" definition of the limit concept to replace Cauchy's intuitive, "kinematic" conception. This was accomplished by Weierstrass when he gave his "static" definition of limit in terms of inequalities involving ε's and δ's – the definition we use today, at least in our formal, rigorous incarnations. It is ironic that inequalities, used in the eighteenth century for estimation, and ε, used by some to indicate error, became in the hands of Weierstrass the tools of supreme precision.

With his ε–δ formulation Weierstrass did away with infinitesimals, used by Cauchy and his predecessors for over two centuries (two millennia, if we consider the Greek contributions). During the next several decades the continuum of real numbers was shown to be logically reducible to the discrete collection of positive integers. The arithmetization of analysis was now complete. To Plato God ever geometrized, while to Jacobi He ever arithmetized. The logical supremacy of arithmetic, however, was not lasting. In the 1880s Dedekind and Frege undertook a reconstruction of arithmetic based on ideas from set theory and logic. But that is another story.

4.6.4 Didactic Observation

The teaching of calculus – what should be taught and how it should be taught – is an issue under continuing debate. We make general remarks – desiderata – guided by the historical account.

As we pointed out before, calculus involves algorithms, theory, and applications. At some point, then, students should be exposed to its technical power, its logical harmony, and its usefulness. Calculus is also the answer to a 2000-year-old quest for describing continuity and variability; it is an *intellectual* accomplishment of the first rank. The spirit of these thoughts should animate our teaching of the subject. Central ideas should stand out among the hundreds of formulas and techniques.

Hilbert noted that every mathematical theory goes through three periods of development: the naïve, the formal, and the critical. In the case of calculus, the naïve period occurred in the seventeenth century, the formal in the eighteenth, and the critical in the nineteenth. The evolution of a mathematical idea often proceeds in four stages: discovery (or invention), use, understanding, and justification (cf. the discussion of the derivative in Sect. 4.6.2). It is important to keep the *order* of these stages in mind in discussing *any* concept or theory.

The question of rigor in the teaching of calculus is of ongoing concern. G. F. Simmons provides good advice [60, p. ix]:

> Mathematical rigor is like clothing: in its style it ought to suit the occasion, and it diminishes comfort and restricts freedom of movement if it is either too loose or too tight.

The notion of rigor is not absolute. Mathematicians' views of what constitutes an acceptable proof have evolved, and so should students' (see Chaps. 7–10). To begin a calculus course with a definition of limit may be logically constructive but pedagogically destructive. In general, rigor for rigor's sake will defeat the students. They must be *convinced* of the usefulness, the importance, of having rigorous definitions and proofs of what appear to be intuitive concepts and results. For example, there is little point giving a formal definition of continuity if one offers only the "naïve" examples of traceability with pen on paper; nor of giving a rigorous proof of the Mean Value Theorem if the only available notion of continuity is the naïve. To demonstrate the *need* for higher standards of rigor one should give *counterexamples* to plausible and widely-held notions, for example, that a continuous function is differentiable with the possible exception of finitely many points. In their absence, it is legitimate, and it may be desirable, to give heuristic, tentative – but not sloppy! – definitions "to suit the occasion," and to revise them when (*if*) the need arises. That is unmistakably the lesson of history.

4.7 The Twentieth Century: The Nonstandard Analysis of Robinson

4.7.1 Introduction

About a century after Weierstrass had banished infinitesimals "for good" – so we all thought until 1960 – they were brought back to life as genuine and rigorously defined mathematical objects in the "nonstandard analysis" conceived by the mathematical logician Abraham Robinson. (Another example of the resurrection of a previously banished concept is divergent series, outlawed, as we recall, by Cauchy and Abel in the early nineteenth century but rigorously reintroduced as asymptotic series by Poincaré and Stieltjes in the latter part of that century.)

While "standard analysis" – the calculus we inherited from Cauchy, Weierstrass, and others – is based on the complete and ordered, hence Archimedean, field $\mathbf{R}$ of real numbers, nonstandard analysis is grounded in the ordered, *but not complete*, field $\mathbf{R}^*$ of "hyperreal" numbers. $\mathbf{R}^*$ is an extension field of $\mathbf{R}$ in which one can rigorously define infinitesimals: $\varepsilon \in \mathbf{R}^*$ is *infinitesimal* if $-a < \varepsilon < a$ for all positive $a \in \mathbf{R}$. Thus the only *real* infinitesimal is zero. The inverse of a nonzero infinitesimal is an *infinite* (hyperreal) number.

Nonstandard analysis was "in part, inspired," says Robinson, "by the so-called non-standard models of Arithmetic whose existence was first pointed out by Skolem [in 1934]" [55, p. vii]. Both Skolem's and Robinson's works were part of the newly emerging subfield of mathematical logic called *model theory*. Here is how Robinson puts it [55, p. vii]:

> In the fall of 1960 it occurred to me that the concepts and methods of contemporary Mathematical Logic are capable of providing a suitable framework for the development of the Differential and Integral Calculus by means of infinitely small and infinitely large numbers.

It is ironic that infinitesimals were excluded from calculus in the nineteenth century because they proved to be logically unsatisfactory, and they were rendered mathematically respectable in the twentieth century thanks to logic. Robinson was very gratified that it was mathematical logic that made nonstandard analysis possible. Gödel valued Robinson's work because it made logic and mathematics come together in such a fundamental way, and the contemporary mathematician Simon Kochen echoed him: "Robinson, via model theory, wedded logic to the mainstream of mathematics" [16, p. 195].

Robinson was also guided in his work in nonstandard analysis by a sense of history. He saw it as being in the tradition of Leibniz, Euler, and Cauchy. In fact, he argued that "Leibniz' ideas can be fully vindicated" by his own rigorous theory of infinitesimals [55, p. 2]. More on this later.

Leibniz, as we have seen, tried to justify his work with infinitesimals on essentially two grounds: the pragmatic (that it yielded correct results) and the logical (that it could be made rigorous by the method of exhaustion). He also attempted to rationalize his handling of infinitesimals with a rather vague *principle of continuity*: that (in our language) properties of the reals also hold for the hyperreals (but clearly not all properties – for example, not the Archimedean property; see Chap. 9). Robinson observed that

> What was lacking at the time [of Leibniz] was a formal language which would make it possible to give a precise expression of, and delimitation to, the laws which were supposed to apply equally to the finite numbers and to the extended system including infinitely small and infinitely large numbers [55, p. 266].

4.7.2 Hyperreal Numbers

It is the working out of this program for which Robinson is to be credited. More specifically, one needed

(a) To define (construct) a non-Archimedean field of hyperreal numbers containing the reals which would provide for a rigorous definition of infinitesimals.
(b) To formulate a "transfer principle" which would give formal expression to Leibniz' principle of continuity and thus render precise those properties which are transferable from the reals to the hyperreals.

Keisler, who through his textbooks was instrumental in bringing Robinson's infinitesimals into the calculus classroom, observed that

> The reason Robinson's work was not done sooner is that the Transfer Principle for the hyperreal numbers is a type of axiom that was not familiar in mathematics until recently [40, p. 904].

There are two approaches to the above program – the axiomatic and the constructive. In the former one *postulates*

1. The existence of an ordered, proper extension field $\mathbf{R}^*$ of $\mathbf{R}$; *this implies the existence of infinitesimals in* $\mathbf{R}^*$.
2. A *transfer principle* which enables one to carry over from $\mathbf{R}$ to $\mathbf{R}^*$ all "elementary" statements (roughly, statements which quantify over elements and not over subsets of $\mathbf{R}$). More precisely, all finitary operations and relations on $\mathbf{R}$ must be extendable canonically to $\mathbf{R}^*$ and the truth of an elementary statement in $\mathbf{R}$ must imply its truth in $\mathbf{R}^*$.

For example, the function $f(x) = \sqrt{1 - x^2}$, where $x \in \mathbf{R}$ and $-1 \leq x \leq 1$, is extendable to a function $f^*(a) = \sqrt{1 - \mathrm{a}^2}$ in $\mathbf{R}^*$ (that is, $a \in \mathbf{R}^*$), and since $1-x^2 \geq 0$ if and only if $-1 \leq x \leq 1$, hence $1-a^2 \geq 0$ if and only if $-1 \leq a \leq 1$.

The statement "for all $a, b \in \mathbf{R}$ with $b > 0$, there exists a positive integer n such that $nb > a$" is not elementary, hence not extendable to $\mathbf{R}^*$. Robinson specified a formal language – in the sense of modern logic – such that those and only those properties which are expressible in that language are transferable between $\mathbf{R}$ and $\mathbf{R}^*$.

Just as one can develop standard analysis from an axiomatic description of $\mathbf{R}$ as a complete ordered field, so one can develop nonstandard analysis from the axiomatic description of $\mathbf{R}^*$ given above. In fact, one can derive all results in *standard* calculus by nonstandard means using infinitesimals, which is, of course, what Leibniz, Euler, and others had done. The basic idea is as follows:

If you wish to prove a theorem over $\mathbf{R}$ (an ordinary calculus theorem), translate it into a statement over $\mathbf{R}^*$ using the transfer principle, prove it by nonstandard methods – which is usually easier to do, since one can employ infinitesimal arguments – and restrict it back to $\mathbf{R}$. To paraphrase a statement of Hadamard, the shortest path between two truths in the real domain passes through the hyperreal domain.

The last step is accomplished via the so-called *standard part theorem*: for every finite hyperreal number a, there exists exactly one real number "infinitely close" to it, denoted by st(a). Two hyperreal numbers are *infinitely close* if their difference is an infinitesimal. For example, it can be shown that for any (standard) function $f(x)$, the usual definition of the derivative is equivalent to $f'(x) = \mathrm{st}\{[f(x+\varepsilon) - f(x)]/\varepsilon\}$, where ε is an infinitesimal.

How does this compare with Leibniz's definition of slope? Leibniz defines the slope of the curve $y = f(x)$ at (x, y) as $\mathrm{d}y/\mathrm{d}x$, where $\mathrm{d}x$ is the differential of x and $\mathrm{d}y = f(x + \mathrm{d}x) - f(x)$ the corresponding differential of y. Thus, according to Leibniz, $f'(x)[f(x + \mathrm{d}x) - f(x)]/\mathrm{d}x$. The modern nonstandard formulation is, as noted, $f'(x) = \mathrm{st}[f(x + \mathrm{d}x) - f(x)]/\mathrm{d}x$. For example, in the final step of the computation of the derivative of $f(x) = x^2$, Leibniz would *identify* $2x + \mathrm{d}x$ with $2x$, while Robinson would write $\mathrm{st}(2x + \mathrm{d}x) = 2x$. It is this (implicit) identification in Leibniz' calculus of hyperreal numbers with their standard parts, and in particular of infinitesimals with zero, that was the cause of its logical difficulties. Robinson's standard-part apparatus replaces the need for limits.

How did Robinson know that the axioms for the hyperreal numbers are consistent? By constructing a "model" for them and deducing the postulates as theorems. The construction is analogous to that of the real numbers as equivalence classes of Cauchy sequences of rationals. Robinson defined the hyperreal numbers as equivalence classes of arbitrary sequences – *all* sequences – of reals, where the equivalence relation is given in terms of "ultraproducts." His construction involved rather sophisticated notions of mathematical logic. Subsequently various simplifications were offered. See [17, 41].

4.7.3 Wider Implications

What has nonstandard analysis accomplished and how has it been received by the mathematical community? Robinson introduced its methods into topology, differential geometry, measure theory, complex analysis, and Lie group theory. They have also been applied in functional analysis, differential equations, probability, areas of mathematical physics, and economics. These inroads of the subject, in such a short time-span, are indeed impressive. Detractors have argued that there is nothing essentially new in all this since, by the transfer and standard-part principles, every result provable by the methods of nonstandard analysis has, at least in theory, a standard proof. One can also claim that every geometric result provable by synthetic geometry can be proved analytically, but does that make synthetic geometry superfluous? A key point is that new results have been *discovered* or *first* proved via nonstandard analysis. New ways of looking at a given idea should be encouraged.

4.7.4 Robinson and Leibniz

Although Robinson's infinitesimals are in the spirit of Leibniz', Robinson (as we noted) viewed nonstandard analysis as a *vindication* of Leibniz' (and Euler's) calculus. In fact, he maintained that it called for a rewriting of the history of calculus.

Robinson's reconstruction would argue that seventeenth- and eighteenth-century analysis was not merely based on a bold and imaginative notion of infinitesimal – for how could that yield such powerful results? – but that contemporary analysts worked with a sense of confidence that infinitesimals could be rigorously justified (by such a theory, say, as Robinson developed).

Such a rehabilitation of the infinitesimals of the seventeenth and eighteenth centuries has been forcefully rejected by many historians of mathematics. Indeed, which of Robinson's sophisticated notions of nonstandard analysis are present in Leibniz' infinitesimals? For that matter, which Weierstrassian ε–δ ideas concerning limits appear in Newton's ultimate ratios? How does Eudoxus' definition of the multiplication of ratios serve, as some have argued, as a precursor of group theory?

Historical reconstructions should be treated with extreme care. "Hindsight sees much to which foresight was blind," observed the mathematician and historian of mathematics E.T. Bell [3, p. 136].

Setting aside the idea of "reconstruction," what light does nonstandard analysis shed on Leibniz' (and others') reflections about the *existence* of infinitesimals? Of course, mathematical existence has an entirely different meaning for us than what it had for Leibniz' contemporaries. *We* (or at least most of us) would accept the existence of infinitesimals by virtue of the existence of $\mathbf{R}^*$: infinitesimals are simply elements in $\mathbf{R}^*$. But such a notion of existence of mathematical objects would have been alien to pre-nineteenth-century and, in fact, to most nineteenth-century mathematicians, as it is to some mathematicians today. See [16, 17, 46].

The imaginary $\sqrt{-1}$, an ideal point (a point at infinity), Kummer's ideal numbers – all were viewed at the time they were introduced as "ideal" (unreal?) objects needed to bring about desired goals. For us, these objects have shed their metaphysical connotation. Can infinitesimals not be viewed in the same vein? Is Leibniz' conception of infinitesimals as "fictions useful to abbreviate and to speak universally" inconsistent with the above picture?

4.7.5 Didactic Observation

Should we teach calculus via the methods of nonstandard analysis? It depends to a large extent on one's objectives. In general, desirable features of a mathematical theory are the clarity of its concepts, their intuitive appeal, the beauty and simplicity of the theory, and the manipulative ease of its technical apparatus. On these ground nonstandard analysis has much to recommend it.

At an informal level, infinitesimals would seem to have at least as much appeal as limits. Physicists and engineers have been using infinitesimals for that reason long after they were formally banished from mathematics, and mathematicians still use them informally, with confidence that they have rigorous backing. The notion of something so small that it can be neglected is familiar. Formally, the notion of infinitesimal is grounded in the hyperreals, as the notion of limit is in the reals. And formally, the reals are hardly more "real" than the hyperreals. Of course, the considerable intuitive appeal of the reals comes from their geometric model – the points on a line. But geometric models have also been proposed for the "hyperreal line" [40]. See also [36].

Some have strongly opposed the teaching of nonstandard calculus. The constructivist Errett Bishop calls such attempts "a debasement of meaning," adding that "the real damage lies in [the] obfuscation and devitalization of those wonderful ideas [of standard calculus]" [16, p. 189].

The historical context would seem to provide sufficient grounds for recommending that teachers be at least familiar with the rudimentary ideas and techniques of nonstandard calculus, and that they might convey these to students at some point in the latter's mathematical education. For over two millennia infinitesimal methods have been used with great success by mathematicians such as Archimedes, Leibniz, Newton, Euler, and Cauchy. Robinson's nonstandard analysis is a fitting culmination, if not a vindication, of these ideas. It is also testimony that (according to Lynn Steen) "the epistemological foundation of mathematical analysis is far from settled" [62, p. 92].

References

1. K. Andersen, Cavalieri's method of indivisibles, *Arch. Hist. Exact Sciences* 31 (1985) 291–367.
2. M. Baron, *The Origins of the Infinitesimal Calculus*, Dover, 1987.
3. E. T. Bell, *The Development of Mathematics*, McGraw-Hill, 1945.
4. J. L. Bell, Infinitesimals, Synthese 75 (1988) 285–315.
5. H. J. M. Bos, Differentials, higher-order differentials and the derivative in the Leibnizian calculus, *Arch. Hist. Exact Sciences* 14 (1974) 1–90.
6. U. Bottazzini, *The Higher Calculus: A History of Real and Complex Analysis from Euler to Weierstrass*, Springer, 1986.
7. C. B. Boyer, *The History of the Calculus and its Conceptual Development*, Dover, 1959.
8. C. B. Boyer, Cavalieri, limits and discarded infinitesimals, *Scripta Mathematica* 8 (1941) 79–91.
9. D. Bressoud, *A Radical Approach to Lebesgue's Theory of Integration*, Math. Assoc. of Amer., 2008.
10. D. Bressoud, *A Radical Approach to Real Analysis*, Math. Assoc. of Amer., 1994.
11. F. Cajori, Indivisibles and "ghosts of departed quantities" in the history of mathematics, *Scientia* 37 (1925) 301–306.
12. F. Cajori, Grafting of the theory of limits on the calculus of Leibniz, *Amer. Math. Monthly* 30 (1923) 223–234.
13. F. Cajori, Discussion of fluxions: from Berkeley to Woodhouse, *Amer. Math. Monthly* 24 (1917) 145–154.
14. R. Calinger (ed.), *Vita Mathematics: Historical Research and Integration with Teaching*, Math. Assoc. of Amer., 1996.
15. A. – L. Cauchy, *Cours d'Analyse*, translated into English, with annotations, by R. Bradley and E. Sandifer, Springer, 2009 (orig. 1821).
16. J. W. Dauben, Abraham Robinson and nonstandard analysis: history, philosophy, and foundations of mathematics. In W. Aspray and P. Kitcher (eds.), *History and Philosophy of Modern Mathematics*, Univ. of Minnesota Press, 1988, pp. 177–200.
17. M. Davis and R. Hersh, Nonstandard analysis, *Scient. Amer.* 226 (1972) 78–86.
18. R. Dedekind, *Essays on the Theory of Numbers*, Dover, 1963 (orig. 1872 & 1888).
19. U. Dudley (ed.), *Readings for Calculus*, Math. Assoc. of Amer., 1993.
20. W. Dunham, *The Calculus Gallery: Masterpieces from Newton to Lebesgue*, Princeton Univ. Press, 2005.
21. W. Dunham, *Euler: The Master of Us All*, Math. Assoc. of Amer., 1999.
22. W. Dunham, *Journey Through Genius: The Great Theorems of Mathematics*, Wiley, 1990.
23. C. H. Edwards, *The Historical Development of the Calculus*, Springer, 1979.
24. L. Euler, *Inrtoductio in Analysin Infinitorum*, vols. 1 and 2, English tr. by J. Blanton, Springer, 1989 (orig. 1748).
25. H. Eves, *Great Moments in Mathematics*, 2 vols., Math. Assoc. of Amer., 1983.
26. J. Fauvel (ed.), The use of history in teaching mathematics, Special Issue of *For the Learning of Mathematics*, vol. 11, no. 2, 1991.

27. J. Fauvel (ed.), *History in the Mathematics Classroom: The IREM Papers*, vol. 1, The Mathematical Association (London), 1990.
28. J. Fauvel and J. van Maanen (eds.), *History in Mathematics Education*, Kluwer, 2000.
29. C. Fraser, The calculus as algebraic analysis: some observations on mathematical analysis in the 18th century, *Arch. Hist. Exact Sciences* 39 (1989) 317–335.
29a. A. Gardiner, *Infinite Processes: Background to Analysis*, Springer, 1979.
30. J. W. Grabiner, Who gave you the epsilon? Cauchy and the origins of rigorous calculus, *Amer. Math. Monthly* 90 (1983) 185–194.
31. J. W. Grabiner, The changing concept of change: the derivative from Fermat to Weierstrass, *Math. Mag.* 56 (1983) 195–206.
32. J. W. Grabiner, *The Origins of Cauchy's Rigorous Calculus*, MIT Press, 1981.
33. I. Grattan-Guinness (ed.), *From the Calculus to Set Theory, 1630–1910*, Princeton Univ. Press, 2000.
34. I. Grattan-Guinness, *The Development of the Foundations of Mathematical Analysis from Euler to Riemann*, MIT Press, 1970.
35. E. Hairer and G. Wanner, *Analysis by its History*, Springer, 1996.
36. V. Harnik, Infinitesimals from Leibniz to Robinson: time to bring them back to school, *Math. Intelligencer* 8 (2) (1986) 41–47, 63.
37. K. Kalman, Six ways to sum a series, *College Math. Jour.* 24 (1993) 402–421.
38. V. J. Katz, *A History of Mathematics*, 3rd ed., Addison-Wesley, 2009.
39. V. J. Katz (ed.), *Using History to Teach Mathematics: An International Perspective*, Math. Assoc. of America, 2000.
40. J. Keisler, *Elementary Calculus: An Infinitesimal Approach,* 2nd ed., Prindle, Weber & Schmidt, 1986.
41. J. Keisler, *Foundations of Infinitesimal Calculus*, Prindle, Weber & Schmidt, 1976.
42. P. Kitcher, *The Nature of Mathematical Knowledge*, Oxford Univ. Press, 1983.
43. P. Kitcher, Fluxions, limits, and infinite littlenesse: a study of Newton's presentation of the calculus, *Isis* 64 (1973) 33–49.
44. M. Kline, Euler and infinite series, *Math. Mag.* 56 (1983) 307–314.
45. M. Kline, *Mathematical Thought from Ancient to Modern Times*, Oxford Univ. Press, 1972.
46. I. Lakatos, Cauchy and the continuum: the significance of non-standard analysis for the history and philosophy of mathematics, *Math. Intelligencer* 1 (1978) 151–161.
47. R. E. Langer, Fourier series: the genesis and evolution of a theory, *Amer. Math. Monthly* 54 (S) (1947) 1–86.
48. R. Laubenbacher and D. Pengelley, *Mathematical Expeditions: Chronicles by the Explorers*, Springer, 1999.
49. D. Laugwitz, On the historical development of infinitesimal mathematics, I, II, *Amer. Math. Monthly* 104 (1997) 447–455, 660–669.
50. N. MacKinnon (ed.), Use of the history of mathematics in the teaching of the subject, Special Issue of *Mathematical Gazette*, Vol. 76, no. 475, 1992.
51. P. Mancosu, The metaphysics of the calculus: a foundational debate in the Paris Academy of Sciences, 1700–1706, *Hist. Math.* 16 (1989) 224–248.
52. K. O. May, History in the mathematics curriculum, *Amer. Math. Monthly* 81 (1974) 899–901.
53. NCTM, *Historical Topics for the Mathematics Classroom*, 2nd ed., National Council of Teachers of Mathematics, 1989.
54. D. Pimm, Why the history and philosophy of mathematics should not be rated X, *For the Learning of Mathematics* 3 (1982) 12–15.
55. A. Robinson, *Non-Standard Analysis*, North-Holland, 1966.
56. R. Roy, The discovery of the series formula for π by Leibniz, Gregory, and Nilakantha, *Math. Mag.* 63 (1990) 291–306.
57. A. Shell-Gellasch (ed.), *Hands on History: A Resource for Teaching Mathematics*, Math. Assoc. of Amer., 2007.
58. A. Shell-Gellasch and R. Jardine (eds.), *From Calculus to Computers: Using the Last 2000 Years of Mathematical History in the Classroom*, Math. Assoc. of Amer., 2005.

59. G. F. Simmons, *Calculus Gems: Brief Lives and Memorable Mathematics*, McGraw-Hill, 1992.
60. G. F. Simmons, *Differential Equations*, McGraw-Hill, 1972.
61. S. Stahl, *Real Analysis: A Historical Approach*, Wiley, 1999.
62. L. A. Steen, New models of the real-number line, *Scient. Amer.* 225 (1971) 92–99.
63. J. Stillwell, *Mathematics and its History*, 2nd ed., Springer-Verlag, 2002.
64. D. J. Struik, *A Concise History of Mathematics*, 4th ed., Dover, 1987.
65. F. Swetz et al (eds.), *Learn from the Masters!*, Math. Assoc. of Amer., 1995.
66. O. Toeplitz, *The Calculus: A Genetic Approach*, The Univ. of Chicago Press, 1963.
67. N. Y. Vilenkin, *In Search of Infinity*, transl. from the Russian by A. Shenitzer, Birkhäuser, 1995.
68. D. T. Whiteside, *The Mathematical Papers of Isaac Newton*, 8 vols., Cambridge Univ. Press, 1967–1981.
69. R. L. Wilder, History in the mathematics curriculum: its status, quality and function, *Amer. Math. Monthly* 79 (1972) 479–495.
70. R. M. Young, *Excursions in Calculus*, Math. Assoc. of Amer., 1992.

Chapter 5
A Brief History of the Function Concept

5.1 Introduction

The evolution of the concept of function goes back 4000 years; 3,700 of these consist of anticipations. The idea evolved for close to 300 years in intimate connection with problems in calculus and analysis. In fact, a one-sentence definition of analysis as the study of properties of various classes of functions would not be far off the mark. Moreover, the concept of function is one of the distinguishing features of "modern" as against "classical" mathematics. W. L. Schaaf goes a step further [26, p. 500]:

> The keynote of Western culture is the function concept, a notion not even remotely hinted at by any earlier culture. And the function concept is anything but an extension or elaboration of previous number concepts – it is rather a complete emancipation from such notions.

The evolution of the function concept can be seen initially as a tug of war between two elements, two mental images: the geometric, expressed in the form of a curve, and the algebraic, expressed as a formula – first finite and later allowing infinitely many terms, the so-called "analytic expression" [9, p. 256]. Subsequently, a third element enters, the "logical" definition of function as a correspondence, with a mental image of an input–output machine. In the wake of this development, the geometric conception of function is gradually abandoned. A new tug of war soon ensues – and is, in one form or another, still with us today – between this novel "logical" ("abstract," "synthetic," "postulational") conception of function and the old "algebraic" ("concrete," "analytic," "constructive") conception.

In this chapter, we will elaborate on these points and try to give the reader a sense of the excitement and the challenge that some of the best mathematicians of all time confronted in trying to come to grips with the basic conception of function that we now accept as commonplace.

I. Kleiner, *Excursions in the History of Mathematics*,
DOI 10.1007/978-0-8176-8268-2_5, © Springer Science+Business Media, LLC 2012

5.2 Precalculus Developments

The notion of function in explicit form did not emerge until the beginning of the eighteenth century, although implicit manifestations of the concept date back to about 2000 BC. The main reasons that the function concept did not emerge earlier were:

- Lack of algebraic prerequisites – coming to terms with the continuum of real numbers, and the development of symbolic notation.
- Lack of motivation. Why define an abstract notion of function unless one had many examples from which to abstract?

In the course of about 200 years (c. 1450–1650) there occurred a number of developments that were fundamental to the rise of the function concept:

- Extension of the concept of number to embrace real and to some extent complex numbers (Bombelli, Stifel, et al.).
- The creation of a symbolic algebra (Viète, Descartes, et al.).
- The study of motion as a central problem of science (Kepler, Galileo, et al.).
- The wedding of algebra and geometry (Fermat, Descartes, et al.).

The seventeenth century witnessed the emergence of modem mathematized science and the invention of analytic geometry. Both of these developments suggested a dynamic, continuous view of the functional relationship as against the static, discrete view held by the ancients.

In the blending of algebra and geometry, the key elements were the introduction of *variables* and the expression of the relationship between variables by means of *equations*. The latter provided a large number of examples of curves – potential functions – for study and set the final stage for the introduction of the function concept. What was lacking was the identification of the independent and dependent variables in an equation [2, p. 348]:

> Variables are not functions. The concept of function implies a unidirectional relation between an 'independent' and a 'dependent' variable. But in the case of variables as they occur in mathematical or physical problems, there need not be such a division of roles. And as long as no special independent role is given to one of the variables involved, the variables are not functions but simply variables.

See [7, 17, 29] for details.

The calculus developed by Newton and Leibniz had not the form that students see today. In particular, it was not a calculus of *functions*. The principal objects of study in seventeeenth-century calculus were curves. For example, the cycloid was introduced geometrically and studied extensively well before it was given as an equation. Seventeenth-century analysis originated as a collection of methods for solving problems about curves, such as finding tangents to curves, areas under curves, lengths of curves, and velocities of points moving along curves. Since the problems that gave rise to calculus were geometric and kinematic, and since Newton and Leibniz were preoccupied with exploiting the marvelous tool that they had created, time and reflection would be required before calculus could be recast in algebraic form.

The variables associated with a curve were geometric – abscissas, ordinates, subtangents, subnormals, and the radii of curvature of a curve. In 1692 Leibniz introduced the word "function" to designate a geometric object associated with a curve [27, p. 272]. For example, Leibniz asserted that "a tangent is a function of a curve." [14, p. 85].

Newton's "method of fluxions" applies to "fluents," not functions. Newton calls his variables "fluents" – the image (as in Leibniz) is geometric, of a point "flowing" along a curve. Newton's major contribution to the development of the function concept was his use of power series. These were important for the subsequent development of that concept (see Chap. 4 and Sect. 6.4).

As increased emphasis came to be placed on the formulas and equations relating the functions associated with a curve, attention was focused on the role of the symbols appearing in the formulas and equations, and thus on the relations holding among these symbols, independent of the original curve. The correspondence between Leibniz and Johann Bernoulli (1694–1698) traces how the lack of a general term to represent quantities dependent on other quantities in such formulas and equations brought about the use of the term "function" as it appears in Bernoulli's definition of 1718 [25, p. 72] (see also [3, p. 9], and [29, p. 57]):

> One calls here Function of a variable a quantity composed in any manner whatever of this variable and of constants.

This was the first formal definition of function, although Bernoulli did not explain what "composed in any manner whatever" meant. See [3, 7, 14, 29] for details on this section.

5.3 Euler's *Introductio in Analysin Infinitorum*

In the first half of the eighteenth century, we witness a gradual separation of the seventeenth-century analysis from its geometric origin and background. This process of "degeometrization of analysis" [2, p. 345] saw the replacement of the concept of variable, applied to geometric objects, with the concept of function as an algebraic formula. This trend was embodied in Euler's classic *Introductio in Analysin Infinitorum* of 1748, intended as a survey of the concepts and methods of analysis and analytic geometry needed for a study of the calculus. See [8].

Euler's *Introductio* was the first work in which the concept of function plays an explicit and central role. In the preface, Euler claims that mathematical analysis is the general science of variables and their functions. He begins by defining a function as an "analytic expression" (that is, a "formula") [25, p. 72]:

> A function of a variable quantity is an analytical expression composed in any manner from that variable quantity and numbers or constant quantities.

Euler does not define the term "analytic expression," but he tries to give it meaning by explaining that admissible "analytic expressions" involve the four algebraic operations, roots, exponentials, logarithms, trigonometric functions, derivatives, and integrals. (The term "analytic expression," which will appear often throughout this chapter, was formally defined only in the late nineteenth century (see Sect. 5.8).) Euler classifies functions as being algebraic or transcendental, single-valued or multivalued, and implicit or explicit. The *Introductio* contains one of the earliest treatments of trigonometric functions as numerical ratios ([15]), as well as the earliest algorithmic treatment of logarithms as exponents. The entire approach is algebraic. Not a single picture or drawing appears (in Vol. 1). See [8].

Expansions of functions in power series play a central role in this treatise. In fact, Euler claims that any function can be expanded in a power series: "If anyone doubts this, this doubt will be removed by the expansion of every function" [3, p. 10]. (Youschkevitch claims that "because of power series the concept of function as analytic expression occupied the central place in mathematical analysis" [29, p. 54].) This remark was in keeping with the spirit of mathematics in the eighteenth century. Hawkins summarizes Euler's contribution to the emergence of function as an important concept [12, p. 3]:

> Although the notion of function did not originate with Euler, it was he who first gave it prominence by treating the calculus as a formal theory of functions.

Euler's view of functions was soon to evolve, as we shall see in the next section. See [2,3,7,29] for details of the above.

5.4 The Vibrating-String controversy

Of crucial importance for the subsequent evolution of the concept of function was the Vibrating-String problem:

An elastic string having fixed ends (0 and l, say) is deformed into some initial shape and released to vibrate. The problem is to determine the function that describes the shape of the string at time t (Fig. 5.1).

The controversy centered around the meaning of "function." Grattan-Guinness suggests that in the controversy over various solutions of this problem,

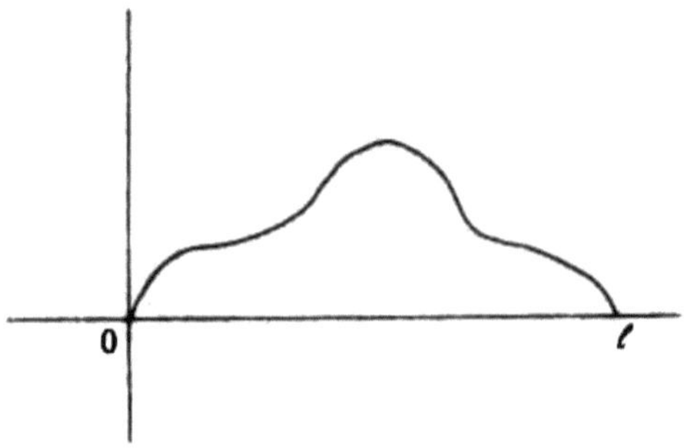

Fig. 5.1 An initial shape of an elastic string released to vibrate

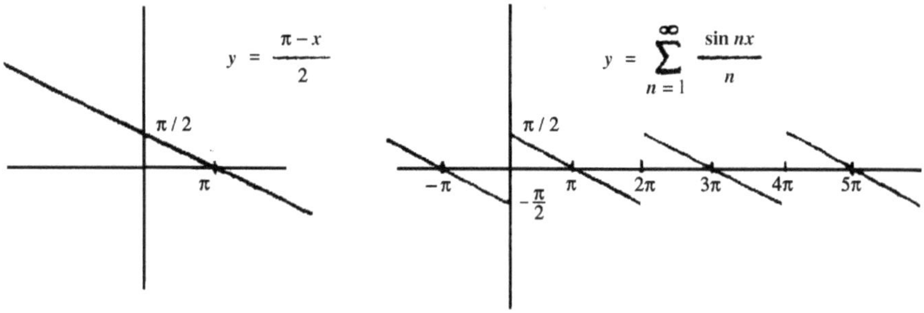

Fig. 5.2 Two analytic expressions that agree on the interval $(0, 2\pi)$ but nowhere else

> the whole of eighteenth-century analysis was brought under inspection: the theory of functions, the role of algebra, the real line continuum and the convergence of series.... [11, p. 2].

To understand the debates that surrounded the Vibrating-String problem, we must first mention an "article of faith" of eighteenth-century mathematics: *If two analytic expressions agree on an interval, they agree everywhere.*

This was not an unnatural assumption, given the type of functions (analytic expressions) considered at the time. On this view, the whole course of a curve given by an analytic expression is determined by any small part of the curve. This implicitly assumes that the independent variable in an analytic expression ranges over the whole domain of real numbers, without restriction.

In view of this, it is baffling – to us – that as early as 1744 Euler wrote to Goldbach stating that $(\pi - x)/2 = \sum_{n=1}^{\infty} (\sin nx)/n$ [29, p. 67]. Here, indeed, is an example of two analytic expressions that agree on the interval $(0, 2\pi)$, but nowhere else. Euler must surely have recognized this (Fig. 5.2), but, according to Youschkevitch,

> This is not the only occasion on which EULER knew examples which did not comply with his conceptions but which he may have considered to be insignificant exceptions from the general rule [29, p. 67]. See also [21].

In 1747, d'Alembert solved the Vibrating-String problem by showing that the motion of the string is governed by the partial differential equation

$$\partial^2 y/\partial t^2 = a^2(\partial^2 y/\partial x^2) \text{ (}a\text{ is a constant)},$$

the so-called *wave equation*. Using the boundary conditions $y(l,t) = 0$, and the initial conditions $y(x,0) = f(x)$ and $\partial y/\partial t|_{t=0} = 0$, he solved the partial differential equation to obtain $y(x,t) = [\varphi(x + at) + \varphi(x - at)]/2$ as the "most general" solution of the Vibrating-String problem, φ being an "arbitrary" function. It follows readily that

$$y(x,0) = f(x) = \varphi(x) \text{ on } (0,l)$$

$$\varphi(x + 2l) = \varphi(x) \quad \text{and}$$

$$\varphi(-x) = \varphi(x)$$

Fig. 5.3 Euler would allow this shape as a possible initial shape of a vibrating string

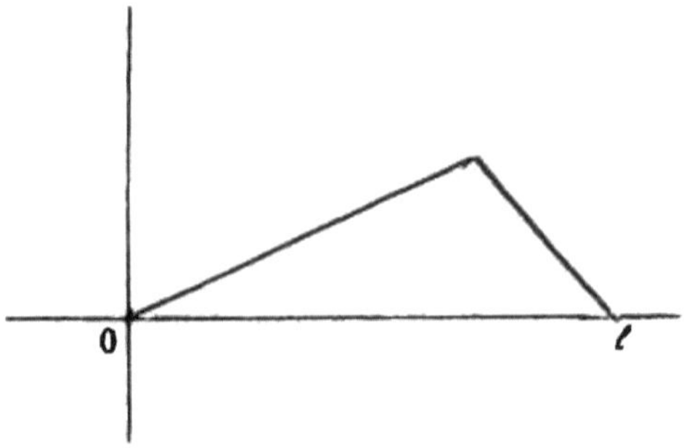

Thus, φ is determined on $(0, l)$ by the initial shape of the string, and is continued, by the "article of faith," as an odd periodic function of period $2l$.

D'Alembert believed that the function $\varphi(x)$ (and hence $f(x)$) must be an "analytic expression" – that is, it must be given by a formula. To d'Alembert, these were the only permissible functions. Moreover, since this analytic expression satisfies the wave equation, it must be twice differentiable.

In 1748 Euler wrote a paper on the same problem, in which he agreed completely with d'Alembert concerning the solution but differed from him on its *interpretation*. Euler contended that d'Alembert's solution was not the "most general," as the latter had claimed. Having himself solved the problem mathematically, Euler claimed his experiments showed that the solution $y(x,t) = [\varphi(x + at) + \varphi(x - at)]/2$ gives the shapes of the string for different values of t even when the initial shape is not given by a (single) formula. From physical considerations, Euler argued that the initial shape of the string can be given by

(a) Several analytic expressions in different subintervals of $(0, l)$, say, circular arcs of different radii in different parts of $(0, l)$ or, more generally.
(b) A curve drawn free-hand.

But according to the "article of faith" prevalent at the time, neither of these two types of initial shapes could be given by a single analytic expression, since such an expression determines the shape of the entire curve by its behavior on any interval, no matter how small. Thus, d'Alembert's solution could not be the most general.

It is interesting to note that Euler called functions of types (a) and (b) "discontinuous," reserving the word "continuous" for functions given by a single analytic expression. Thus, he regarded the two branches of a hyperbola as a (single) continuous function! [19, p. 301]. This conception of "continuity" persisted until 1821, when Cauchy gave the definition used nowadays.

D'Alembert, who was much less interested in the vibrations of the string than in the mathematics of the problem, claimed that Euler's argument was "against all rules of analysis." Euler believed that it is admissible to apply certain of the operations of analysis to arbitrary curves. His, but not d'Alembert's, "rules of analysis" would allow him to admit, for example, the triangular-shaped curve (Fig. 5.3) as the initial shape of a vibrating string. For, Euler would argue that one could change the shape of the curve at the "top" by an infinitely small amount and thus "smooth" it out. Since infinitesimal changes were ignored in analysis, this would have no effect on the solution. Langer explains the differing views of Euler and d'Alembert concerning the Vibrating-String problem in terms of their general approach to mathematics [18, p. 17]:

> Euler's temperament was an imaginative one. He looked for guidance in large measure to practical considerations and physical intuition, and combined with a phenomenal ingenuity an almost naive faith in the infallibility of mathematical formulas and the results of manipulations upon them. D'Alembert was a more critical mind, much less susceptible to conviction by formalisms. A personality of impeccable scientific integrity, he was never inclined to minimize shortcomings that he recognized, be they in his own work or in that of others.

Daniel Bernoulli entered the picture in 1753 by giving yet another solution of the Vibrating-String problem. Bernoulli, who was essentially a physicist, based his argument on the physics of the problem and the known facts about musical vibrations, discovered earlier by Rameau and others. It was generally recognized at the time that musical sounds and, in particular, vibrations of a "musical" string, are composed of fundamental frequencies and their harmonic overtones. This physical evidence, and some "loose" mathematical reasoning, convinced Bernoulli that the solution to the Vibrating-String problem must be given by

$$y(x,t) = \sum_{n=1}^{\infty} b_n \sin(n\pi x/l)\ \cos(n\pi a t/l).$$

This, of course, meant that an arbitrary function $f(x)$ can be represented on $(0,l)$ by a series of sines,

$$y(x,0) = f(x) = \sum_{n=1}^{\infty} b_n \sin(n\pi x/l).$$

Note that Bernoulli was only interested in solving a physical problem, and did not give a definition of function. By an "arbitrary function" he meant an "arbitrary shape" of the vibrating string.

Both Euler and d'Alembert, as well as other mathematicians of that time, found Bernoulli's solution absurd. Relying on the eighteenth-century "article of faith," they argued that since $f(x)$ and the sine series agree on $(0,l)$, they must agree everywhere. But this implies the manifestly absurd conclusion that an "arbitrary" function $f(x)$ is odd and periodic. (Since Bernoulli's initial shape of the string was given by an analytic expression, Euler rejected Bernoulli's solution as being the most general solution.) Bernoulli retorted that d'Alembert's and Euler's solutions constitute "beautiful mathematics but what has it to do with vibrating strings?" [24, p. 78].

Joined later by Lagrange, the debate lasted for several more years, and died down without being resolved. Ravetz [24, p. 81] characterized the essence of the debate as one between d'Alembert's mathematical world, Bernoulli's physical world, and Euler's "no-man's land" between the two. The debate did, however, have important consequences for the evolution of the function concept. Its major effect was to extend that concept to include:

(a) Functions defined piecewise by analytic expressions in different intervals. Thus, $f(x) = \begin{cases} x, & x \geq 0 \\ -x, & x < 0 \end{cases}$ was now, for the first time, considered to be a bona fide function.

(b) Functions drawn freehand and possibly not given by any combination of analytic expressions.

As Lützen put it [20]:

> D'Alembert let the concept of function limit the possible initial values, while Euler let the variety of initial values extend the concept of function. We thus see that this extension of the concept of function was *forced* upon Euler by the physical problem in question.

To see how Euler's view of functions evolved over a period of several years, compare the definition given in his 1748 *Introductio* with the following definition, given in 1755, in which the term "analytic expression" does not appear [25, pp. 72–73]:

> If, however, some quantities depend on others in such a way that if the latter are changed the former undergo changes themselves then the former quantities are called functions of the latter quantities. This is a very comprehensive notion and comprises in itself all the modes through which one quantity can be determined by others. If, therefore, x denotes a variable quantity then all the quantities which depend on x in any manner whatever or are determined by it are called its functions. . . .

Euler's view of functions was reinforced later in that century by work in partial differential equations [24, p. 86]:

> The work of Monge in the 1770s, giving a geometric interpretation to the integration of partial differential equations, seemed to provide a conclusive proof of the fact that functions 'more general than those expressed by an equation' were legitimate mathematical objects. . . .

See [3, 5, 11, 18, 19, 21, 24, 29] for details on this section.

5.5 Fourier Series

Fourier's work on heat conduction, submitted to the Paris Academy of Sciences in 1807 but published only in 1822 in his classic *Analytic Theory of Heat*, was a revolutionary step in the evolution of the function concept. Fourier's main result of 1822 was the following:

Theorem

Any function $f(x)$ defined over $(-l, l)$ is representable over this interval by a series of sines and cosines,

$$f(x) = a_0/2 + \sum_{n=1}^{\infty} [a_n \cos(n\pi x/l) + b_n \sin(n\pi x/l)],$$

where the coefficients a_n and b_n are given by

$$a_n = 1/l \int_{-l}^{l} f(t)\cos(n\pi t/l)\mathrm{d}t,\ b_n = 1/l \int_{-l}^{l} f(t)\sin(n\pi t/l)\mathrm{d}t.$$

Fourier's announcement of this result met with incredulity, for it upset several tenets of eighteenth-century mathematics. The result was known to Euler and Lagrange, among others, but only for certain functions. Fourier, of course, claimed that it is true for *all* functions, where the term "function" was given the most general contemporary interpretation [25, p. 73]:

> In general, the function $f(x)$ represents a succession of values or ordinates each of which is arbitrary. An infinity of values being given to the abscissa x, there are an equal number of ordinates $f(x)$. All have actual numerical values, either positive or negative or null. We do not suppose these ordinates to be subject to a common law; they succeed each other in any manner whatever, and each of them is given as if it were a single quantity.

Fourier's proof of his theorem was loose even by the standards of the early nineteenth century. In fact, it was formalism in the spirit of the eighteenth century – "a play upon symbols in accordance with accepted rules but without much or any regard for content or significance" [18, p. 33]. To convince the skeptical mathematical community of the reasonableness of his claim, Fourier needed to show that:

(a) The coefficients of the Fourier series can be calculated for *any* $f(x)$.
(b) *Any* function $f(x)$ can be represented by its Fourier series in $(-l, l)$. (Fourier was among the first to highlight the issue of convergence of series, which was of little concern to mathematicians of the eighteenth century.)

He showed this by:

(a′) Interpreting the coefficients a_n and b_n in the Fourier series expansion of $f(x)$ as areas, which made sense for "arbitrary" functions $f(x)$, not necessarily given by analytic expressions.
(b′) Calculating a_n and b_n (for small values of n) for a great variety of functions $f(x)$, and noting the close agreement in $(-l, l)$, but not outside that interval, between the initial segments of the resulting Fourier series and the functional values of $f(x)$.

Fourier accomplished all these using mathematical reasoning that would be clearly unacceptable to us today. However, as Langer put it so perceptively,

> It was, no doubt, partially because of his very disregard for rigor that he was able to take conceptual steps which were inherently impossible to men of more critical genius [18, p. 33].

Fourier's work raised the analytic (algebraic) expression of a function to at least an equal footing with its geometric representation (as a curve). His work, moreover, had a fundamental and far-reaching impact on subsequent developments in mathematics.

For example, it forced mathematicians to reexamine the notion of integral, and was the starting point of the researches that led Cantor to his creation of the theory of sets. As for its impact on the evolution of the function concept, Fourier's work:

- Did away with the "article of faith" held by eighteenth-century mathematicians. Thus, it was now clear that two functions given by different analytic expressions can agree on an interval without necessarily agreeing outside the interval.
- Showed that Euler's concept of "discontinuous" was flawed. Some of Euler's discontinuous functions were shown to be representable by a Fourier series – an analytic expression – and were thus continuous in Euler's sense.
- Gave renewed emphasis to analytic expressions.

All this forced a reevaluation of the function concept, as we shall see. Consult [3,7, 9,11,18,21] for details.

To summarize: The period 1720–1820 was characterized by a development and exploitation of the tools of the calculus bequeathed by the seventeenth century. These tools were employed in the solution of important "practical" problems, for example the Vibrating-String problem and the Heat-Conduction problem. These problems, in turn, clamored for attention to important "theoretical" concepts, for example function, continuity, convergence. A new subject – analysis – began to take form, in which the concept of function was central. But both the subject and the concept were still in their formative stages. It was a period of "formalism" in analysis – formal manipulations dictated the "rules of the game," with little concern for rigor. The concept of function was in a state of flux – an analytic expression (an "arbitrary" formula), then a curve (drawn freehand), then again an analytic expression, but this time a "specific" formula, namely a Fourier series. Both the subject of analysis – certainly its basic notions – and the concept of function were ripe for a reevaluation and a reformulation. This is the next stage in our story.

5.6 Dirichlet's Concept of Function

Dirichlet was one of the early exponents of the critical spirit in mathematics ushered in by the nineteenth century (others were Gauss, Abel, and Cauchy). He undertook a careful analysis of Fourier's work to make it mathematically respectable. The task was not simple: "To make sense out of what he [Fourier] did took a century of effort by men of 'more critical genius,' and the end is not yet in sight" [5, p. 263].

Fourier's result that any function can be represented by its Fourier series was, of course, incorrect. In a fundamental paper of 1829, Dirichlet gave sufficient conditions for such representability:

Theorem

If a function f has only finitely many discontinuities and finitely many maxima and minima in $(-l, l)$, then f may be represented by its Fourier series on $(-l, l)$. The Fourier series converges pointwise to f where f is continuous, and to $[f(x+) + f(x-)]/2$ at each point x where f is discontinuous.

For a mathematically rigorous proof of this theorem, one needed

(a) Clear notions of continuity, convergence, and the definite integral.
(b) Clear understanding of the function concept.

Cauchy contributed to the former, Dirichlet to the latter. We first turn briefly to Cauchy's contribution.

Cauchy was one of the first mathematicians to usher in a new spirit of rigor in analysis. In his famed *Cours d'Analyse* of 1821 and subsequent works, he rigorously defined the concepts of continuity, differentiability, and integrability of a function in terms of limits [4]. (Bolzano had done much of this earlier, but his work went unnoticed for 50 years.) It should be noted, however, that standards of rigor have changed in mathematics (not always from less rigor to more), and that Cauchy's rigor is not ours (cf. Chaps. 7–10). Kitcher [16] suggests that Cauchy's motivation in rigorizing the basic concepts of calculus came from work in Fourier series. See also [10] for background to Cauchy's work in analysis.

In dealing with continuity, Cauchy addresses himself to Euler's conceptions of "continuous" and "discontinuous" (Sect. 5.4). He shows that the function

$$f(x) = \begin{cases} x, & x \geq 0 \\ -x, & x < 0 \end{cases},$$

which Euler considered discontinuous, can also be written as $f(x) = \sqrt{x^2}$, and $f(x) = 2/\pi \int_0^\infty [x^2/(x^2 + t^2)]\mathrm{d}t$, which means that $f(x)$ is also continuous in Euler's sense. This paradoxical situation, Cauchy claims, cannot happen when *his* definition of continuity is used.

Cauchy's conception of function is not very different from that of his predecessors [3, p. 104]:

> When the variable quantities are linked together in such a way that, when the value of one of them is given, we can infer the values of all the others, we ordinarily conceive that these various quantities are expressed by means of one of them which then takes the name of *independent variable;* and the remaining quantities, expressed by means of the independent variable, are those which one calls the *functions* of this variable.

Although Cauchy gives a rather general definition of function, his subsequent comments suggest that he had in mind something more limited (see [10, p. 10]). He classifies functions as "simple" and "mixed." The "simple functions" are $a + x, a - x, ax, a/x, x^a, a^x, \log x, \sin x, \cos x, \arcsin x, \arccos x$; the "mixed functions" are composites of the "simple" ones, for example, $\log(\sin x)$. See [3, 7, 10, 11, 14, 16] for Cauchy's contribution.

Fig. 5.4 Lejeune Dirichlet (1805–1859)

Now let us consider Dirichlet's definition of function [21]:

> y is a function of a variable x, defined on the interval $a < x < b$, if to every value of the variable x in this interval there corresponds a definite value of the variable y. Also, it is irrelevant in what way this correspondence is established.

The novelty in Dirichlet's conception of function as an arbitrary correspondence lies not so much in the definition as in its application. Mathematicians from Euler through Fourier to Cauchy had paid lip service to the "arbitrary" nature of functions, but in practice they thought of them as analytic expressions or curves. Dirichlet was the first to take seriously the notion of function as an arbitrary correspondence (but see [3, p. 201]). This is made abundantly clear in his 1829 paper on Fourier series, at the end of which he gives an example of a function, now known as the *Dirichlet function*,

$$D(x) = \begin{cases} c, & x \text{ rational} \\ d, & x \text{ irrational} \end{cases} \qquad (c \neq d),$$

that does not satisfy the hypothesis of his theorem on the representability of a function by a Fourier series [12, p. 15]. The Dirichlet function:

- Was the first explicit example of a function that was not given by an analytic expression, or by several such, nor was it a curve drawn freehand.
- Was the first example of a function that is discontinuous *everywhere* (in our, not Euler's sense).
- Illustrated the concept of function as an arbitrary pairing.

Another important point is that Dirichlet was among the first to restrict explicitly the domain of the function to an interval; in the past, the independent variable was allowed to range over all real numbers. See [3, 6, 11, 12, 17, 20, 29] for details about Dirichlet's work.

5.7 "Pathological" Functions

With his example of the function $D(x)$, Dirichlet let the genie escape from the bottle. A flood of "pathological" functions, and classes of functions, followed in the succeeding half century. Certain functions were introduced to test the domain of applicability of various results; for example, the Dirichlet function was introduced in connection with the representability of a function by a Fourier series. Certain classes of functions were introduced in order to extend various concepts or results; for example, functions of bounded variation were introduced to test the domain of applicability of the Riemann integral.

The character of analysis began to change. Since the seventeenth century, the processes of analysis were assumed to be applicable to "all" functions, but it now turned out that they are restricted to particular *classes* of functions. In fact, the investigation of various classes of functions – such as continuous functions, semicontinuous functions, differentiable functions, functions with nonintegrable derivatives, integrable functions, monotonic functions, continuous functions that are not piecewise monotonic – became a principal concern of analysis. (One example is Dini's study of continuous nondifferentiable functions, for which he defined the so-called Dini derivatives.) Whereas mathematicians had formerly looked for order and regularity in analysis, they now took delight in discovering exceptions and irregularities. The towering personalities connected with these developments were Riemann and Weierstrass, although many others made important contributions, for example, du Bois Reymond and Darboux.

The first major step in these developments was taken by Riemann in his *Habilitationsschrift* of 1854, which dealt with the representation of functions by Fourier series. As we recall, the coefficients of a Fourier series are given by integrals. Cauchy had defined the integral only for continuous functions, but his ideas could be extended to functions with finitely many discontinuities. Riemann extended Cauchy's integral and thus enlarged the class of functions representable by Fourier series. This extension, known today as the Riemann integral, applies to functions of bounded variation, a much broader class of functions than Cauchy's continuous functions. Thus, a function can have infinitely many discontinuities, which can be dense in any interval, and still be Riemann-integrable. (There are, of course, restrictions on the discontinuities of a Riemann-integrable function. As we now know, following Lebesgue, a function is Riemann-integrable if and only if its discontinuities form a set of Lebesgue measure zero.) Riemann gave the following example in his *Habilitationsschrift* (published in 1867):

$$f(x) = 1 + (x)/1^2 + (2x)/2^2 + (3x)/3^2 + \dots,$$

where for any real number a the function (a) is defined as 0 if $a = 1/2 + k$ (k an integer), and a minus the nearest integer when $a \neq 1/2 + k$. This function is discontinuous for all $x = m/2n$, where m is an integer relatively prime to $2n$ [7, p. 325]. In contrast to Dirichlet's function $D(x)$, this one is given by an analytic expression and is Riemann-integrable.

Riemann's work may be said to mark the beginning of a theory of the mathematically discontinuous, although there are isolated examples in Fourier's and Dirichlet's works. It planted the discontinuous firmly upon the mathematical scene. The importance of this development can be inferred from the following statement of Hawkins [12, p. 3]:

> The history of integration theory after Cauchy is essentially a history of attempts to extend the integral concept to as many discontinuous functions as possible; such attempts could become meaningful only after existence of highly discontinuous functions was recognized and taken seriously.

In 1872 Weierstrass startled the mathematical community with his famous example of a continuous nowhere-differentiable function

$$f(x) = \sum_{n=1}^{\infty} b^n \cos(a^n \pi x),$$

where a is an odd integer, b a real number in (0,1), and $ab > 1 + 3\pi/2$ [14, p. 387]. (Bolzano had given such an example in 1834, but it went unnoticed.) This example was contrary to all geometric intuition. In fact, up to about 1870 most books on the calculus "proved" that a continuous function is differentiable except possibly at a finite number of points! [12, p. 43]. Even Cauchy believed that.

The malaise in the understanding and use of the function concept around this time can be gathered from the following account by Hankel (in 1870) concerning the function concept as it appears in the "better textbooks of analysis" (Hankel's phrase [20]; see also [3, p. 198]):

> One [text] defines function in the Eulerian manner; the other that y should change with x according to a rule, without explaining this mysterious concept; the third defines them as Dirichlet; the fourth does not define them at all; but everyone draws from them conclusions that are not contained therein.

Weierstrass' example began the disengagement of the continuous from the differentiable in analysis. His work, and others' in this period, necessitated a reexamination of the foundations of analysis and led to the so-called "arithmetization of analysis," in which process Weierstrass was a prime mover (see Sect. 4.6.3). As Birkhoff notes [3, p. 71]:

> Weierstrass demonstrated the need for higher standards of rigor by constructing *counterexamples* to plausible and widely held notions.

Counterexamples play an important role in mathematics. They illuminate relationships, clarify concepts, and often lead to the creation of new mathematics. (An interesting case study of the role of counterexamples in mathematics can be found in the book *Proofs and Refutations* by I. Lakatos.) The impact of the developments we have been describing was, as we have already noted, to change the character of analysis. A new subject was born – the theory of functions of a real variable. Hawkins gives a vivid description of the state of affairs [12, p. 119]:

> The nascent theory of functions of a real variable grew out of the development of a more critical attitude, supported by numerous counterexamples, towards the reasoning of earlier mathematicians. Thus, for example, continuous nondifferentiable functions, discontinuous series of continuous functions, and continuous functions that are not piecewise monotonic were discovered. The existence of exceptions came to be accepted and more or less expected. And the examples of nonintegrable derivatives, rectifiable curves for which the classical integral formula is inapplicable, nonintegrable functions that are the limit of integrable functions, Harnack-integrable derivatives for which the Fundamental Theorem II is false, and counterexamples to the classical form of Fubini's Theorem appear to have been received in this frame of mind. The idea, as Schoenflies put it in his report,... was to proceed, as in human pathology, to discover as many exceptional phenomena as possible in order to determine the laws according to which they could be classified.

Not everyone however was pleased with these developments, as the following quotations from Hermite (in 1893) and Poincaré (in 1899), respectively, attest [17, p. 973]:

> I turn away with fright and horror from this lamentable evil of functions which do not have derivatives (Hermite).
>
> Logic sometimes makes monsters. For half a century we have seen a mass of bizarre functions which appear to be forced to resemble as little as possible honest functions which serve some purpose. More of continuity, or less of continuity, more derivatives, and so forth. Indeed, from the point of view of logic, these strange functions are the most general; on the other hand those which one meets without searching for them, and which follow simple laws appear as a particular case which does not amount to more than a small comer.
> In former times when one invented a new function it was for a practical purpose; today one invents them purposely to show up defects in the reasoning of our fathers and one will deduce from them only that.
> If logic were the sole guide of the teacher, it would be necessary to begin with the most general functions, that is to say with the most bizarre. It is the beginner that would have to be set grappling with this teratologic museum (Poincaré).

The effect of the events we have been describing on the function concept can be summarized as follows. Stimulated by Dirichlet's conception of function and his example $\mathrm{D}(x)$, the notion of function as an arbitrary correspondence is given free rein and gains general acceptance; the geometric view of function is given little consideration. Riemann's and Weierstrass' functions could certainly not be "drawn," nor could many of the other examples of functions given during this period. After Dirichlet's work, the term "function" acquired a clear meaning independent of the term "analytic expression." During the next half century, mathematicians introduced many examples of functions in the spirit of Dirichlet's broad definition, and the time was ripe for an effort to determine which functions were describable by means of "analytic expressions," a vague term in use during the previous two centuries. See [3, 12, 16, 17] for details of this period.

5.8 Baire and Analytically Representable Functions

The question whether every function in Dirichlet's sense is representable analytically was first posed by Dini in 1878 [6, p. 31]. In his doctoral thesis of 1898 Baire undertook to give an answer. The very notion of analytic representability had to be clarified, since it was used in the past in an informal way. Dini himself used it vaguely, asking "if every function can be expressed analytically, for all values of the variable in the interval, by a finite or infinite series of operations ('operations du calcul') on the variable" [6, p. 32].

The starting point for Baire's scheme was the Weierstrass Approximation Theorem (published in 1885): *Every continuous function f(x) on an interval* $[a, b]$ *is a uniform limit of polynomials on* $[a, b]$. Baire called the class of continuous functions *class* 0. He then defined the functions of *class* 1 to be those that are not in class 0 but are (pointwise) limits of functions of class 0. In general the functions of *class m* are those functions which are not in any of the preceding classes but are representable as limits of sequences of functions of class $m-1$. This process is continued by transfinite induction to all ordinals less than the first uncountable ordinal Ω. Since the Baire functions thus constructed are closed under limits, nothing new results if this process is repeated. This classification into Baire classes α ($\alpha < \Omega$) is called the *Baire classification,* and the functions which constitute the union of the Baire classes are called *Baire functions* (see [26a]).

Baire called a function *analytically representable* if it belonged to one of the Baire classes. Thus a function is analytically representable in Baire's sense if it can be built up from a variable and constants by a finite or denumerable set of additions, multiplications, and passages to pointwise limits.

The collection of analytically representable functions (Baire functions) is very encompassing. For example, discontinuous functions representable by Fourier series belong to class 1. Thus functions representable by Fourier series constitute only a "small" part of the totality of analytically representable functions. (Recall Fourier's claim that *every* function can be represented by a Fourier series!) As another example, Baire showed that the "pathological" Dirichlet function $D(x)$ is of class 2, since

$$D(x) = \begin{cases} c, & x \text{ rational} \\ d, & x \text{ irrational} \end{cases} = (c-d) \lim_{n\to\infty} (\cos n!\pi x)^{2m} + d.$$

Moreover, any function obtained from a variable and constants by an application of the four algebraic operations and the operations of analysis (such as differentiation, integration, expansion in series, use of transcendental functions) – the kind of function known in the past as an "analytic expression" – was shown to be analytically representable.

Lebesgue pursued these studies and showed (in 1905) that each of the Baire classes is nonempty, and that the Baire classes do not exhaust all functions.(In fact, there are (Lebesgue-measurable) functions which are not Baire functions. At the same time, Lebesgue showed that to every measurable function f there corresponds a Baire function which differs from f only on a set of measure zero.) Thus Lebesgue established that there are functions which are not analytically representable in

Baire's sense. This he did by actually exhibiting a function outside the Baire classification, "using a profound but extremely complex method" [21]. The construction is quite "messy" and uses the axiom of choice. Using nonconstructive methods, one can show by a counting argument that the Baire functions have cardinality c. Since the set of all functions has cardinality 2^c, there are uncountably many functions which are not analytically representable in Baire's sense.

Thus not all functions in the sense of Dirichlet's conception of function as an arbitrary correspondence are analytically representable in Baire's sense, although it is (apparently) very difficult to produce a specific function that is not. Do such functions, which are not analytically representable, "really" exist? This is part of our story in the next section.

5.9 Debates About the Nature of Mathematical Objects

Function theory was characterized by some at the turn of the twentieth century as the branch of mathematics which deals with counterexamples. This view was not universally applauded, as the earlier quotations from Hermite and Poincaré indicate (see Sect. 5.7). In particular, Dirichlet's general conception of function began to be questioned. Objections were raised against the phrase in his definition that "it is irrelevant in what way this correspondence is established" (see Sect. 5.6). Subsequently, the arguments for and against this point linked up with the arguments for and against the axiom of choice, explicitly formulated by Zermelo in 1904, and broadened into a debate over whether mathematicians are free to create their objects at will.

There was a famous exchange of letters in 1905 among Baire, Borel, Hadamard, and Lebesgue concerning the contemporary logical state of mathematics (see [6, 22, 23] for details). Much of the debate was about function theory – the critical question being whether the definition of a mathematical object, say a number or a function, however given, legitimizes the existence of that object; in particular, whether Zermelo's axiom of choice is a legitimate mathematical tool for the definition or construction of functions. In this context, Dirichlet's conception of function was found to be too broad by some, for example Lebesgue, and devoid of meaning by others, for example Baire and Borel, but was acceptable to yet others, for example Hadamard. Baire, Borel, and Lebesgue supported the requirement of a definite "law" of correspondence in the definition of a function. The "law," moreover, had to be reasonably explicit – that is, understood by and communicable to anyone who wanted to study the function.

To illustrate the point, Borel compares the number π, whose successive digits can be unambiguously determined, and which he therefore regards as well defined, with the number obtained by carrying out the following "thought experiment." Suppose we lined up infinitely many people and asked each of them to name a digit at random. Borel claims that, unlike π, this number is not well defined since its digits are not related by any law. This being so, two mathematicians discussing the number will never be certain that they are talking about the *same* number. Put briefly, Borel's position is that without a definite law of formation of the digits of an infinite decimal, one cannot be certain of its identity.

Hadamard had no difficulty in accepting as legitimate the number resulting from Borel's thought experiment. By way of illustration, he alluded to the kinetic theory of gases, where one speaks of the velocities of molecules in a given volume of gas although no one knows them precisely. Hadamard felt that "the requirement of a law that determines a function... strongly resembles the requirement of an *analytic expression* for that function, and that this is a throwback to the eighteenth century" [21].

The issues described here were part of broad debates about various ways of doing analysis – synthetic vs. analytic, or idealist vs. empiricist. These debates, in turn, foreshadowed subsequent "battles" between proponents and opponents of the various philosophies of mathematics, for example, formalism and intuitionism, dealing with the nature and meaning of mathematics. And, of course, even now the issue has not been resolved. (There has been a renewed interest in recent decades, by computer scientists, among others, in Brouwer's "intuitionistic mathematics." The revival, in the form of "constructive mathematics," was led by E. Bishop, and is highlighted in an article by M. Mandelkern, "Constructive Mathematics," *Math. Mag.* 58 (1985) 272–280.). See [5, 6, 21–23, 26a] for details.

The period 1830–1910 witnessed an immense growth in mathematics, both in scope and in depth. New mathematical fields were formed, for example complex analysis, algebraic number theory, noneuclidean geometry, abstract algebra, mathematical logic, and older ones were deepened, for example real analysis, probability, analytic number theory, calculus of variations. Mathematicians felt free to create their systems (almost) at will, without finding it necessary to seek motivation from or applications to concrete (physical) settings. At the same time there was throughout the nineteenth century a reassessment of gains achieved, accompanied by a concern for the foundations of various branches of mathematics. These trends are reflected in the evolution of the notion of function. The concept unfolds from its modest beginnings as a formula or a geometric curve (eighteenth and early nineteenth centuries) to an arbitrary correspondence (Dirichlet). This latter idea is exploited throughout the nineteenth century by way of the construction of various "pathological" functions. Toward the end of the century, there is a reevaluation of past accomplishments (Baire classification, controversy relating to use of the axiom of choice), much of it in the broader context of debates about the nature and meaning of mathematics.

5.10 Recent Developments

Here we touch briefly on three more recent developments relating to the function concept.

(a) *L_2 Functions*. The set $L_2 = \{f(x): f^2(x)$ is Lebesgue-integrable$\}$ forms a "Hilbert space" – a fundamental object in functional analysis. Two functions in L_2 are considered to be the same if they agree everywhere except possibly on a set of Lebesgue measure zero. Thus, in L_2 Function Theory, one can always work with representatives in an equivalence class rather than with individual functions. These notions, Davis and Hersh observed,

> involve a further evolution of the concept of function. For, an element in L_2 is not a function, either in Euler's sense of an analytic expression, or in Dirichlet's sense of a rule or mapping associating one set of numbers with another. It is function-like in the sense that it can be subjected to certain operations normally applied to functions (adding, multiplying, integrating). But since it is regarded as unchanged if its values are altered on an arbitrary set of measure zero, it is certainly not just a rule assigning values at each point in its domain [5, p. 269].

(b) *Generalized Functions (Distributions)*. The concept of a distribution or generalized function is a very significant and fundamental extension of the concept of function. The theory of distributions arose in the 1930s and 1940s. It was created to give mathematical meaning to the differentiation of nondifferentiable functions – a process which the physicists had employed (unrigorously) for some time. Thus Heaviside (in 1893) "differentiated" the function

$$f(x) = \begin{cases} 1, & x > 0 \\ 1/2, & x = 0, \\ 0, & x < 0 \end{cases}$$

to obtain the impulse "function"

$$\delta(x) = \begin{cases} 0, & x \neq 0 \\ \infty, & x = 0 \end{cases}.$$

The following is a heuristic argument. Approximate $f(x)$ by a sequence of differentiable funcitons $f_\varepsilon(x)$ as in the diagram (Fig. 5.5); then $f'_\varepsilon(x) \to \delta(x)$ as $\varepsilon \to 0$. In 1930 Dirac introduced $\delta(x)$ as a convenient notation in the mathematical formulation of quantum theory.

Formally, a *distribution* is a continuous linear functional on a space D of infinitely differentiable functions, called "test functions," that vanish outside some interval [a, b]. To any continuous (or locally integrable) function F, there corresponds a distribution $\Phi_F : D \to C$ given by $\Phi_F(x) = \int_{-\infty}^{\infty} F(t)x(t)dt$. However, not every

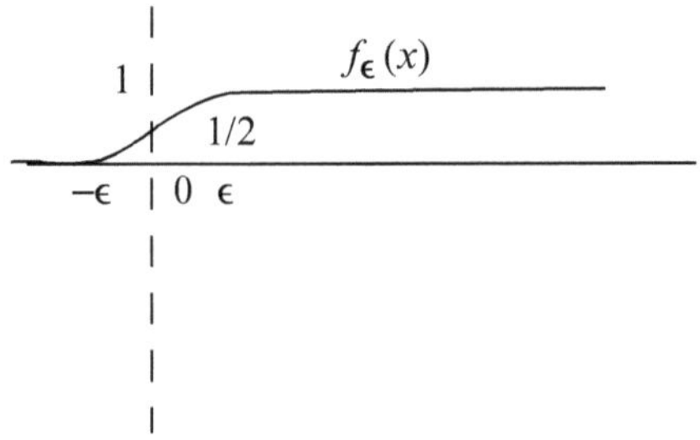

Fig. 5.5 A heuristic argument for obtaining the Dirac δ-function

distribution comes from such a function: The distribution $\delta : D \to C$ given by $\delta(x) = x(0)$ corresponds to the "Dirac $\delta-$function" mentioned above, and does not arise from any function F in the way described above. See [5, 19, 28].

A basic property of distributions is that each distribution has a derivative that is again a distribution. In particular, every continuous function is "differentiable," that is, has a distribution as its "derivative." In fact Laurent Schwartz, one of the creators of the theory of distributions, claimed that he had introduced distributions to enable differentiation of continuous functions. Lützen [19, p. 305] asserts that "the theory of distributions probably constitutes the closest approximation to Euler's vision of a generalized calculus," a vision that Euler tried to put into practice in his solution of the Vibrating-String problem.

Treves put it thus [28, p. 338]:

> The enduring merit of distribution theory has been that the basic operations of analysis, differentiation and convolution, and the Fourier/Laplace transforms and their inversion, which demanded so much care in the classical framework, could now be carried out without qualms by obeying purely algebraic rules.

(c) *Category Theory.* The notion of a function as a mapping between arbitrary sets gradually became dominant in the mathematics of the twentieth century. Algebra had a major impact on this development, in which the concept of function was placed in the general framework of the concept of mapping from one set into another. Thus linear transformations of vector spaces (principally $\mathbf{R}^n$ and $\mathbf{C}^n$), were dealt with throughout much of the nineteenth century. Homomorphisms of groups and automorphisms of fields were introduced in the latter part of that century. As early as 1887, Dedekind gave a fairly modern definition of the term "mapping" [25, p. 75]:

> By a mapping of a system [set] S a law is understood, in accordance with which to each determinate element s of S there is associated a determinate object, which is called the image of s and is denoted by $\varphi(s)$; we say too, that $\varphi(s)$ corresponds to the element s, that $\varphi(s)$ is caused or generated by the mapping φ out of s, that s *is* transformed by the mapping φ into $\varphi(s)$.

Analysis too played a major role in the extension of the domain and range of definition of a function to arbitrary sets. (Recall that Dirichlet's definition of function was as an arbitrary correspondence between (real) *numbers.*) Thus, Euler and others in the eighteenth century treated (informally) functions of several variables.

In 1887, considered the year of birth of functional analysis, Volterra defined the notion of a "functional" which he called a "function of functions." (A *functional* is a function whose domain is a set of functions and whose range is the real or complex numbers.) In the first two decades of the twentieth century, the notions of metric space, topological space, Hilbert space, and Banach space were introduced; functions (operators, linear operators) between such spaces play a prominent role. See [17] for details.

In 1939 Bourbaki gave the following definition of a function [3, p. 7]:

> Let E and F be two sets, which may or may not be distinct. A relation between a variable element x of E and a variable element y of F is called a *functional relation* in y if, for all $x \in$ E, there exists a unique $y \in$ F which is in the given relation with x.
> We give the name of *function* to the operation which in this way associates with every element $x \in$ E the element $y \in$ F which is in the given relation with x; y *is* said to be the *value* of the function at the element x, and the function is said to be *determined* by the given functional relation. Two equivalent functional relations determine the *same* function.

Bourbaki then gave the definition of a function from E to F as a certain subset of the Cartesian product E X F. This is, of course, the definition of function as a set of ordered pairs.

All these "modern" definitions of function were given in terms of sets, and hence their logic must receive the same scrutiny as that of set theory. ("Naïve" set theory was developed by Cantor during the last three decades of the nineteenth century.)

In category theory, which arose in the late 1940s to give formal expression to certain aspects of homology theory, the concept of function assumes a fundamental role. It can be described as an "association" from an "object" A to another "object" B. The "objects" A and B need not have any elements, that is, they need not be sets in the usual sense. In fact, the objects A and B can be entirely dispensed with. A "category" can then be defined as consisting of functions (or "maps"), *which are taken as undefined (primitive) concepts* satisfying certain relations or axioms. In 1966 Lawvere outlined how category theory can replace set theory as a foundation for mathematics. See [13] for details.

In the recent developments outlined in this section, we have seen the function concept modified (L_2 functions), generalized (distributions), and finally "generalized out of existence" (category theory). Have we come full circle?

References

1. G. Birkhoff, *A Source Book in Classical Analysis*, Harvard Univ. Press, 1973.
2. H. Bos, Mathematics and rational mechanics. In *Ferment of Knowledge*, ed. by G. S. Rousseau and R. Porter, Cambridge Univ. Press, 1980, pp. 327–355.
3. U. Bottazzini, *The Higher Calculus: A History of Real and Complex Analysis from Euler to Weierstrass,* Springer, 1986.
4. A.– L. Cauchy, *Cours d'Analyse*, translated into English, with annotations, by R. Bradley and E. Sandifer, Springer, 2009 (orig. 1821).
5. P. J. Davis and R. Hersh, *The Mathematical Experience,* Birkäuser, 1981.

6. P. Dugac, Des fonctions comme expressions analytiques aux fonctions representables analytiquement. In *Mathematical Perspectives*, ed. by J. Dauben, Academic Press, 1981, pp. 13–36.
7. C. H. Edwards, *The Historical Development of the Calculus,* Springer, 1979.
8. L. Euler, *Introductio in Analysin Infinitorum*, Vols 1 & 2, translated into English by J. Blanton, Springer, 1989 (orig. 1748).
9. A. Gardiner, *Infinite Processes,* Springer-Verlag, 1982.
10. J. V. Grabiner, *The Origins of Cauchy's Rigorous Calculus*, M.I.T. Press, 1981.
11. I. Grattan-Guinness, *The Development of the Foundations of Mathematical Analysis from Euler to Riemann*, M.I.T. Press, 1970.
12. T. Hawkins, *Lebesgue's Theory of Integration—Its Origins and Development,* Chelsea, 1975.
13. H. Herrlich and G. E. Strecker, *Category Theory,* Allyn & Bacon, 1973.
14. R. F. lacobacci, *Augustin-Louis Cauchy and the Development of Mathematical Analysis,* Ph.D. Dissertation, New York Univ., 1965. (University Microfilms, no. 65–7298, 1986.)
15. V. J. Katz, Calculus of the trigonometric functions, *Hist. Math.* 14 (1987) 311–314.
16. P. Kitcher, *The Nature of Mathematical Knowledge*, Oxford Univ. Press, 1983.
17. M. Kline, *Mathematical Thought from Ancient to Modern Times*, Oxford Univ. Press, 1972.
18. R. E. Langer, Fourier series: the genesis and evolution of a theory. The First Herbert Ellsworth Slaught Memorial Paper, *Math. Assoc. of America,* 1947. (Suppl. to Vol. 54 of the *Amer. Math. Monthly*, pp. 1–86.)
19. J. Lützen, Euler's vision of a generalized partial differential calculus for a generalized kind of function, *Math. Mag*. 56 (1983) 299–306.
20. J. Lützen, The development of the concept of function from Euler to Dirichlet (in Danish), *Nordisk. Mat. Tidskr.* 25/26 (1978) 5–32.
21. N. Luzin, Function (in Russian), *The Great Soviet Encyclopedia*, Vol. 59 (ca. 1940), pp. 314–334.
22. A. F. Monna, The concept of function in the 19th and 20th centuries, in particular with regard to the discussion between Baire, Borel and Lebesgue, *Arch. Hist. Ex. Sci*. 9 (1972/73) 57–84.
23. G. H. Moore, *Zermelo's Axiom of Choice: Its Origins, Development, and Influence,* Springer, 1982.
24. J. R. Ravetz, *Vibrating strings and arbitrary functions*. In *The Logic of Personal Knowledge: Essays Presented to M. Polanyi on his Seventieth Birthday,* The Free Press, 1961, pp. 71–88.
25. D. Rüthing, Some definitions of the concept of function from Joh. Bernoulli to N. Bourbaki, *Math. Intelligencer* 6:4 (1984) 72–77.
26. W. L. Schaaf, Mathematics and world history, *Math. Teacher* 23 (1930) 496–503.
26a. J. Stillwell, *Roads to Infinity: The Mathematics of Truth and Proof,* A K Peters, 2010.
27. D. J. Struik, *A Source Book in Mathematics, 1200–1800,* Harvard Univ. Press, 1969.
28. F. Treves, Review of *The Analysis of Linear Partial Differential Operators* by L. Hormander, *Bull. Amer. Math. Soc.* 10 (1984) 337–340.
29. A. P. Youschkevitch, The concept of function up to the middle of the 19th century, *Arch. Hist. Ex. Sci*. 16 (1976) 37–85.

Chapter 6
More on the History of Functions, with Remarks on Teaching

6.1 Introduction

The notion of function is central in both mathematics and mathematics education. Textbook definitions or descriptions of function have varied with time, context, and level of presentation. A function has been viewed as a formula, a rule, a correspondence, a relation between variables, a table of values, a graph, a mapping, a transformation, an operation, a set of ordered pairs (see, e.g., [14, 19, 22]). These ideas reflect the historical evolution of the function concept. We will briefly trace some aspects of this evolutionary process, and following the historical account in each section (except for the first and last), draw some pedagogical morals. Further discussion of pedagogical issues can be found in [5, 8, 13, 14, 16, 17, 25, 30, 32].

Recorded mathematical history goes back almost 4,000 years. During its first 3,500 years mathematicians developed the elements of algebra, deductive geometry, trigonometry, and even aspects of analytic geometry and the integral calculus. Yet the concept of function, perhaps surprisingly, is not part of that mathematics (see Sect. 6.3 for some reasons). The concept originated in the early eighteenth century, well into the so-called modern period in the evolution of mathematics. And although the concept of function now permeates all areas of mathematics, it had its origins in calculus and analysis.

In dealing with the historical origin of mathematical ideas the "big bang" theory rarely applies. Mathematical concepts usually develop gradually, in response to mathematical needs. While the function concept dates back about 300 years, the "instinct for functionality" may be said to be about 4,000 years old. We now briefly describe the expression of that "instinct," the prehistory of the notion of function.

I. Kleiner, *Excursions in the History of Mathematics*,
DOI 10.1007/978-0-8176-8268-2_6,

6.2 Anticipations of the Function Concept

The notion of function as a dependence of one quantity on others is all-pervasive. It is implicit in ancient mathematics in the form of tables, curves, physical laws, and relationships between geometric quantities.

6.2.1 *Babylonian Mathematics*

The Babylonians, as long ago as 1800 BC, were avid "tablemakers." To facilitate arithmetical and algebraic calculations they constructed tables of reciprocals, squares, cubes, square roots, cube roots, and others. The following is a transcribed table of reciprocals [15, p. 7]:

igi 2 gál-bi 30	igi 8 gál-bi 7, 30
igi 3 gál-bi 20	igi 9 gál-bi 7, 40
igi 4 gál-bi 15	....
igi 6 gál-bi 10	igi 27 gál-bi 2, 13, 20

The table says that (for example) the reciprocal of 2 is 30/60, and the reciprocal of 8 is $7/60 + 30/60^2$ (60 was the Babylonian number base).

The Babylonians also tabulated astronomical observations, namely the positions of the sun, moon, and planets at various times. They then used what we would call linear interpolation to compute values not included among the original observations. See [15, 35].

6.2.2 *Greek Mathematics*

Mathematical relationships which would nowadays be expressed by means of equations, and thus viewed as functional relations, were described by the Greeks as proportions. Thus the Greek counterpart of the equation $A = \pi r^2$ for the area of a circle is stated in Euclid's *Elements* (c. 300 BC) as: the areas of circles are to each other as the squares on their radii. The Greek discovery of what was perhaps the first law of mathematical physics, namely the relationship between the lengths of plucked strings and the musical sounds they emitted, was also expressed in terms of proportions; for example, a string half the length of a given one produces a tone one octave higher.

The Greeks represented sections of a cone algebraically by means of "symptoms," which have been interpreted as equations of the conics [15, p. 92]. They also considered the spiral, the quadratrix, the cissoid, and the conchoid. These curves, the first two now known to be transcendental, were defined kinematically. For example, the spiral was given as the locus of a point which moves at a uniform rate along a straight line which revolves about one of its points at a uniform rate.

The Greeks, especially Ptolemy in his *Almagest* (c.150 AD), also developed the elements of trigonometry. Ptolemy computed quite accurate tables of chords of a circle, similar to later tables of the sine function. See [33].

For further details about Greek contributions see [3, 4, 10, 15, 33, 35].

6.2.3 The Latitude of Forms

In the thirteenth and fourteenth centuries Paris and Oxford became the seats of two major schools of mathematical philosophy whose declared aim was the study of natural phenomena using mathematics as a tool. The resulting theory, known as the "latitude of forms," dealt for the first time with *nonuniform* motion. In particular, the case of uniformly accelerated motion was investigated and was represented by the scholastic philosopher Oresme as a graph. This was the first graphical representation of a physical law. See [10, 15, 35] for details.

6.2.4 Precalculus Developments

The major developments relating to functions during the late sixteenth and early seventeenth centuries were the emergence of mathematized science with Kepler, Galileo and others, and the invention of analytic geometry by Descartes and Fermat. The former gave mathematical expression, in terms of curves and equations (or proportions), to such physical problems as the determination of the motions of a pendulum, of a freely falling body, and of the planets; of the paths of a projectile, and of a point on a circle rolling along a line; and of the shape of a rope suspended from two fixed points. See [4, 15] for details.

Analytic geometry was most important for the development of the function concept. One could now represent known curves, defined in the past kinematically or geometrically, by means of equations, and conversely, one could obtain a curve simply by writing down an equation connecting two variables x and y. Before this time only about a dozen curves were known, but now it was possible to create an infinity of curves, hence an infinity of potential functions. In fact, in the early seventeenth century Fermat introduced the infinite family of so-called parabolas and hyperbolas of Fermat, namely $y = kx^n$, with $n > 0$ and $n < 0$, respectively. See [3, 4, 35].

Yet the notion of function did not arise explicitly at this time. For equations between variables are not considered as functions unless an identification is made of their independent and dependent variables. And there was no compelling reason at that time (the 1630s) to make such identification. See [35] and Chap. 5.

6.2.5 *The Calculus of Newton and Leibniz*

Among the basic ideas of the calculus today are function, limit, continuity, derivative, and definite integral. None of these was explicitly present in the calculus created by Newton and Leibniz in the last third of the seventeenth century. In particular, theirs is *not* a calculus of functions. It is, rather, a calculus of curves represented by equations. For, the major problems which gave rise to the calculus were geometric or kinematic, involving curves, such as finding the tangent to a curve, the area under a curve, the length of a curve, and the instantaneous velocity of a point moving along a curve. The algorithms – a calculus – for dealing with such problems were based on the representation of curves as equations rather than as functions. For example, to find the tangent at a point (x, y) to the conic $x^2 + 2xy = 5$, replace x and y by $x + \mathrm{d}x$ and $y + \mathrm{d}y$, respectively ($(x + \mathrm{d}x, y + \mathrm{d}y)$ represented a point on the conic "infinitely close" to (x, y)). Then $(x + \mathrm{d}x)^2 + 2(x + \mathrm{d}x)(y + \mathrm{d}y) = 5 = x^2 + 2xy$. Simplifying and discarding $(\mathrm{d}x)(\mathrm{d}y)$ and $(\mathrm{d}x)^2$, which are negligible in comparison with $\mathrm{d}x$ and $\mathrm{d}y$, yields $2x\mathrm{d}x + 2x\mathrm{d}y + 2y\mathrm{d}x = 0$. Dividing by $\mathrm{d}x$ and solving for $\mathrm{d}y/\mathrm{d}x$ (considered as a quotient of two differentials), we get $\mathrm{d}y/\mathrm{d}x = -x - y/x$. This is what we would get by writing $x^2 + 2xy = 5$ as $y = (5 - x^2)/2x$ and differentiating this *functional* relation. See [10, 15, 24, 35], and Chaps. 4 and 5 for details.

6.2.6 *Remark on Teaching*

The above account of the prehistory of the function concept has been very brief. Each of the topics in this sketchy survey can be expanded by students into a historical project, discussing more fully the various topics. Another interesting historical project, which cuts across several of the above sections, is to consider the following three stages in the evolution of the idea of function: proportion, equation, and function. For this, the article by Boyer [3] will be very useful.

6.3 The Emergence and Consolidation of the Function Concept

During the early decades of the eighteenth century the calculus gradually became detached from its geometric origin. The powerful algebraic (algorithmic, analytic) apparatus developed by Newton and Leibniz was augmented and exploited by their successors to solve problems not directly related to the geometry of curves. The formulas connecting the variables and their differentials began to take on a life of their own, independent of the geometric objects they represented. Leibniz and Johann Bernoulli groped for a concept to express this new reality and eventually came up with the idea of a function (see [35, pp. 56–60] for details about their correspondence on this issue). Although Leibniz was the first to *use* the word "function," Bernoulli supplied (in 1718) the first formal definition [27, p. 72]:

> One calls here Function of a variable a quantity composed in any manner whatever of this variable and of constants.

Bernoulli did not specify what "composed in any manner whatever" meant, although it is apparent from the context that "function" to him meant an algebraic expression. The concept of variable, applied to geometric objects, was thus replaced with the concept of function as an algebraic formula.

Why did the concept of function arise so late in the evolution of mathematics? Put simply, there was no clear need for it earlier. One introduces an abstract concept, such as that of function, only when one has many concrete examples from which to abstract. But, as we noted, only a handful of examples of functions, mostly in the form of curves, were available before the seventeenth century. With the advent of analytic geometry and mathematized science the store of potential functions increased dramatically. The calculus lent them significance. The time was now ripe for the abstract concept of function to emerge. Since the calculus of Newton and Leibniz was motivated by, and had its main applications to, geometry and physics, it took several more decades for the calculus to be recast in algebraic terms with the function concept as its centerpiece.

The recasting was done largely by Euler, through his influential textbooks of the mid-eighteenth century. Euler turned the seventeenth-century calculus of variables and equations into a calculus of functions. His definition of function, given in his famous 1748 text *Introductio in Analysin Infinitorum* (see [2, 11]), is as follows [27, p. 72]:

> A function of a variable quantity is an analytic expression composed in any manner from that variable quantity and numbers or constant quantities.

An "analytic expression" was to Euler an algebraic formula generated from algebraic and transcendental functions (i.e., polynomials, trigonometric, inverse trigonometric, exponential, and logarithmic functions) by the four algebraic operations plus the composition of functions, and the taking of n-th roots. We now call such functions *elementary*. For example, $\log[(\sin x^2)/x + 1] + \sqrt{x}$ is an analytic expression. It is important to note that a function with a "split domain," such as

$$f(x) = \begin{cases} x^2, & x > 0 \\ x, & x \geq 0 \end{cases},$$

was *not* considered a bona fide function by mathematicians of the mid-eighteenth century: the algorithms of the calculus applied at that time only to functions given by a *single* analytic expression. Also, the domain of a function was not restricted; it was usually taken to be all real numbers (possibly with a finite number of exceptions, as in $f(x) = \frac{1}{x}$). See [10, 13, 35] for further details.

6.3.1 Remarks on Teaching

The historical account of the early evolution of the function concept suggests several pedagogical questions: Should one teach calculus without functions? When should one introduce the notion of function? What definition of function should one give to beginning students?

It is certainly possible to teach *elementary* versions of the calculus without functions, with the emphasis placed on *curves*: finding tangents to, and areas under, curves, and discussing applications. The curve is a much more "natural" object for students than a function, and generations of students have been taught the calculus of curves rather than of functions. Is this a desirable approach? It may be in some instances, although this is perhaps a heretical view. But it is not unlike suggesting that ideas of linear algebra be taught without introducing the notion of a vector space – another heretical view; such ideas were first developed by mathematicians in the seventeenth century, if not earlier, but the concept of a vector space was introduced only in the last decades of the nineteenth century. Many important ideas of linear algebra can be taught without introducing vector spaces.

When should one introduce the notion of function? *Only* when there is a manifest need for it. For example, although plotting graphs is a frequent mathematical activity, it need not be related to functions. It is well to remember that mathematical concepts were introduced to meet mathematical needs, and this should also be an overriding principle in pedagogy. Do not bring a cannon onto the stage unless you are prepared to fire it, exhorted Chekhov.

What definition of function should we give to beginning students? One that suits the occasion – that is, their needs. It may be function as a formula, or as a rule, but it should most certainly *not* be function as a set of ordered pairs. Giving this latter definition and proceeding to discuss only linear and quadratic functions makes little pedagogical sense.

The historical account should alert teachers to the fact that defining functions on a "split-domain," or restricting the domain of definition to an interval, are not "natural" ideas and should be introduced with care. In fact, it is perfectly legitimate and reasonable to define a function as a single formula, without specifying a domain of definition. Subsequently, when the need arises, one extends the definition to include functions given by two or more formulas, or functions whose domains are intervals or more general subsets of the reals. Giving *tentative* definitions of mathematical concepts and, moreover, telling students that they *are* tentative and will require revision with changing circumstances is sound pedagogical practice.

6.4 Functions Repesented by Power Series

Descartes classified curves into "geometric" and "mechanical." In his analytic geometry he dealt only with geometric curves, since he believed that only with such curves could one associate an equation. The mechanical curves, he argued, were not amenable to his method. Descartes' "geometric" curves are our *algebraic* curves and his "mechanical" curves are our transcendental curves. Algebraic curves are curves defined by equations of the form $p(x, y) = 0$, where $p(x, y)$ is a polynomial. Curves that are not algebraic are called *transcendental*. Examples of transcendental curves are the sine, cosine, logarithm, spiral, and cycloid – curves originally defined kinematically or geometrically rather than by equations. For example, the logarithm was defined in terms of motions of points along two lines, and was later represented as the area under a hyperbola; and the sine (of a central angle in a circle) was given in terms of the length of the chord subtended by the angle (see [4, 10, 24, 33] for details). One of Newton's and Leibniz' major achievements was their application of the analytic apparatus of the calculus to transcendental curves. The major tool they employed for this purpose was power series.

A *power series* is an expression of the form $a_0 + a_1x + a_2x^2 + a_3x^3 + \cdots$, with a_i real or complex numbers – an "infinite polynomial," if you will. Among Newton's early discoveries was the extension of the binomial theorem to fractional and negative exponents. This enabled him to integrate algebraic functions, for example $\sqrt{(1-x^2)}$, by expressing them as power series and integrating term by term: $\int \sqrt{(1-x^2)}\,dx = \int[1 - (1/2)x^2 - (1/8)x^4 - (1/16)x^6 - \cdots]\,dx = x - (1/6)x^3 - (1/40)x^5 - \cdots$. The evaluation of this integral baffled Newton's predecessors. Of course the question of whether it is permissible to integrate a power series – an infinite sum – term by term as if it were a finite sum must be dealt with, but this was not an explicit concern of Newton and his contemporaries.

Newton and others also obtained power-series expansions of transcendental functions, notably the sine, cosine, logarithm, inverse tangent, and exponential functions. For example, the power series expansion of the logarithmic function was obtained as follows:

Expand $1/(1+x)$ in a power series (by the binomial expansion of $(1+x)^{-1}$ or by long division): $1/(1+x) = 1 - x + x^2 - x^3 + \cdots$. Integrate both sides to get the *Mercator series* (obtained independently by Newton):

$$\log(1+x) = x - (x^2/2) + (x^3/3) - (x^4/4) + \cdots. \tag{6.1}$$

Term-by-term integration requires justification.

The power series for $\arctan x$ (derived by James Gregory in 1668) can be obtained similarly: $1/(1+x^2) = 1 - x^2 + x^4 - x^6 + \cdots$; integrating both sides gives

$$\arctan x = x - (x^3/3) + (x^5/5) - (x^7/7) + \cdots. \tag{6.2}$$

Again, justification is called for.

Substitution of $x = 1$ (with proper justification) into formulas (6.1) and (6.2) yields two interesting results: $\log 2 = 1 - (1/2) + (1/3) - (1/4) + \cdots$ and $\pi/4 = 1 - (1/3) + (1/5) - (1/7) + \cdots$. The latter famous formula is due to Leibniz [10, p. 247].

Given the power-series expansions of $\log(1+x)$ and $\arctan x$, one can integrate these functions by integrating their power series term by term. Although these functions can now be integrated by more standard procedures, there are functions which cannot, and for which expansion in a power series is the key method of integration. Thus it was *proved* in the nineteenth century that such functions as e^{-x^2}, e^x/x, $\sqrt{(1-x^3)}$, $\cos x^2$, $\sin x/x$ cannot be integrated in finite terms – that is, although the integrals exist, they are not elementary functions [23]. To evaluate such $\int f(x)dx$, one obtains the power-series expansion of $f(x)$ and integrates term by term (again, with proper justification).

For example, $e^{-x^2} = 1 - x^2 + x^4/2! - x^6/3! + \cdots$ (one gets this from the power-series expansion of e^x; see below). Hence $\int e^{-x^2}dx = x - x^3/3 + x^5/5 \cdot 2! - x^7/7 \cdot 3! + \cdots$. One can now evaluate the definite integral: $\int_0^1 e^{-x^2}dx = 1 - 1/3 + 1/5 \cdot 2! - 1/7 \cdot 3! + \cdots$, and approximate this integral to any desired degree of accuracy. See [10, 15, 28] for details.

Power series – sometimes called the infinite decimals of analysis – continued to be a major tool of the calculus in the eighteenth century. In fact, Euler claimed that *every* function can be expanded in a power series, and challenged mathematicians to prove him wrong! (see Chap. 5). (He allowed for the possibility of nonintegral exponents in a power series.) Indeed, the functions considered in eighteenth-century calculus were for the most part expandable in power series. Moreover, the most frequent method of differentiation and integration during that century was by the use of power series, as described above. Thus Euler showed that the derivative of $\sin x$ is $\cos x$ by first expanding these functions in power series. (Newton had given these expansions earlier, but Euler's was the first analytic derivation [15].) Thus, given

$$\sin x = x - x^3/3! + x^5/5! - \cdots, \text{ and } \cos x = 1 - x^2/2! + x^4/4! - \cdots,$$

Euler differentiates the power series of $\sin x$ term by term and obtains the power series of $\cos x$, so that $(d/dx)(\sin x) = \cos x$.

An important advantage of viewing transcendental functions as power series is that one can, following Euler, readily extend their definition to *complex* values of the variable. Thus $\sin z$, $\cos z$, and e^z are defined for any complex z as:

$$\sin z = z - z^3/3! + z^5/5! - \cdots, \quad \cos z = 1 - z^2/2! + z^4/4! - \cdots,$$

$$e^z = 1 + z + z^2/2! + z^3/3! + \cdots.$$

Nowadays, the convergence of these series would have to be considered, though it was not in Euler's time.

To set the record straight, we make several elementary observations about power series.

(a) A power series $\sum_0^\infty a_n x^n$ does not, in general, converge for all x: it converges for $|x| < r$ and diverges for $|x| > r$, where r is a nonnegative real number or ∞, called the *radius of convergence* of the power series. In the extreme cases, the power series may not converge for any $x \neq 0$, or it may converge for all x, in which cases we formally designate r as 0 or ∞, respectively.

(b) If $f(x) = \sum_0^\infty a_n x^n$ for $|x| < r$, one can differentiate and integrate the power series term by term, infinitely often. All the resulting power series also converge for $|x| < r$. Thus if $f(x)$ has a power-series representation (for $|x| < r$), it must be infinitely differentiable (for $|x| < r$). Differentiating both sides of $f(x) = \sum_0^\infty a_n x^n$ repeatedly and setting each time $x = 0$ (which is permissible), we get $a_n = f^{(n)}(0)/n!$, hence $f(x) = f(0) + f'(0)x + (f''(0)/2!)x^2 + (f'''(0)/3!)x^3 + \cdots$, the *Taylor expansion of* $f(x)$ ($f^{(n)}$ is the n-th derivative of f).

(c) It follows from the above that infinite differentiability of a function is a necessary condition for its representability by a power series. But it is not sufficient, as the classic example

$$f(x) = \begin{cases} e^{-1/x^2}, & x \neq 0 \\ 0, & x = 0 \end{cases},$$

given by Cauchy in the 1820s shows. It is easy to prove that $f(x)$ is infinitely differentiable for all real numbers x (including 0), and that $f^{(n)}(0) = 0$ for all n. But $f(x) = f(0) + f'(0)x + (f''(0)/2!)x^2 + \cdots$ is impossible, except for $x = 0$, since the right-hand side is zero for all x while the left-hand side is zero only for $x = 0$. See [28] for details.

Power series are nevertheless a powerful tool in analysis; they are central in complex analysis. See Sect. 6.6 for the use of power series in the solution of differential equations.

6.4.1 Remarks on Teaching

It would be instructive to introduce power series *early* in the study of calculus. To do that one would need to know the derivative of x^n, the binomial theorem, and little else (see [10]). Power series can:

1. Encourage students to think of analogies with polynomials – what the two have in common and how they differ. Analogy is a potent tool for mathematical discovery, but one which ought to be treated with caution. For example, the range of values for which polynomials can be differentiated and integrated term by term is all of **R**, whereas the corresponding range for power series is their interval of convergence.
2. Give students the tools for numerical computations and approximations. For example, the Gregory series $\arctan x = x - x^3/3 + x^5/5 - \cdots$ was used, with $x = 1/\sqrt{3}$, to get $\pi/6 = 1/\sqrt{3}(1 - 1/3 \cdot 3 + 1/3^2 \cdot 5 - 1/3^3 \cdot 7 + \cdots)$, and thus to calculate π to 72 decimal places. One can, in fact, put the arctan series to better use to get series approximations of π which converge more rapidly, and these ideas invite students to think of convergence and of rates of convergence. See [1, p. 144].

3. Exhibit unexpected relations among functions. For example, the famous and important Euler-Cotes formula $e^{i\theta} = \cos\,\theta + i\sin\,\theta$ relating the exponential and trigonometric functions is readily obtained from the power-series expansions of $\sin\,\theta$, $\cos\,\theta$, and e^{θ}. Putting $\theta = \pi$ in the formula yields $e^{i\pi} + 1 = 0$, the famous equality relating the five most important numbers in mathematics. Setting $\theta = \pi/2$ in the Euler-Cotes formula gives Euler's remarkable result $i^{i} = e^{-\pi/2}$. In the 1930s it was shown that this implies that e^{π} is transcendental. See [26, p. 134].
4. Help derive interesting results not directly related to calculus. Two such results were obtained above.

Teachers of calculus courses should be encouraged to give such nonroutine, interesting applications of the ideas of calculus which are not part of the prescribed curriculum. Such "gems" help decompartmentalize mathematics and can enliven a mathematics class.

6.5 Functions Defined by Integrals

We often define the log function in terms of an integral: $\log x = \int_1^x (1/t)dt$. All the usual properties of logarithms can be recovered from this definition. More generally, given any continuous function $f(x)$, one can define another function $F(x)$ by setting $F(x) = \int_a^x f(t)dt$ (the integral of a continuous function always exists), and proceed to study its properties (e.g., $F(x)$ is differentiable). Of course, to merit special attention such functions should be of some intrinsic interest. In the eighteenth and nineteenth centuries functions of the type $F(x) = \int R(x, \sqrt{P(x)})dx$, where $P(x)$ is a polynomial of degree 3 or 4 and $R(x, \int R(x, \sqrt{P(x)})$ is a rational function (a quotient of polynomials) of x and $\sqrt{P(x)}$, were singled out for study; $\int \sqrt{(1-x^3)}dx$, and $\int [x/\sqrt{(1-x^4)}]dx$ are examples. Such integrals are called *elliptic integrals* since they arise in finding the length of an arc of an ellipse [28]. It was shown by Liouville in the first half of the nineteenth century that such integrals cannot be evaluated in terms of elementary functions [23]. They are examples of new types of transcendental functions, the so-called *higher transcendental functions* or *special functions*. See [29].

Various properties of functions defined by elliptic integrals were obtained in the eighteenth century, but the crucial developments, due to Abel and Jacobi, came in the early nineteenth century. The major idea of Abel and Jacobi was to "invert" these functions, namely to focus not on the functions $F(x) = \int R(x, \sqrt{P(x)})dx$ but on their inverses. These inverse functions, now called *elliptic functions,* yielded more interesting properties than the original elliptic integrals. For example, let $F(x) = \int_0^x dt/\sqrt{(1-t^2)}$, $-1 < x < 1$. This is the arcsine function, and its inverse is a branch of the sine function.

The elliptic functions were, in turn, generalized by Poincaré in the late nineteenth century to the so-called *automorphic functions* [15]. The important change in point of view was to study the elliptic functions, and later the automorphic functions, as functions of a *complex* variable. In fact, complex function theory (the calculus of complex functions) was one of the most beautiful and important inventions of nineteenth-century mathematics, and the theory of elliptic functions became an important branch of that vast subject, proving to have applications in algebra and number theory.

Another important higher transcendental function given by an integral is the *gamma function*, $\Gamma(x) = \int_0^\infty t^{x-1}e^{-t}dt$, x positive. It was defined by Euler in the 1730s as an extension of the factorial to noninteger values. Euler showed that $\Gamma(x+1) = x\Gamma(x)$ (try to show this using integration by parts), from which it follows by induction that $\Gamma(n+1) = n!$ for positive integral n. It can also be shown, for example, that $\Gamma(1/2) = \sqrt{\pi}$ [29, p. 236]. The gamma function has turned out to be important in analysis, geometry, number theory, probability, and physics [15].

Two more examples of important higher transcendental functions defined by integrals are:

(a) $F(x) = [1/\sqrt{2\pi}]\int_{-\infty}^{x} e^{-t^2/2}dt$, the *normal distribution function* arising in probability, introduced by De Moivre in the eighteenth century and used widely by Laplace and Gauss in the early nineteenth century.
(b) $G(x) = \int_2^x dt/\log t$, introduced and studied extensively in the nineteenth century in connection with the problem of the distribution of primes among the integers [15, p. 830]. It was shown that $\int_2^x dt/\log t \sim x/\log x$ ($f(x) \sim g(x)$ means that $\lim_{x\to\infty} f(x)/g(x) = 1$). At the end of the nineteenth century it was proved that $x/\log x \sim \pi(x)$, where $\pi(x)$ denotes the number of primes $\leq x$. This is the justly famous *prime number theorem.*

6.5.1 Remarks on Teaching

Teaching integration techniques without indicating that some indefinite integrals cannot be expressed in finite terms, that is, by means of elementary functions, does not serve students well. The other important point to bear in mind is that the inability to integrate some functions in finite terms had very positive consequences: it provided mathematicians with a useful method for creating important new functions, such as the elliptic functions. These functions have been tabulated and studied extensively. Practiced mathematicians feel no less at home with elliptic functions than students do with circular (trigonometric) functions. See [15, 29].

6.6 Functions Defined as Solutions of Differential Equations

A *differential equation* is an equation involving an unknown function and one or more of its derivatives; more precisely, it is an equation of the form $F(x, y, y', y'', \ldots, y^{(n)}) = 0$, for some function F, where y is a function of x, and $y', y'', \ldots, y^{(n)}$

are its first, second, ..., $n-$th derivatives. To be exact, this is an *ordinary* differential equation, in contrast to a partial differential equation, to be considered in the next section. Examples are $y'' - xy = e^x$ and $y' = y^2/(1 - xy)$. Differential equations serve as mathematical models for various physical phenomena [29], and it can thus be vitally important to find the function $y(x)$ that satisfies a given differential equation. How to do this comprises the vast and rich branch of analysis known as the theory of differential equations, begun modestly by Newton and Leibniz and continuing with great vigor to this day.

The simplest differential equation is $y' = f(x)$. Its solution is, of course, $y = \int f(x)dx$. We have considered some functions y of this type in the previous section. Another simple example of a differential equation is $y' = ky$; it is a mathematical model of growth and decay. We can readily solve this equation: $y'/y = k$, hence $\int (y'/y)dx = \int k dx$. This gives $\log y = kx + k'$, hence $y = ce^{kx}$ for some constant c. This is an ad hoc procedure not suitable for solving most types of differential equations. We will now solve the same equation by the use of power series, which constitutes a general method applicable to a wide variety of differential equations.

We assume that the equation $y' = ky$ has a power-series solution $y = a_0 + a_1x + a_2x^2 + a_3x^3 + \cdots$, and try to determine the coefficients. We have $y' = a_1+2a_2x+3a_3x^2+\cdots$ (as we noted earlier, a power series can be differentiated term by term within its interval of convergence). Since $y' = ky$, $a_1+2a_2x+3a_3x^2+\cdots = k(a_0 + a_1x + a_2x^2 + a_3x^3 + \cdots)$. Comparing coefficients we get $a_1 = ka_0, 2a_2 = ka_1 = k^2a_0$, hence $a_2 = k^2a_0/2, a_3 = k^3a_0/3 \cdot 2, \ldots, a_n = k^n a_0/n!, \ldots$. Hence $y = a_0 + a_1x + a_2x^2 + \cdots = a_0(1 + kx + ((kx)^2/2!) + (kx)^3/3!) + \cdots)$. Thus, if the differential equation $y' = ky$ has a power-series solution, the power series must be of the indicated form. Conversely, we readily verify that $y = a_0(1 + kx + ((kx)^2/2!) + ((kx)^3/3!) + \cdots)$ *is* a solution of the differential equation $y' = ky$ and that the power series converges for all real numbers x [28]. We recognize this power series as representing the function a_0e^{kx}, and we have thus recovered the solution obtained above by other means.

Note that we need to know nothing about the exponential function to solve the equation $y' = ky$ in power series. In fact, we can *define* a function $f_k(x)$ as *the* solution of $y' = ky$ satisfying $y(0) = 1$. The existence and uniqueness of a solution in this case are easy to show [22]. We can then show that $f_k(x)$ satisfies the usual properties of an exponential function. For example, $f_k(x + a) = f_k(x)f_k(a)$, because both sides are solutions of $y' = ky$ with the same value $f_k(a)$ at $x = 0$. We can then *define* e^x to be $f_1(x)$.

This is the reverse of the approach usually taken in texts, but it is logically just as valid. It can also be used to *define* the log function as the solution, by means of power series, of the differential equation $xy' = 1$, as well as to define the sin and cos functions as power-series solutions of the differential equation $y'' + y = 0$. See [17] for details.

The importance of the power-series method for solving differential equations lies, however, in its application to the solution of differential equations which *cannot* be solved by simple integrations which lead to elementary functions. Thus, although it is known that solutions of the differential equation $y'' + P(x)y' + Q(x)y = 0$ exist, it is also known that such solutions cannot, in general, be obtained in terms of elementary functions, unless $P(x)$ and $Q(x)$ are constants (just as we know that $\int e^{x^2}dx$ exists but cannot be expressed as an elementary function). In such cases recourse to power-series solutions has been fundamental. Among the important differential equations solved in this manner, some in the eighteenth but most in the nineteenth century, are:

The *Bessel equation* $x^2y'' + xy' + (x^2 - a^2y) = 0$
The *Legendre equation* $(1 - x^2)y'' - 2xy' + a(a + 1)y = 0$
The *Hermite equation* $y'' - 2xy' + 2ay = 0$
The *hypergeometric equation* $x(1 - x)y'' + [c - (a + b + 1)x]y' - aby = 0$

(the equations depend on the parameters a, b, c). The power-series solutions of these equations define new and important classes of higher transcendental functions: *Bessel functions, Legendre functions, Hermite functions,* and *hypergeometric functions.* For example, $y(x) = 1 - (2a/2!)x^2 + [(2^2a(a-2)/4!]x^4 - [(2^3a(a-2)(a-4)/6!]x^6 + \cdots$ is a Hermite function. See [22].

Perhaps the most important of these functions is the hypergeometric function,

$$F(a,b,c,x) = 1+[ab/1\cdot c]x+[a(a+1)b(b+1)/1\cdot 2c(c+1)]x^2 + [a(a+1)(a+2)b(b+1)(b+2)/1\cdot 2\cdot 3c(c+1)(c+2)]x^3+\cdots,$$

studied in the early nineteenth century by Gauss, who showed that the series converges for $|x| < 1$ [25]. This was likely the earliest satisfactory treatment of the convergence of an infinite series. Note that $F(1,b,b,x) = 1 + x + x^2 + \cdots$, thus the hypergeometric function (series) is an extension of the geometric series, hence its name. Moreover, a number of the major functions of analysis are related to the hypergeometric function:

$$\begin{aligned}(1+x)^a &= F(-a,b,b,-x), \log(1+x) = xF(1,1,2,-x),\\ \sin^{-1}x &= xF(1/2,1/2,3/2,x^2),\ e^x = \lim_{b\to\infty} F(a,b,a,x/b),\\ \sin x &= x\lim_{a\to\infty} F(a,a,3/2,-x^2/4a^2),\\ \cos x &= \lim_{a\to\infty} F(a,a,1/2,-x^2/4a^2).\end{aligned}$$

The higher transcendental functions mentioned above are among scores of "special functions" considered of sufficient interest in the eighteenth and nineteenth centuries to merit individual study. Although the utility of these functions is undiminished, they are now also studied collectively via a group-theoretic approach. See [15, 29].

6.6.1 Remark on Teaching

In the last few sections we have indicated some of the different ways in which functions arise in analysis. The idea was to show the rich and multifaceted nature of functions. Students might undertake as a project a comparison of various definitions of (say) the log function – in terms of continuous motions of points along straight lines (Napier's definition), as the solution of a differential equation, as the inverse of the exponential function, and as an integral – and discuss advantages and disadvantages of the different approaches. See [6a, 10, 13, 15, 29].

6.7 Partial Differential Equations and the Representation of Functions by Fourier Series

Partial differential equations, especially the famous (one-dimensional) *wave equation* $\partial^2 y/\partial t^2 = a^2\partial^2 y/\partial x^2$ and *heat equation* $\partial y/\partial t = a^2\partial^2 y/\partial x^2$, have been of fundamental importance in the evolution of the function concept. The wave equation is a mathematical model for the vibrations of an elastic string fixed at both ends (e.g., a violin string), with $y(x, t)$ representing the displacement of the string from its original position at time t. This equation was solved in the mid-eighteenth century by d'Alembert, Euler, and D. Bernoulli. Their respective solutions differed, however, a fact which caused considerable debate and controversy among them. The debate centered on admissible initial forms of the string and resulted in the extension of the then-held view of functionality.

Recall that Euler defined a function as an analytic expression, namely a *single* algebraic formula. As a result of the debate over the vibrating-string problem, the concept of function was extended to include

(a) Expressions given by several formulas, for example

$$f(x) = \begin{cases} x^2, & x \geq 0 \\ x, & x < 0 \end{cases},$$

(b) Freely drawn curves

This extension of the function concept was not, however, universally accepted in the eighteenth century, since contemporary practice in the calculus could not be readily extended to such functions. See [21, 35], and Chap. 5 for details.

The heat equation $\partial y/\partial t = a^2\partial^2 y/\partial x^2$ is a mathematical model for the distribution of temperature through a body. Fourier solved the equation in the early nineteenth century and concluded, incorrectly as it turned out (see below), that *any* function $f(x)$ defined on an interval $(-c, c)$ can be represented by an infinite series of sines and cosines ($f(x)$ was to represent the initial temperature distribution $y(x, 0)$ of the body): $f(x) = a_0/2 + \sum_1^\infty (a_n \cos n\pi x/c + b_n \sin n\pi x/c)$, where the coefficients a_n and b_n are given by $a_n = 1/c \int_{-c}^{c} [f(t)\cos(n\pi t/c)]\mathrm{d}t$, $b_n = 1/c \int_{-c}^{c} [f(t)\sin(n\pi t/c)]\mathrm{d}t$. See [10, 15, 21, 29] for details.

Fig. 6.1 Joseph Fourier (1768–1830)

Fourier's result took the mathematical community by surprise since it ran counter to fundamental eighteenth-century tenets. In particular, mathematicians questioned

(a) How a function given by two or more distinct expressions (formulas) can equal a function given by a single expression, namely, the Fourier series of the function.
(b) How a sum, albeit infinite, of periodic functions can equal a function which need not be periodic, the equality, admittedly, being only on an interval.
(c) How a sum of "smooth" functions such as the sine and cosine can equal a function with (say) corners (a nondifferentiable function), or worse, a function with breaks (a discontinuous function).

Although Fourier's result that *any* function can be represented by a Fourier series turned out to be incorrect, a large class of functions *can* be so expressed. In particular, Dirichlet showed in 1829 that any function $f(x)$ with finitely many discontinuities and finitely many maxima and minima in an interval $(-c, c)$ can be represented in that interval by a Fourier series, which converges to $f(x)$ at the points of continuity of $f(x)$, and to $1/2(f(x_0^+) + f(x_0^-))$ at points x_0 of discontinuity of $f(x)$. The class of such functions is quite broad, including essentially all functions studied in elementary calculus. In particular, there are functions which confirm the possibilities raised in (a), (b), and (c) above. Indeed, the discontinuous (hence nondifferentiable) nonperiodic function

$$f(x) = \begin{cases} -1, & x < 0 \\ 0, & x = 0 \\ 1, & x > 0 \end{cases}$$

given by three separate expressions can be represented in $(-\pi, \pi)$ by a single expression, which is the sum of differentiable (hence continuous) periodic functions,

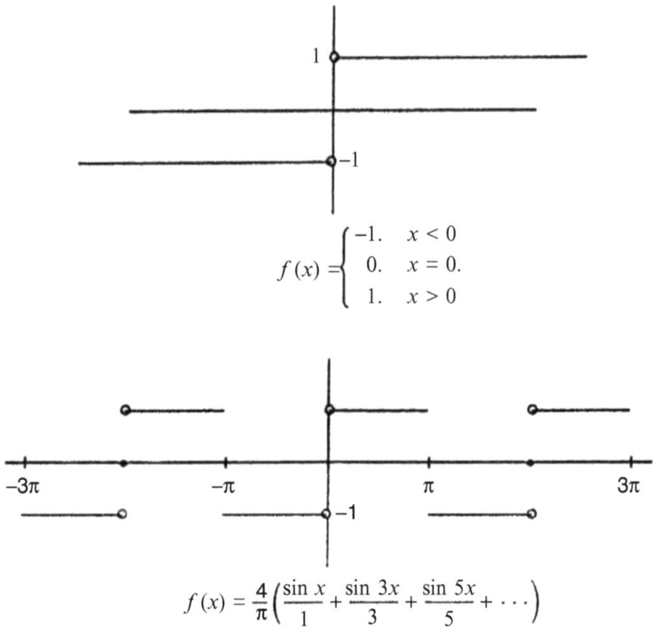

Fig. 6.2 A function given by three separate expressions is represented in $(-\pi, \pi)$ by a single expression, its Fourier series

namely the Fourier series $(4/\pi)[(\sin x)/1+(\sin 3x)/3+(\sin 5x)/5+\cdots]$ of $f(x)$. (Compute the coefficients a_n and b_n by the above formula and note that the $a_n = 0$; see [13, pp. 8–11] for details.) See Fig. 6.2 for the graphs of the two functions.

Fourier's work necessitated a careful examination of the basic concepts of calculus, in particular continuity, convergence, and the definite integral. In fact, none of these ideas was rigorously formulated before the nineteenth century (see [10, 15]). Fourier's work also focussed attention on the meaning of function, and especially on the *domain* of a function as an important ingredient of its makeup. Indeed, the function $f(x) = x$ can be represented by the Fourier series $g(x) = \pi - 2\sum_1^\infty (\sin nx)/n$ for $0 < x < 2\pi$, but $f(x)$ and $g(x)$ differ outside the interval $(0, 2\pi)$ since $g(x)$ is a periodic function while $f(x)$ is not. Thus two functions can agree on an interval but nowhere outside it. That would have been unacceptable to eighteenth-century mathematicians, since to them equality of functions on an interval implied their equality on the entire real line. See [35] and Chap. 5.

We note that Fourier series offer a much more comprehensive way of representing functions than do power series (Sect. 6.4). To be representable by a power series in an interval, a function must be infinitely differentiable there, and this is only a necessary condition, while to be representable by a Fourier series in an interval, a function need not even be continuous there. The subject of Fourier series has inspired diverse and important mathematical discoveries, including Cantor's set theory (see [34]). It is still intensively investigated.

6.7.1 *Remarks on Teaching*

The following pedagogical points, drawn from the above historical account, are worthy of note:

(a) It is well known that mathematics has been widely applied in the physical (and other) sciences. It is probably little known that, conversely, physical problems have had an important impact on the development of mathematics. The vibrating-string problem and the heat-conduction problem are excellent examples of the latter phenomenon. Teachers should become aware of this important insight and exploit it whenever possible.
(b) Fourier's work was significant as much for the questions it raised as for the answers it provided: questions about the nature of functions, continuity, convergence, and integration. Questions and problems are the lifeblood of mathematics, and they should, whenever possible, be starting points for work in the mathematics classroom.
(c) We have noted the unnaturalness, for beginning students, of restricting the domain of definition of a function to a subset of the reals, and of defining functions on a split domain. Both of these issues came to the fore in Fourier's work. By showing that two functions may be identical on an interval but differ outside it, Fourier *forced* mathematicians to reckon with functions defined on intervals. More elementary, "natural" examples which impose a restricted domain of definition on functions are the *inverses* of the exponential and trigonometric functions. For example, while the exponential function is defined for all reals, the log function is not. Other examples are functions represented by power series: it is clear that, for example, $f(x) = 1 + x + x^2 + \cdots$ is not defined for $x \geq 1$ (the prohibition $x \leq -1$ is more difficult for students to contend with).

Fourier also showed that the distinction between functions defined by a single formula and those given by two or more formulas is irrelevant: the latter can also be given by a single formula. An elementary example, due to Cauchy, is

$$f(x) = \begin{cases} x, & x \geq 0 \\ -x, & x < 0 \end{cases},$$

which can also be represented as $f(x) = \sqrt{x^2}$ or $f(x) = 2/\pi \int_0^\infty [x^2/(x^2 + t^2)]\mathrm{d}t$.

(d) Fourier's and Cauchy's examples point to the important conceptual distinction between a function and its description(s): different formulas may describe the same function. These ideas also highlight the distinction between a function and its values: two functions are equal, even if they do not "look" alike, provided they have the same values.

6.8 Functions and Continuity

Although the notion of continuity is nowadays fundamental in calculus, it did not arise, at least in the way we understand it, until the nineteenth century, about 150 years after the invention of calculus by Newton and Leibniz. In the eighteenth century, Euler did define a notion of "continuity" to distinguish between functions as analytic expressions and the new types of functions which emerged from the vibrating-string debate. Thus a continuous function was one given by a single analytic expression, while functions given by several analytic expressions or freely drawn curves were considered discontinuous. For example, to Euler the function

$$f(x) = \begin{cases} x^2 & x > 0 \\ x & x \leq 0 \end{cases},$$

was discontinuous, while the function comprising the two branches of a hyperbola was considered continuous (!) since it was given by the single analytic expression $f(x) = 1/x$.

The work on Fourier series showed the untenability of the eighteenth-century notion of continuity. Indeed, a function such as

$$f(x) = \begin{cases} -1, & -\pi < x < 0 \\ 0, & x = 0 \\ 1, & 0 < x < \pi \end{cases}$$

could be represented (as we have seen) by a single analytic expression, namely its Fourier series, hence it was *both* continuous *and* discontinuous in the eighteenth-century sense of that concept. So was a freely drawn curve. (Fourier gave a heuristic argument to show that functions given by such curves have Fourier-series expansions.)

In his important *Cours d'Analyse* of 1821 Cauchy initiated a reappraisal and reorganization of the foundations of eighteenth-century calculus. In this work he defined continuity essentially as we understand it, although he used the then-prevailing language of infinitesimals rather than the now-accepted $\varepsilon - \delta$ formulation given by Weierstrass in the 1850s (see [7, 10] and Chap. 5). The shift in point of view from Euler's to Cauchy's conception of continuity was fundamental: from continuity as a global property – that is, a function defined by a single expression on the whole real line, to continuity as a local property – a function defined at each point of an interval. Nevertheless the concept proved to be subtle, and was not completely understood even by Cauchy and his contemporaries in the early nineteenth century. For example:

(a) Continuity was sometimes identified with the geometric notion of traceability with an uninterrupted motion of (say) pen on paper. This identification is incorrect: not every continuous function can be drawn, as Weierstrass showed in the 1860s (see (c) below). Continuity was also identified with the Intermediate

Value Property (IVP), namely that a function defined on a closed interval takes on all values intermediate between those at the endpoints. This identification, too, is incorrect: although a continuous function *on a closed interval* has the IVP, the function

$$f(x) = \begin{cases} \sin(1/x), & x \neq 0 \\ 0, & x = 0 \end{cases},$$

has the IVP in any interval $[a, b]$ with $a < 0, b > 0$, but is not continuous at $x = 0$ (see [10, 28]). This example was given by Darboux c. 1870.

(b) Cauchy "proved" that an infinite sum (a convergent series) of continuous functions is a continuous function. This is incorrect of course, although it is true for finite sums. A counterexample was given by Abel in the 1820s – it is essentially the series $\sum_0^\infty \sin(2n+1)x/(2n+1)$ we encountered earlier (Sect. 6.7), which is discontinuous at $x = k\pi, k = 0, \pm 1, \pm 2, \ldots$ [13, p. 269]. The error in Cauchy's proof resulted from his failure to distinguish between convergence and uniform convergence of a series of functions. (A *uniformly* convergent series of continuous functions is indeed continuous.) Cauchy also failed to distinguish between pointwise and uniform continuity of a function – a fundamental, albeit subtle, distinction [15, 28].

(c) Euler's continuous functions were, in practice, differentiable, except possibly at isolated points. So were Cauchy's – at least this is what Cauchy and his contemporaries believed, and what some of them "proved." It was therefore an astonishing event when Weierstrass gave (in the 1860s) an example of a continuous function which is *nowhere* differentiable. (Bolzano had given an example c. 1830 which went unnoticed.) Although such functions can be constructed by elementary means, the constructions entail the use of a limiting process and the functions cannot be drawn. The following graph (Fig. 6.3) represents one stage of the limiting process (the dots represent the results of subdivisions), yielding Bolzano's example of a continuous nowhere-differentiable function [30, p. 35]. See [13, 21] for other examples.

These examples showed that the concept of continuity is considerably broader than that of differentiability, and thus established continuity as an important concept of investigation in its own right. The examples also showed the limitations of intuitive geometric reasoning in analysis – the continuous nowhere-differentiable functions, that is, continuous functions with tangents at no point, are clearly entirely nonintuitive – and thus the need for careful, analytic formulations of basic notions.

Such "pathological" functions, however, were not greeted with universal sanction, as Hermite's statement indicates: "I turn away with fright and horror from this regrettable plague of [continuous] functions which do not have derivatives" [15, p. 973]. (Cf. mathematicians' opposition to the introduction of negative, irrational, and complex numbers [15].) Such functions are now, however, commonplace in mathematics, and, in fact, are the most elementary examples of "fractals" [20].

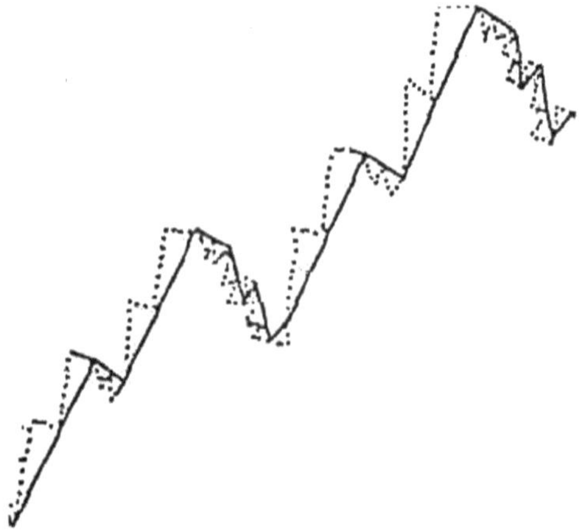

Fig. 6.3 A stage in the limiting process representing Bolzano's example of a continuous nowhere-differentiable function

(d) Cauchy was the first to give (in 1823) a rigorous definition of the definite integral (as a limit of a sum), such an integral being viewed in the past as an area, or as an antiderivative evaluated at upper and lower limits [10]. He then proved the existence, on a closed interval, of the definite integral of a *continuous* function. The idea of proving the existence of a mathematical entity which one may not be able to evaluate or construct was also novel. The notion that *highly discontinuous* functions can have an "area" – that is, an integral – is due to Riemann, who in the 1850s extended Cauchy's concept of integral. He gave an example of a function with a dense set of discontinuities in any interval, no matter how small, which has an integral in his extended sense. (A set of points is "dense" if there exists a point of the set between any two others; for example, the rational numbers are a dense set.) One such example (not Riemann's; for his example see [10] or [15]) is

$$f(x) = \begin{cases} 1/|q|, & \text{if } x = p/q \text{ with } (p,q) = 1 \\ 0, & \text{if } x \text{ is irrational} \end{cases}.$$

This function, which gives another indication of the subtlety of the notion of continuity, is continuous at the irrationals and discontinuous at the rationals (try to show that $\lim_{x\to a} f(x) = 0$ for all real a). By a result of Lebesgue, its (Riemann) integral exists over any interval $[a, b]$. On the other hand, the famous *Dirichlet function*,

$$D(x) = \begin{cases} 1, & x \text{ rational} \\ 0, & x \text{ irrational} \end{cases},$$

given in Dirichlet's 1829 paper on Fourier series, is everywhere discontinuous and does not have a Riemann integral. In one of the great achievements of early-twentieth-century mathematics, Lebesgue extended still further the notion of integral, giving meaning in particular to the integral of such everywhere-discontinuous functions as $D(x)$.

6.8.1 *Remarks on Teaching*

Which definition of continuity should we give students? It depends, of course, on their level of mathematical sophistication. We advocate tentative definitions of continuity, just as we advocated tentative definitions of function. Defining continuity as traceability with an uninterrupted motion of the chalk on the board is entirely acceptable for beginning students *provided* they are told that this does not represent "the whole story." It does represent mathematical honesty on the part of the teacher, which is always desirable. Mathematical honesty should not, however, be confused with mathematical rigor, which is *not* always desirable.

Counterexamples are wonderful devices for illustrating the subtleties of mathematical concepts. The above examples can be used in the classroom to show that (1) the notion of continuity is subtle; (2) the notion of area is subtle; and (3) geometric intuition, though fundamental, *can* be misleading. Formal, rigorous mathematics is with us to stay. See also Chap. 8.

6.9 Conceptual Aspects of Functions

In the last few sections we have described various methods introduced in the eighteenth and nineteenth centuries for generating and representing functions. But what was the nature of the function *concept* during that period? Three major elements or threads can be discerned in its evolution: the algebraic – a formula, an analytic expression; the geometric – a curve; and the abstract – a rule, a correspondence. At various times, one or another of these views of functions dominated.

As we noted, the number and variety of curves increased dramatically in the early seventeenth century due to the invention of analytic geometry and the rise of mathematized science. Curves and their equations were the main objects of study of seventeenth-century calculus. Gradually, the algebraic-algorithmic aspects of the calculus evolved without reference to curves. The concept of function arose in the early eighteenth century to facilitate this development. The concept was algebraic. For much of the eighteenth century a function was thought of as a single formula, the so-called analytic expression. Following the vibrating-string debate in the mid-eighteenth century, the concept was extended to include functions given by two or more analytic expressions in different parts of the real line, as well as curves drawn freehand. This extended notion of function, however, was not dominant in the eighteenth century. See Chap. 5.

The early decades of the nineteenth century saw the emergence of the concept of function as an arbitrary rule or correspondence. This change of viewpoint, too, was forced on mathematicians by developments in calculus, especially those due to Fourier and Cauchy [7, 10, 15]. Nineteenth-century analysis differed in fundamental ways from that of the eighteenth century. The eighteenth century's largely global, algebraic perspective gave way in the nineteenth century to a local, analytic view. The study of the behavior of a function over the whole real line was replaced by a study of its properties at points of an interval on the real line. For example, in the eighteenth century the derivative of a function was computed from its algebraic representation, often as a power series (Sect. 6.4), while in the nineteenth century it was computed at a *point*, relying on the limit process. The notion of function changed to reflect this new perspective: from a universally valid formula to a rule focusing on a correspondence between numbers. The so-called Dirichlet definition of function in the 1820s is indicative of this new view [18, p. 264]:

> y is a function of a variable x, defined in the interval $a < x < b$, if to every value of the variable x in this interval there corresponds a definite value of the variable y. Also, it is irrelevant in what way this correspondence is established.

The Dirichlet function

$$D(x) = \begin{cases} 1, & x \text{ rational} \\ 0, & x \text{ irrational} \end{cases},$$

reflected the emerging spirit of functionality: It is not a function given by an analytic expression, nor can it be represented by a curve. It was a new type of function, the first of many "pathological" functions. See Chap. 5.

The domain and range in the Dirichlet definition of function are sets of *real numbers*. The notion of function with arbitrary sets as domain and range evolved gradually in the nineteenth century. (The concept of function with non-numerical domain is *implicit* much earlier: maps of the globe are functions of the sphere into the Euclidean plane; the derivative, as an operator, is a function with domain the set of differentiable functions and range the set of all functions; truth tables are functions with domain a set of statements and range the set $\{T, F\}$.) Functions arose as transformations in geometry, as homomorphisms in algebra, and as operators in analysis, with domains Euclidean spaces, groups or rings, and sequences or function spaces, respectively. When set theory was developed during the last decades of the nineteenth century, such examples of functions became subsumed under the general notion of a function (or mapping) between arbitrary sets. In 1917 Carathéodory defined a function as a rule of correspondence from an arbitrary set to the set of real numbers. In the 1930s Bourbaki gave his well-known definition of a function from a set A to a set B as a special kind of binary relation between A and B, and also as a set of ordered pairs (a subset of $A \times B$). (The idea of a function as a set of ordered pairs is essentially already present in Hausdorff's classic 1914 book on set theory.) See Chap. 5.

Dirichlet's broad conception of function as an arbitrary correspondence prevailed for much of the nineteenth century, but signs of dissatisfaction began to appear toward that century's end. Some claimed that Dirichlet's definition gave mathematicians too broad a license to create functions. Such misgivings were part of the broader context of the rise of the intuitionist school of mathematics, which began to question various nineteenth-century concepts and practices, including the notion of mathematical existence. The phrase "there exists" occurs frequently in mathematics, but what does it mean? Does it signify merely the absence of a logical contradiction, or some transcendent reality? This was a major issue in debate in the early twentieth century between the intuitionist and formalist schools of mathematical philosophy. As applied to functions, if, for example, we define $f(x)$ by

$$f(x) = \begin{cases} 1, & \text{if } x \text{ is a positive integer and there are } x \text{ successive zeros} \\ & \text{in the decimal expansion of } \pi \\ 0, & \text{otherwise} \end{cases},$$

does $f(x)$ exist? Is it well defined? While the formalists would respond in the affirmative, the intuitionists would take the opposite view. To them $f(x)$ is not a bona fide function since we cannot determine its values for all values x in the domain. For example, what is $f(99)$? We do not know if $f(99) = 1$ or 0 since we do not know, and may never know, if there are 99 consecutive zeros in the decimal expansion of π. Hence to the intuitionists such a function does not exist. The debate has not been resolved to date. See Chap. 5.

6.9.1 *Remarks on Teaching*

At some point in their mathematical studies, students should come to realize that functions are not just analytic expressions or curves, although the two are prototypes of dominant currents of mathematical thought – the algebraic (or analytic) and the geometric (or synthetic). They should also come to understand local vs. global and existential vs. constructive notions. And they should come to recognize what may seem to them a startling phenomenon: mathematicians can have fundamental and irreconcilable differences. See Chap. 10.

6.10 Analytically Representable Functions

Despite the abstract definition of function as a rule or correspondence, interest persisted in "concrete," analytic representations of functions. Which functions are so representable? That is, which functions can be given by "formulas"?

In the first half of the eighteenth century a function was defined to be an analytic expression, hence every function was representable analytically by virtue of its definition, the power series being the "universal mode" of such representation. (We recall that to Euler and his contemporaries every function was representable by a power series.) In the latter part of the eighteenth century the notion of function was extended to include functions defined on split domains and freely drawn curves. In the early nineteenth century Fourier showed, albeit nonrigorously, that such functions too can be represented analytically, namely by Fourier series.

The "universal mode" of analytic representability of functions now became the trigonometric series. Subsequent decades of the nineteenth century saw the rise of various "pathological" functions, often defined by rules of correspondence and thus, presumably, not analytically representable. The Dirichlet function

$$D(x) = \begin{cases} 1, & x \text{ rational} \\ 0, & x \text{ irrational} \end{cases}$$

was one such example.

At the end of the nineteenth century Baire, and later Lebesgue, took up again the question of analytic representability of functions. Baire's "universal mode" of analytic representability was given by denumerable limits of continuous functions. That is, a function $f(x)$ was (to Baire) analytically representable if $f(x) = \lim_{n_q \to \infty} \dots \lim_{n_2 \to \infty} \lim_{n_1 \to \infty} f_{n_1, n_2, \dots, n_q}(x)$ is a continuous function (see Chap. 5). (By a previous result of Weierstrass the continuous functions were representable as limits of polynomials.) Using his scheme, Baire showed that the "pathological" Dirichlet function is, in fact, a "tame," analytically representable function, namely $D(x) = \lim_{m \to \infty} \lim_{n \to \infty} (\cos m!\pi x)^n$.

Lebesgue showed that Baire's analytically representable functions include "almost all" functions which appear in practice. (They are coextensive with the Borel-measurable functions.) On the other hand, one can show by a "counting argument" that the set of all analytically representable functions (à la Baire) has the cardinality c of the continuum, while the set of *all* functions (from the reals to the reals) has cardinality 2^c. Thus there are (in theory) uncountably many functions which are not analytically representable, although we hardly ever encounter one in practice! See [35] and Chap. 5 for details.

6.10.1 *Remark on Teaching*

In this section we have introduced a new mode of representing functions, namely by the use of limits, and have thus extended the range of functions representable by formulas. Note that already in the eighteenth century Euler used limits in describing functions, when he showed that $e^x = \lim_{n \to \infty}(1 + x/n)^n$; and we have used them in this chapter since, for example,

$$\sin x = x - x^3/3! + x^5/5! - \cdots = \lim_{n \to \infty}[x - x^3/3! \\ + x^5/5! - \cdots + (-1)^n x^{2n+1}/(2n+1)!].$$

But what *is* a formula? This "simple" question has no simple answer; in fact, it has no definitive answer, as the historical tale has shown. The notion of formula has evolved, and is likely still evolving. This speaks to the changing nature of mathematics and argues for tentative definitions of mathematical concepts, as we have tried to suggest, to respond to pedagogical circumstances and needs.

6.11 Conclusion

It is perhaps surprising that such a simple, elementary concept as function has such a rich and varied history. On the other hand, the centrality of the function concept in analysis and, more generally, in mathematics as a whole, makes plausible the depth and complexity of its history.

What about the set-theoretic, ordered-pair definition of function? It is very broad and general, and when expressed in the form $f = \{(a, b) \in A \times B : \ldots\}$, brings into sharp relief the separate existence of f, apart from $f(x)$. But it does not capture the rich and creative history of the function concept. It may be satisfying logically, but hardly psychologically.

In general, too broad a concept may suggest weakness rather than strength. Functions as ordered pairs possess too few common properties to warrant serious study. It is continuous functions, differentiable functions, integrable functions, functions representable by power series or by Fourier series, and so on, which are objects worthy of consideration in analysis. In algebra, it is homomorphisms (structure-preserving mappings) rather than arbitrary mappings which are of the essence. In geometry, the transformations of interest are linear, or projective, or distance-preserving.

The ordered-pair definition of function may be useful in indicating what to include in, and perhaps more importantly what to exclude from, the stock of examples we call functions. It may be useful in courses in topology and functional analysis. And it should be presented to students, at some point in their studies, as a "theoretical construct" rather than as a working definition – just as students ought to see at some point the definition of the real numbers in terms of Dedekind cuts or Cauchy sequences, and of the integers in terms of the Peano axioms. In the final analysis, though, it is not so much what a function is as what you can do with functions which is of the essence. (Just as the question of what a number is, or a matrix, or... is a much less weighty issue than what you can do with numbers, or matrices, or....)

References

1. P. Beckmann, *A History of* π, St. Martin's Press, 1971.
2. C. B. Boyer, The foremost textbook of modern times, *Amer. Math. Monthly* 58 (1951) 223-226.
3. C. B. Boyer, Proportion, equation, function: three steps in the development of a concept, *Scripta Math.* 12 (1946) 5-13.
4. C. B. Boyer, Historical stages in the definition of curves, *National Math. Mag.* 19 (1945) 294-310.
5. R. C. Buck, Functions, in E. Begle (ed.), *Mathematics Education,* Yearbook of the Nat. Soc. Stud. Educ., Vol. 69, 1970, pp. 236-259.
6. F. Cajori, *A History of Mathematical Notations*, Vol. 2, Open Court Publ., 1929.

6a. F. Cajori, History of exponential and logarithmic concepts, *Amer. Math. Monthly* 20 (1913) 5-14, 35-47, 75-84, 107-117, 148-151, 173-182, 205-210.

7. A.–L. Cauchy, *Cours d'Analyse*, translated into English, with annotations, by R. Bradley and E. Sandifer, Springer, 2009 (orig. 1821).
8. R. B. Deal, Some remarks on "A dilemma in definition," *Amer. Math. Monthly* 74 (1967) 1258-1259.
9. T. Dreyfus and T. Eisenberg, Intuitive functional concepts: a baseline study on intuitions, *Jour. Res. Math. Educ.* 13 (1982) 360-380.
10. C. H. Edwards, *The Historical Development of the Calculus,* Springer-Verlag, 1979.
11. L. Euler, *Introductio in Analysin Infinitorum*, Vols. 1 & 2, translated into English by J. Blanton, Springer, 1989 (orig. 1748).
12. G. Ferraro and M. Panza, Developing into series and returning from series: a note on the foundations of eighteenth-centuiry analysis, *Hist. Math.* 30 (2003) 17-46.
13. A. Gardiner, *Infinite Processes: Background to Analysis*, Springer-Verlag, 1982.
14. D. W. Hight, Functions: dependent variables to pickle pickers, *Math. Teacher* 61 (1968) 575-579.
15. M. Kline, *Mathematical Thought from Ancient to Modern Times*, Oxford Univ. Press, 1972.
16. M. Kronfellner, Ein genetischer Zugang zum Funktionsbegriff, *Math. Didactica* 10 (1987) 81-100.
17. R. E. Langer, Functions. In *Insights into Modern Mathematics,* 23rd Yearbook of the NCTM, National Council of Teachers of Mathematics, 1957, pp. 241-272.
18. N. Luzin, Function, Parts I and II, *Amer. Math. Monthly* 105 (1998) 59-67 (Part I) and 263-270 (Part II).
19. M. A. Malik, Historical and pedagogical aspects of the definition of function, *Int. Jour. Math. Educ. Sci. Technol.* 11 (1980) 489-492.
20. B. Mandelbrot, *The Fractal Geometry of Nature*, W. H. Freeman, 1977.
21. J. H. Manheim, *The Genesis of Point-Set Topology,* Macmillan, 1964.
22. Z. Markovits, B. – S. Eylon, and M. Bruckheimer, Functions today and yesterday, *For the Learning of Mathematics* 62 (2) (1986) 18-24, 28.
23. D. G. Mead, Integration, *Amer. Math. Monthly* 68 (1961) 152-156.
24. National Council of Teachers of Mathematics, *Historical Topics for the Mathematics Classroom,* National Council of Teachers of Mathematics, 1969.
25. C. P. Nicholas, A dilemma in definition, *Amer. Math. Monthly* 73 (1966) 762-768.
26. I. Niven, *Irrational Numbers*, Math. Assoc. of America, 1956.
27. D. Rüthing, Some definitions of the concept of function from Joh. Bernoulli to N. Bourbaki, *Math. Intell.* 6 (4) (1984) 72-77.
28. G. F. Simmons, *Calculus with Analytic Geometry,* McGraw-Hill, 1985.
29. G. F. Simmons, *Differential Equations, with Applications and Historical Notes,* McGraw-Hill, 1972.
30. D. G. Tahta, A startling discovery, *Math. Teaching* 45 (1968) 33-37.
31. D. Tall and S. Vinner, Concept images and concept definition in mathematics with particular reference to limits and continuity, *Educ. Stud. Math.* 12 (1981) 151-169.
32. H. A. Thurston, A reply to "A dilemma in definition", *Amer. Math. Monthly* 74 (1967) 1259-1261.
33. G. Van Brummelen, *The Mathematics of the Heaavens and the Earth: The Early History of Trigonometry*, Princeton Univ. Press, 2008.
34. E. B. Van Vleck, The influence of Fourier's series upon the development of mathematics, *Science* 39 (1914) 113-124.
35. A. P. Youschkevitch, The concept of function up to the middle of the 19th century, *Arch. Hist. Exact Sci.* 16 (1976) 37-85.

Part C
Proof

Chapter 7
Highlights in the Practice of Proof: 1600 BC–2009

7.1 Introduction

> Mathematical rigor is like clothing: in its style it ought to suit the occasion, and it diminishes comfort and restricts freedom of movement if it is either too loose or too tight [63, p. ix].

The above observation, by G.F. Simmons, is sound pedagogical advice. It also reflects mathematical practice and its historical evolution. Standards of rigor have changed in mathematics, and not always from less rigor to more. Mathematicians' views of what constitutes an acceptable proof have evolved. In this chapter we will give examples pointing to that evolution. For further examples see Chaps. 8 and 9. For discussions of the nature and role of proof see [15, 21, 23, 27, 46, 48, 62, 74], and Chap. 10.

Several themes emerge:

(a) The validity of a proof is a reflection of the overall mathematical climate at any given time.
(b) The causes of transition from less rigor to more rigor, or vice versa, were, in general, not aesthetic or epistemological; there were good *mathematical* reasons for such changes.
(c) Every tightening or relaxation of the standards of rigor created new problems having to do with rigor – a familiar theme in mathematics: each time a problem is solved, new ones emerge.
(d) There was frequently no agreement among contemporary mathematicians about what makes for a satisfactory proof.

7.2 The Babylonians

Babylonian mathematics is the most advanced and sophisticated of pre-Greek mathematics, but it lacks proof. There are no general statements in Babylonian mathematics, and there is no attempt at deduction of the results or even at

I. Kleiner, *Excursions in the History of Mathematics*,

DOI 10.1007/978-0-8176-8268-2_7,

explanation of their validity. Mathematics without proof – a paradox? Their mathematics deals with specific problems, and the solutions are prescriptive – do this and that and you will get the answer. The following is an example (c. 1600 BC) of one such problem and its solution [39, p. 24]:

> I summed the area and two-thirds of my side-square and it was 0;35 [35/60 in sexagesimal notation]. [What is the side of my square?]

Solution. You put down 1, the projection. Two-thirds of 1, the projection, is 0;40. You combined its half, 0;20 and 0;20. You add [the result] 0;06,40 to 0;35 and [the result] 0;41,40 squares 0;50. You take away 0;20 that you combined from the middle of 0;50, and the square-side is 0;30.

In modem notation the problem is to solve the equation $x^2 + (2/3)x = 35/60$. The instructions for its solution can be expressed as:

$$\begin{aligned} x &= \sqrt{(0;40/2)^2 + 0;35} - 0;40/2 \\ &= \sqrt{0;06,40 + 0;35} - 0;20 \\ &= \sqrt{0;41,40} - 0;20 \\ &= 0;50 - 0;20 \\ &= 0;30 \end{aligned}$$

For us, these instructions amount to the use of the formula

$$x = \sqrt{(a/2)^2 + b} - a/2$$

to solve the equation $x^2 + ax = b$ – a remarkable feat, indeed, although geometry is apparently at the root of the solution [39, p. 24].

Note that this is a "fun" problem without practical utility – mathematics for its own sake c. 1600 BC. See [39, p. 27].

Many similar examples appear in Babylonian mathematics (see, for example, [61, 76]). Indeed, the accumulation of example after example of the same type of problem indicates the existence of some form of justification of Babylonian mathematical procedures. In any case, as Wilder suggests [81, p. 156]:

> The Babylonians had brought mathematics to a stage where two basic concepts of Greek mathematics were ready to be born – the concept of a *theorem* and the concept of a *proof.*

See [39, 43, 76, 81] for further details on this section.

7.3 Greek Axiomatics

Proof as deduction from explicitly stated postulates was, of course, conceived by the Greeks. The axiomatic method is, without doubt, the single most important contribution of ancient Greece to mathematics. The explicit recognition that mathematics deals with abstractions and that proof by deductive reasoning offers a foundation for mathematical reasoning was, indeed, an extraordinary development. When, how, and why this came about is open to conjecture. Various reasons – both internal and external to mathematics (Wilder [81] calls them "hereditary" and "environmental" stresses, respectively) – have been advanced for the emergence of the deductive method in ancient Greece, the so-called Greek mathematical miracle. Among the suggested reasons are:

(a) The need to resolve the "crisis" engendered by the Pythagoreans' proof of the incommensurability of the diagonal and side of a square (see [20]). This might have provided an important impetus for a critical re-evaluation of the logical foundations of mathematics.
(b) The desire to decide among contradictory results bequeathed to the Greeks by earlier civilizations (see [76, p. 89]). (For example, the Babylonians used the formula $3r^2$ for the area of a circle, the Egyptians $[(8/9) \times 2r)]^2$. There is evidence that the Babylonians also used $3\frac{1}{8}$ as an estimate for π [43, p. 11].) This encouraged the notion of mathematical demonstration, which in time evolved into the deductive method.
(c) The nature of Greek society. Democracy in Greece required the art of argumentation and persuasion, and hence encouraged logical, deductive reasoning. Moreover, the existence of a leisure class, supported by a large slave class, was (probably) at least a necessary condition for mathematical contemplation and abstract thinking. Thus, paradoxically, both democracy and slavery apparently contributed to the emergence of the deductive method. See [52, Chap. 4].
(d) The predisposition of the Greeks to philosophical inquiry in which answers to ultimate questions are of prime concern. In particular, it has been argued that the axiomatic method originated in the Eleatic school of philosophy begun by Parmenides and furthered by his pupil Zeno in the early fifth century BC. Zeno does, in fact, use the indirect method of proof in his famous paradoxes. (See [70], but also [44], in which an alternative thesis is proposed.) In this connection, it is interesting to note the view of A.C. Clairaut, an eighteenth-century mathematician and scientist, of Euclid's proofs of obvious propositions [43, pp. 618–619]:

> It is not surprising that Euclid goes to the trouble of demonstrating that two circles which cut one another do not have a common centre, that the sum of the sides of a triangle which is enclosed within another is smaller than the sum of the sides of the enclosing triangle. This geometer had to convince obstinate sophists who glory in rejecting the most evident truths; so that geometry must, like logic, rely on formal reasoning in order to rebut the quibblers.

(e) The need to teach. This forced the Greek mathematicians to consider the basic principles underlying their subject. There were, in fact, about a dozen compilers of "Elements" before Euclid [44, p. 179]. It is noteworthy that the pedagogical motive in the formal organization of mathematics was also present in the works of later mathematicians (as we shall see), notably Lagrange, Cauchy, Weierstrass, and Dedekind.

The axiomatic method in Greece did not come without costs. It is paradoxical that the very perfection of classical Greek mathematics – the insistence on strict, logical deduction – likely contributed to its eventual decline. For this insistence precluded the use by the Greeks of such "working tools" as irrational numbers and the infinite (Eudoxus' theory of incommensurables and his method of exhaustion notwithstanding), which proved fundamental for the subsequent development of mathematics. Thus a very rigorous period in mathematics brought in its wake a long period of mathematical activity with little attention paid to rigor. Too much rigor may lead to rigor mortis. We should note, however, that the predominance of rigorous thinking in Greek mathematics was, of course, not the only cause of the lack of concern for rigor during the following two millennia. See [43, 76].

See [20, 43, 44, 70, 76, 81, 82] for details on this section.

7.4 Symbolic Notation

We take symbolism in mathematics for granted. In fact, mathematics without a well-developed symbolic notation would be inconceivable to us. We should note, however, that mathematics evolved for at least three millennia with hardly any symbols! The introduction and perfection of symbolic notation occurred largely in the sixteenth and seventeenth centuries and is due mainly to Viète, Descartes, and Leibniz. Symbolic notation proved to be the key to a very powerful method of demonstration. One need only compare Cardano's three-page derivation (in 1545) of the formula for the solution of the cubic [68, p. 63] with the corresponding modern half-page proof [39, p. 402]. Moreover, in the absence of symbols Cardano deals with equations with *numerical* coefficients rather than with literal coefficients which, of course, are required for a general proof.

7.4.1 Leibniz

The pedagogical advantages resulting from symbolic notation are well expressed by C.H. Edwards in his comments on Leibniz' felicitous notation for calculus [17, p. 232]:

> It is hardly an exaggeration to say that the calculus of Leibniz brings within the range of an ordinary student problems that once required the ingenuity of an Archimedes or a Newton.

In addition to being the key to a method of demonstration and an invaluable pedagogical aid, symbolic notation also proved to be vital to a method of discovery. For example, the relation between the roots and coefficients of polynomial equations could surely have been noticed only after symbolic notation for such equations was well in place [29]. The discovery of new results was often a consequence of the intimate relation between content and form that a good notation frequently implies. For instance,

> [Leibniz'] infinitesimal calculus is the supreme example in all of science and mathematics, of a system of notation and terminology so perfectly mated with its subject as to faithfully mirror the basic logical operations and processes of that subject [17, p. 232].

As an illustration, we cite Leibniz' discovery (and "proof") of the product rule for differentiation: $\mathrm{d}(xy) = (x + \mathrm{d}x)(y + \mathrm{d}y) - xy = xy + x\mathrm{d}y + y\mathrm{d}x + \mathrm{d}x\mathrm{d}y - xy = x\mathrm{d}y + y\mathrm{d}x$, since, Leibniz notes, "the quantity $\mathrm{d}x\mathrm{d}y\ldots$ is infinitely small in comparison with the rest [17, p. 255]," hence can be discarded. Although this derivation may seem trivial, it was only after a considerable struggle that Leibniz arrived at the correct rules for the differentiation of products and quotients. His striving for an efficient notation for his calculus was part and parcel of his endeavor to find a "universal characteristic" – a symbolic language capable of mechanizing rational expression.

7.4.2 *Euler*

Euler elevated symbol manipulation to an art. Note his uncanny derivation of the power-series expansion of $\cos x$ [29, p. 355]:

Use the binomial theorem to expand the left-hand side of the identity

$$(\cos z + \mathrm{i}\sin z)^n = \cos nz + \mathrm{i}\sin nz.$$

Equate the real part to $\cos nz$ to obtain

$$\begin{aligned}\cos nz = {} & (\cos z)^n - [n(n-1)/2!](\cos z)^{n-2}(\sin z)^2 \\ & +[n(n-1)(n-2)(n-3)/4!](\cos z)^{n-4}(\sin z)^4 - \cdots (*)\end{aligned}$$

Now let n be an infinitely large integer and z an infinitely small number. Then

$$\cos z = 1, \quad \sin z = z, \quad n(n-1) = n^2, \quad n(n-1)(n-2)(n-3) = n^4, \ldots.$$

The equation (*) thus becomes $\cos nz = 1 - n^2z^2/2! + n^4z^4/4! - \cdots$.
Letting $nz = x$ (Euler claims that nz is finite since n is infinitely large and z infinitely small), we finally get

$$\cos x = 1 - x^2/2! + x^4/4! - \cdots (!).$$

This formal "algebraic analysis," so brilliantly used by Euler and practiced by most eighteenth-century mathematicians, accepted as articles of faith that what is true for convergent series is true for divergent series, what is true for finite quantities is true for infinitely large and infinitely small quantities, and what is true for polynomials is true for power series. An elementary example of the use of some of these principles – descendents of Leibniz' "principle of continuity" – was the deduction, from the "identity" $1/(1+x) = 1-x+x^2-x^3+\cdots$, of the equality $1/2 = 1-1+1-1+\cdots$, obtained by setting $x = 1$. The latter result elicited mathematical, metaphysical, and theological discussion. See [43, p. 485] and Chap. 9.

What made mathematicians put their trust in the power of symbols? First and foremost, the use of such formal methods led to important results. A strong intuition by the leading mathematicians of the time kept errors to a minimum. (Errors *were* made. See, for example, [19, p. 10] for Schwarz' counterexample to Euler's proof of the equality $f_{xy} = f_{yx}$ of the partial derivatives of a function $f(x, y)$. For more recent examples of errors made by mathematicians see [12, p. 260] and [15, p. 272].) Moreover, the methods were often applied to physical problems, and the reasonableness of the solutions "guaranteed" the correctness of the results and, by implication, the correctness of the methods. There was also a belief, held by Newton, among others, that mathematicians were simply uncovering God's grand mathematical design of nature. This belief, however, changed by the end of the eighteenth century. When Laplace gave Napoleon a copy of his *Mécanique Céleste*, Napoleon is said to have remarked [43, p. 621]: "M. Laplace, they tell me you have written this large book on the system of the universe and have never even mentioned its Creator," whereupon Laplace replied: "Sir, I have no need of that hypothesis."

See [5, 17, 22, 28, 29, 43] for further details on this section.

7.5 The Calculus of Cauchy

Concern about foundations was never quite absent from mathematics, but it became a dominant feature of its development in the nineteenth century. This century ushered in a spirit of scrutiny of the concepts and methods in various areas of mathematics, especially in analysis. This spirit is already clearly apparent in Gauss' classic *Disquisitiones Arithmeticae* of 1801. (But even the great Gauss' sense of rigor was relative to his time. Thus Smale notes an "immense gap" in Gauss' proof of the Fundamental Theorem of Algebra – a gap filled only in 1920, over 100 years later [64, p. 4].) Other noteworthy examples of concern for rigor are Abel's work in algebra and analysis, Peacock's work in algebra, and Bolzano's work in analysis. We will focus, however, on Cauchy's seminal contribution, begun in his *Cours d'Analyse* of 1821, of providing a rigorous foundation for calculus.

Cauchy selected a few fundamental concepts, namely limit, continuity, convergence, derivative, and integral, established the limit concept as the one on which to base all the others, and derived by fairly modern means the major results of calculus. That this sounds commonplace to us today is, in large part, a tribute to Cauchy's programme – a grand design, brilliantly executed. In fact, most of the above basic concepts of calculus were either not recognized or not clearly delineated before Cauchy's time. Thus, the concept of limit was only adumbrated in the eighteenth century. Euler defined continuity, but in a sense different from Cauchy's (and ours). The differential rather than the derivative was the dominant concept in eighteenth-century analysis; the integral was viewed as an antiderivative. Convergence was rarely considered before the nineteenth century. Cauchy (along with Abel and others) "banished" divergent series – which Euler found so useful – from analysis.

Fig. 7.1 Augustin-Louis Cauchy (1789–1857)

Those series were formally resurrected as legitimate, rigorous mathematical entities only toward the end of the nineteenth century. See [5, 17, 22, 28, 29, 43] for details.

What impelled Cauchy to make such a fundamental departure from established practice? Several reasons can be advanced.

(a) In 1784, Lagrange proposed to the Berlin Academy the foundations of calculus as a prize problem. His lectures on calculus at the Ecole Polytechnique were published in two influential books, in 1797 and 1799–1801. These works made an impact on both Bolzano and Cauchy. But the methods of Lagrange and Cauchy were diametrically opposed. As Lagrange put it, his books were to contain "the principal theorems of the differential calculus without the use of the infinitely small, or vanishing quantities, or limits and fluxions, and reduced to the art of algebraic analysis of finite quantities [43, p. 430]." Thus, Lagrange's foundation for calculus was based on its reduction to algebra, for "he wanted to gain for the calculus the certainty he believed algebra to possess [25, p. 189]." Cauchy's aim, on the other hand, was to *eliminate* algebra as a basis for calculus and thus to repudiate eighteenth-century practice [40, pp. 247–248]:

> As for my methods, I have sought to give them all the rigor which is demanded in [Euclidean] geometry, in such a way as never to run back to reasons drawn from what is usually given in algebra. Reasons of this latter type, however commonly they are accepted, above all in passing from convergent to divergent series and from real to imaginary quantities, can only be considered, it seems to me, as inductions, apt enough sometimes to set forth the truth, but ill according with the exactitude of which the mathematical sciences boast. We must even note that they suggest that algebraic formulas have an unlimited generality, whereas in fact the majority of these formulas are valid only under certain conditions and for certain values of the quantities they contain.

(b) Fourier startled the mathematical community of the early nineteenth century with his work on what came to be known as Fourier series. He claimed that *any* function f defined over $(-l, l)$ is representable over this interval by a series of sines and cosines:

$$f(x) = a_0/2 + \sum_{1}^{\infty}[a_n \cos(n\pi x/l) + b_n \sin(n\pi x/l)],$$

where a_n, b_n are given by

$$a_n = \frac{1}{l}\int_{-l}^{l} f(t)\cos(n\pi t/l)\mathrm{d}t, b_n = \frac{1}{l}\int_{-l}^{l} f(t)\sin(n\pi t/l)\mathrm{d}t.$$

Euler and Lagrange knew that *some* functions have such representations. The "principle of continuity" of eighteenth- and early-nineteenth-century mathematics (see Chap. 9) suggested that the above could not be true for *all* functions: Since sin and cos are continuous and periodic, the same had to be true of a sum of such terms (recall that finite and infinite sums were viewed analogously). However, to refute Fourier's claim one needed – but lacked – clear notions of continuity, convergence, and the integral. (Needless to say, Fourier's result, properly modified, was and remains one of the profound insights of analysis.) Cauchy rose to the challenge of clearing up the meaning of these basic concepts.

(c) Near the end of the eighteenth century a major social change occurred within the community of mathematicians. While in the past they were often attached to royal courts, most mathematicians after the French Revolution earned their livelihood by teaching. Cauchy was a teacher at the influential Ecole Polytechnique in Paris, founded in 1795. It was customary at that institution for an instructor who dealt with material not in standard texts to write up notes for students on the subject of his lectures. The result, in Cauchy's case, was his *Cours d'Analyse* and two subsequent treatises. Since mathematicians presumably think through the fundamental concepts of the subject they are teaching much more carefully when writing for students than when writing for colleagues, this too might have been a contributing factor in Cauchy's careful analysis of the basic concepts underlying calculus.

(d) The above reasons aside, it seems a "natural" process, at least from a historical perspective, that an exploratory period be followed by reflection and consolidation. Geometry in ancient Greece is a case in point. Similarly in the case of calculus, after close to 200 years of vigorous growth with little thought given to foundations, such foundations as did exist were ripe for reevaluation and reformulation.

See [5, 22, 24, 25, 27, 28, 40, 42] for details on this section.

7.6 The Calculus of Weierstrass

Cauchy's new proposals for the rigorization of calculus generated their own problems and enticed a new generation of mathematicians to tackle them. The two major foundational problems were:

(a) Cauchy's verbal definitions of limit and continuity and his frequent use of infinitesimals.
(b) His intuitive appeals to geometry in proving the existence of various limits.

Cauchy defines the notion of limit as follows [40, p. 247]:

> When the values successively attributed to the same variable approach indefinitely a fixed value, eventually differing from it by as little as one could wish, that fixed value is called the *limit* of all the others.

This is followed by a definition of infinitesimal [40, p. 247]:

> When the successive absolute values of a variable decrease indefinitely in such a way as to become less than any given quantity, that variable becomes what is called an *infinitesimal.* Such a variable has zero for its limit.

Cauchy's definition of continuity is as follows [5, pp. 104–105]:

> Let $f(x)$ be a function of the variable x, and let us suppose that, for every value of x between two given limits, this function always has a unique and finite value. If, beginning from one value of x lying between these limits, we assign to the variable x an infinitely small increment α, the function itself increases by the difference $f(x+\alpha)-f(x)$, which depends simultaneously on the new variable α and on the value of x. Given this, the function $f(x)$ will be a *continuous* function of this variable within the two limits assigned to the variable x if, for every value of x between these limits, the numerical value of the difference $f(x+\alpha)-f(x)$ decreases indefinitely with that of α.

These definitions suggest continuous motion – an intuitive idea. Moreover, Cauchy's formulations blur the crucial distinction between, and the placement of, the universal and existential quantifiers that precede x, ε, and δ in a modem (Weierstrassian) definition of limit and continuity. (Although Cauchy at times used $\varepsilon-\delta$ arguments in proofs of various results, he often resorted to the language of infinitesimals. To him an infinitesimal was a variable with zero limit.) These shortcomings were the source of two major errors: Cauchy failed to distinguish between pointwise and uniform continuity of a function and between pointwise and uniform convergence of an infinite series of functions. Thus he "proved" that a convergent series of continuous functions is a continuous function.

The proof, in which he uses infinitesimals freely, goes as follows:

Let

$$s(x)=\sum_{1}^{\infty}u_i(x),\quad s_n(x)=\sum_{1}^{n}u_i(x),\quad r_n(x)=\sum_{n+1}^{\infty}u_i(x),$$

and let α be an infinitesimal. Then

$$s(x+\alpha) - s(x) = [s_n(x+\alpha) - s_n(x)] + [r_n(x+\alpha) - r_n(x)].$$

Since $u_i(x)$ are continuous, $u_i(x+\alpha) - u_i(x)$ is infinitesimal, hence so is $s_n(x+\alpha) - s_n(x)$ (being a finite sum of such terms). Since $\sum_1^\infty u_i(x)$ converges, $r_n(x)$ is infinitesimal for sufficiently large n; the same holds for $r_n(x+\alpha)$. Hence $r_n(x+\alpha) - r_n(x)$ is infinitesimal and thus so is $s(x+\alpha) - s(x)$. Thus an infinitesimal increment in x produces an infinitesimal increment in $s(x)$, hence $s(x)$ is continuous. The use of infinitesimals in the proof masks the distinction between $(\forall\varepsilon)(\forall x)(\exists N)(|\sum_{n+1}^\infty u_i(x)| < \varepsilon)$ and $(\forall\varepsilon)(\exists N)(\forall x)(|\sum_{n+1}^\infty u_i(x)| < \varepsilon)$, and thus the distinction between pointwise and uniform convergence of the series $\sum_1^\infty u_i(x)$. See [5, p. 110] or [40, p. 254] for further details.

The above result is of course false; this was first pointed out in 1826 by Abel, who showed that the series $\sin x - (\sin 2x)/2 + (\sin 3x)/3 - \cdots$ converges to a function discontinuous at $x = 2n+1$ for all integers n [5, p. 113]. It took another 20 years, however, to determine where Cauchy went wrong! Lakatos argues that it is a false reading of history to view Cauchy's proof as erroneous [50, p. 127]. In [49] he gives a reconstruction of Cauchy's arguments in terms of Robinson's nonstandard analysis (see also [51]). One was dealing with subtle concepts indeed.

Other counterexamples to plausible and widely held notions appeared during the half century following Cauchy's publication of his *Cours* and *Résumé*. Among the most unexpected was Weierstrass' example of a continuous nowhere-differentiable function $f(x) = \sum_1^\infty b^n \cos(a^n \pi x)$, a an odd integer, b a real number in (0,1), and $ab > 1 + 3\pi/2$. Cauchy and his contemporaries believed (and some of them "proved") that a continuous function is differentiable except possibly at isolated points. Given the mathematicians' prevailing geometric conception of continuity (see below) and their notions of function (see Chap. 5), this "result" is not surprising.

Since Cauchy's definitions of the fundamental concepts of calculus were given in terms of limits, proofs of the existence of limits of various sequences and functions were of crucial importance. Thus Cauchy's solutions to the eighteenth century's lack of rigor generated new problems. Cauchy resorted to intuitive geometric arguments to establish a number of the fundamental results of analysis. For example, he claimed that "a remarkable property of continuous functions of a single variable is to be able to be represented geometrically by means of straight lines or continuous curves" [40, p. 261], and he used this "remarkable property" of continuous functions to give a – necessarily intuitive – geometric proof of the Intermediate Value Theorem. The proof amounted to noting that, given a function f, if $f(a)$ and $f(b)$ differ in sign then the graph of f must cross the x-axis, hence $f(c) = 0$ for some c in (a, b). See [40, p. 261].

Other – correct – results that Cauchy accepted on intuitive grounds are that an increasing sequence bounded above has a limit, and that a (so-called) Cauchy sequence converges. He used these results to establish, among other things, the existence of the integral of a continuous function, and to give (in an appendix to the *Cours*) an analytic proof of the Intermediate Value Theorem. See [17, pp. 311, 318], [29, pp. 167, 170], and [40, p. 261].

Weierstrass and Dedekind, among others, determined to remedy this "mixture of algebraic formulation and geometric justification which Cauchy favored [and which] did not provide full comprehension of the major results of function theory" [40, p. 264]. Dedekind's expression of the prevailing state of affairs is revealing [14, pp. 1–2].

> As professor in the Polytechnic School in Zurich I found myself for the first time obliged to lecture upon the elements of the differential calculus and felt more keenly than ever before the lack of a really scientific foundation for arithmetic. In discussing the notion of the approach of a variable magnitude to a fixed limiting value, and especially in proving the theorem that every magnitude which grows continually, but not beyond all limits, must certainly approach a limiting value, I had recourse to geometric evidence. Even now such resort to geometric intuition in a first presentation of the differential calculus I regard as exceedingly useful from the didactic standpoint, and indeed indispensable if one does not wish to lose too much time. But that this form of introduction into the differential calculus can make no claim to being scientific, no one will deny. For myself this feeling of dissatisfaction was so overpowering that I made the fixed resolve to keep meditating on the question till I should find a purely arithmetic and perfectly rigorous foundation for the principles of infinitesimal analysis.

Establishing theorems in a "purely arithmetic" manner implied what came to be known as the "arithmetization of analysis." Since the inception of calculus, and even in Cauchy's time, the real numbers were viewed geometrically, without explicit formulation of their properties. Since the real numbers are in the foreground or background of much of analysis, proofs of theorems were of necessity intuitive and geometric. Dedekind's and Weierstrass' astute insight recognized that a rigorous, arithmetic definition of the real numbers would resolve the major obstacle in supplying a rigorous foundation for calculus. It is noteworthy that both Weierstrass and Dedekind presented their ideas on the rigorization of calculus in *lectures* at universities. As in Cauchy's case, so here too pedagogical considerations seem to have been a motive in the search for careful, rigorous formulations of basic mathematical concepts.

The other remaining task was to give a precise "algebraic" definition of the limit concept to replace Cauchy's intuitive, "kinematic" conception. This was accomplished by Weierstrass when he gave his "static" definition of limit in terms of inequalities involving ε's and δ's – the definition we use today, at least in our formal, rigorous incarnation. (It may seem ironic that inequalities, used in the eighteenth century for estimation, and ε, used by some to indicate error, became in the hands of Weierstrass the very tools of precision.) Weierstrass thereby did away with infinitesimals, which were used freely by Cauchy and his predecessors for about two centuries. However, the story of infinitesimals is similar to that of divergent series: about a century after Weierstrass had banished infinitesimals "for good" – so we all thought until 1960 – they were brought back to life by Abraham Robinson as genuine and rigorously defined mathematical objects.

Looking back at 2,500 years of the use of proof in mathematics, we note that not only have standards of rigor changed but so also have the mathematical tools used to establish rigor. Thus in ancient Greece, a theorem was not properly established until it was geometrized. In the Middle Ages and the Renaissance, geometry continued to be the final arbiter of mathematical rigor, even in algebra. Mathematicians' intuition of space appeared, presumably, more trustworthy than their insight into number – a continuing legacy of the consequences of the "crisis of incommensurability" in ancient Greece. The calculus of the seventeenth and especially the eighteenth century was no longer easily justifiable in geometric terms, and algebra became the major tool of justification, such as there was. There was a mix of the algebraic and geometric in Cauchy's work. With Weierstrass and Dedekind in the latter part of the nineteenth century, arithmetic rather than geometry or algebra had become the language of rigorous mathematics. To Plato, God ever geometrized, while to Jacobi, He ever arithmetized. (The creation of non-Euclidean geometry, and the appearance of geometrically nonintuitive examples such as continuous nowhere-differentiable functions must have accelerated this dethroning of geometry.) The logical supremacy of arithmetic, however, was not lasting. In the 1880s Dedekind and Frege undertook a reconstruction of arithmetic based on ideas from set theory and logic. The ramifications of this event will be considered below.

See [5, 17, 26, 40, 43] for details on this section.

7.7 The Reemergence of the Axiomatic Method

Our emphasis on analysis in the last two sections was due to the fact that the most important strides in rigor in the nineteenth century were made in analysis. But algebra, arithmetic, and geometry were also given careful scrutiny during this period. Moreover, mathematical logic came into being in 1847 with Boole's *The Mathematical Analysis of Logic*. All of this led to a rebirth of the axiomatic method late in the nineteenth century. We describe these developments very briefly.

The abstract concept of a group arose from different sources. Thus polynomial theory gave rise to groups of permutations, number theory to groups of numbers, and of "forms" (n-th roots of unity, integers mod n, equivalence classes of binary quadratic forms), and geometry and analysis to groups of transformations. Common features of these concrete examples of groups began to be noted, which resulted in the emergence of the abstract concept of a group in the last decades of the nineteenth century. Similar observations apply to the emergence of the concepts of ring, field, and (to a lesser extent) vector space. See [41].

The arithmetization of analysis reduced the foundations of the subject to that of the real numbers. These were defined in terms of rational numbers. The reduction of the rationals to the positive integers soon followed. (Note that the historical evolution of the logical foundations of the number system – from the reals to the rationals to the integers – is the reverse of the sequence usually presented in textbooks.) There remained the problem of the foundations of the positive integers, that is, arithmetic. This was addressed in different ways by Dedekind, Peano, and Frege during the last two decades of the nineteenth century. All three, however, used axioms (Dedekind less explicitly than the other two) to define the positive integers. See [14, 43].

One of the consequences of the creation of non-Euclidean geometry was a reexamination of the foundations of Euclidean geometry and, more broadly, of axiomatic systems in general. Pasch, Peano, and Hilbert pioneered the development of the modem axiomatic method late in the nineteenth century through a careful analysis of the foundations of geometry. See [43, 82].

Boole, by virtue of his work in mathematical logic and in (what we call today) Boolean algebra, was among the first to promote the view of the arbitrary nature of axioms allowing for different interpretations. In *The Mathematical Analysis of Logic*, Boole subscribes to what was at that time a very novel point of view [82, p. 116]:

> The validity of the processes of analysis does not depend upon the interpretation of the symbols which are employed, but solely upon the laws of their combination. Every system of interpretation which does not affect the truth of the relations supposed, is equally admissible.

The rise of the axiomatic method was gradual and slow (see for example [41, p. 30]). By the early twentieth century, however, the axiomatic method was well established in a number of major areas of mathematics.

In algebra, there were major works in group theory (1904), field theory (1910), and ring theory (1914), crowned by Emmy Noether's groundbreaking papers of the 1920s. In analysis there were Frechet's thesis of 1906 on function spaces, in which a definition of metric space appears; E.H. Moore's work of the same year on "general analysis," an axiomatic formulation of features common to linear integral equations and infinite systems of linear algebraic equations; Banach's researches on Banach spaces (1922); and von Neumann's axiomatization of Hilbert space (1929). In topology Hausdorff defined a topological space in terms of neighborhoods (1914) and P.S. Alexandroff began to develop homology theory (1928) following conversations with E. Noether. In geometry Hilbert's *Foundations of Geometry* (1899) was most influential; Veblen and Young's two-volume abstract treatment of projective geometry (1910–1919) also made a significant impact. In set theory we had Zermelo's axiomatization of the subject in 1908, followed by Fraenkel's improvements in 1921 and von Neumann's version in 1925. Finally, in mathematical logic there was Russell and Whitehead's prodigious three-volume *Principia Mathematica* (1910–1913). See [6, 41, 43, 82], for details of the above.

The axiomatic method, surely one of the most distinctive features of twentieth-century mathematics, flourished in the early decades of the century. Bourbaki, among its most able practitioners and promoters, gives an eloquent description of the essence of the axiomatic method at what was perhaps the height of its power (in 1950) [6, p. 223]:

> What the axiomatic method sets as its essential aim, is exactly that which logical formalism by itself can not supply, namely the profound intelligibility of mathematics. Just as the experimental method starts from the a priori belief in the permanence of natural laws, so the axiomatic method has its cornerstone in the conviction that, not only is mathematics not a randomly developing concatenation of syllogisms, but neither is it a collection of more or less "astute" tricks, arrived at by lucky combinations, in which purely technical cleverness wins the day. Where the superficial observer sees only two, or several, quite distinct theories,

> lending one another "unexpected support" through the intervention of a mathematician of genius, the axiomatic method teaches us to look for the deep-lying reasons for such a discovery, to find the common ideas of these theories, buried under the accumulation of details properly belonging to each of them, to bring these ideas forward and to put them in their proper light.

In this article Bourbaki presents a panoramic view of mathematics organized around what he calls "mother structures" – algebraic, ordered, and topological, and various substructures and cross-fertilizing structures. It must have been an alluring, even bewitching, view of mathematics to those growing up (mathematically) during this period.

There are significant differences between Euclid's axiomatics and its modem incarnation in the last decades of the nineteenth century and the early decades of the twentieth. Euclid's axioms are idealizations of a concrete physical reality and are thus perceived as self-evident truths – a Platonic view, describing a pre-existing reality. In the modern view, axioms are neither self-evident nor true – they are simply assumptions about the relations among the undefined (primitive) terms of the axiomatic system. As early as 1891, Hilbert highlighted this point in the now classic remark that "It must be possible to replace in all geometric statements the words point, line, plane by table, chair, mug" [79, p. 14]. Thus in a modem axiom system the axioms, and hence also the theorems, are *devoid of meaning*. Moreover, such an axiomatic system need not be categorical; that is, it may admit of essentially different (nonisomorphic) interpretations (models), all of which satisfy the same axioms – a fundamentally novel idea.

The modern axiomatic method is thus a unifying and abstracting device. Moreover, while the chief role played by the axiomatic method in ancient Greece was (probably) that of providing a consistent foundation, it became in the first half of the twentieth century also a tool of research. In addition, the axiomatic method was at times indispensable in clarifying the status of various mathematical methods and results (like the axiom of choice and the continuum hypothesis) to which the mathematicians' intuition provided little guide. The method also came to be the arbiter of rigor and precision in mathematics and beyond. (This was also the case, of course, in ancient Greece. At the same time, there is perhaps no better way to bring out the differences between Greek and modem axiomatics than to compare Euclid's *Elements* with Hilbert's *Foundations of Geometry.* The comparison makes it starkly clear how standards of rigor have evolved.) Thus the sometimes opposed activities of discovery and demonstration coexisted within the axiomatic method. For example, Gray notes that Desarguean and non-Desarguean geometries "could never have been discovered without [the axiomatic] method [30, p. 182]."

The modern axiomatic method was, however, not an unmitigated blessing (as we shall see). Although some, for example, Hilbert, claimed that it is the central method of mathematical thought, others, for instance, Klein, argued that as a method of discovery it tends to stifle creativity. And it has its limitations as a method of demonstration.

See [6, 13, 20, 41, 43, 79, 82] for further details.

7.8 Foundational Issues

7.8.1 Introduction

We are referring here to the three philosophies of mathematics – logicism, formalism, and intuitionism – which arose in the first decades of the twentieth century and which dealt with the nature, meaning, and methods of mathematics, and thus, in particular, with questions of rigor and proof in mathematics. Although, as noted, these were twentieth-century developments, they had deep roots in the mathematics of the nineteenth century.

The nineteenth century witnessed a gradual transformation of mathematics – in fact, a gradual revolution, if that is not a contradiction in terms. Mathematicians turned more and more for the genesis of their ideas from the sensory and empirical to the intellectual and abstract. Although this subtle change had already begun in the sixteenth and seventeenth centuries with the introduction of such nonintuitive concepts as negative and complex numbers, instantaneous rates of change, and infinitely small quantities, these were often used (successfully) to solve physical problems and thus elicited little demand for justification.

In the nineteenth century, however, the introduction of non-Euclidean geometries, noncommutative algebras, continuous nowhere-differentiable functions, space-filling curves, n-dimensional geometries, completed infinities of different sizes, and the like, could no longer be justified by physical utility. Cantor's dictum that "the essence of mathematics lies in its freedom" became a reality – but one to which many mathematicians took strong exception, as the following quotations indicate.

> There is still something in the system [of quaternions] which gravels me. I have not yet any clear view as to the extent to which we are at liberty arbitrarily to create imaginaries and to endow them with supernatural properties [41, p. 155].

The reservations are those of John Graves, who communicated them to his friend Hamilton in 1844, shortly after the latter had invented the quaternions. The "supernatural properties" referred mainly to the noncommutativity of multiplication of the quaternions.

> Of what use is your beautiful investigation regarding π? Why study such problems since irrational numbers are nonexistent? [43, p. 1198]. (But see [18, p. 13].)

This was Kronecker's damning praise of Lindemann, who proved in 1882 that π is transcendental, hence that the circle cannot be squared using straightedge and compass.

> I turn away with fright and horror from this lamentable evil of functions without derivatives [43, p. 973].

> Logic sometimes makes monsters. For half a century, we have seen a mass of bizarre functions which appear to be forced to resemble as little as possible honest functions which serve some purpose [43, p. 973].

> I believe that the numbers and functions of analysis are not the arbitrary product of our minds; I believe that they exist outside of us with the same character of necessity as the objects of objective reality; and we find or discover them and study them as do the physicists, chemists, and zoologists [43, p. 1035].

The above quotations, from Hermite (in 1893), Poincaré (in 1899), and again Hermite (in 1905), respectively, are a reaction to various examples of "pathological" functions introduced during the previous half century: integrable functions with discontinuities dense in any interval, continuous nowhere-differentiable functions, nonintegrable functions that are limits of integrable functions, and others (see Chap. 5).

> Later generations will regard *Mengenlehre* [Set Theory] as a disease from which one has recovered [43, p. 1003].

This is Poincaré again, speaking (in 1908) about Cantor's creation of set theory, in particular in connection with the paradoxes that had arisen in the theory. Compare Poincaré's position with that of Hilbert, the other giant of this period:

> No one shall expel us from the paradise which Cantor created for us [43, p. 1003].

The above sentiments, expressed by some of the leading mathematicians of the period, are suggestive of the impending crisis. Although mathematical controversies had arisen before the nineteenth century, for example, the vibrating-string controversy between d'Alembert and Euler, these were isolated cases. The frequency and intensity of the disaffection expressed in the nineteenth century was unprecedented and could no longer be ignored. The result was a split among mathematicians concerning the way they viewed their subject. Its formal expression was the rise in the early twentieth century of three schools of mathematical thought, three philosophies of mathematics – logicism, formalism, and intuitionism. This was the first *formal* expression by mathematicians of what mathematics is about and, in particular, of what proof in mathematics is about. (The "crises" in ancient Greece following Zeno's paradoxes and the proofs of incommensurability might have given rise to similar debates and subsequent formal resolutions, but we have little evidence of that.) The notion of proof – its scope and limits – became a subject of study *by mathematicians.*

7.8.2 Logicism

The logicist thesis, expounded in the monumental *Principia Mathematica* of Russell and Whitehead, held that mathematics is part of logic. Mathematical concepts are expressible in terms of logical concepts; mathematical theorems are tautologies, true by virtue of their form rather than of their factual content. This thesis was motivated, in part, by the paradoxes in set theory, by the work of Frege on mathematical logic and the foundations of arithmetic, and by the espousal of mathematical logic by Peano and his school. Its broad aim was to provide a foundation for mathematics.

Although the logicist thesis was important philosophically and inspired subsequent work in mathematical logic, it was not embraced by the mathematical community. For one thing, it did not grant reality to mathematics other than in terms of logical concepts. For another, it took "forever" to obtain results of any consequence; for example, it is only on p. 362 of the *Principia* that Russell and Whitehead show that $1 + 1 = 2$(!); see [13, p. 334]. "If the mathematical process were really one of strict, logical progression," observe De Millo et al, "we would still be counting on our fingers" [15, p. 272]. There were, moreover, serious technical difficulties in the implementation of the logicist thesis. See [40, 80].

7.8.3 Formalism

The most serious debate within the mathematical community – still unresolved – has been between the adherents of the formalist and intuitionist schools. The formalist thesis, with Hilbert as its main exponent, entails viewing mathematics as a study of axiomatic systems. Both the primitive terms and the axioms of such a system are considered to be strings of symbols to which no meaning is to be attached. These are to be manipulated according to established rules of inference to obtain the theorems of the system.

At the time Hilbert advanced his thesis (the 1920s), the axiomatic method, as we noted, had embraced much of algebra, arithmetic, analysis, set theory, and mathematical logic. Even though Zermelo's axiomatization of set theory in 1908 seemed to have avoided the paradoxes of set theory, there was no assurance that they would not reemerge in one form or another. Hilbert felt that this possibility, and the denial of meaning to the primitive terms and postulates of axiomatic systems, made it imperative to undertake a careful analysis of such systems in order to establish their consistency. The methods by which this was to be accomplished were acceptable also to the intuitionists. These methods came to be known as "metamathematics" or "proof theory." For recent developments in proof theory, see [21, 37, 62, 66a].

The formalists have been accused of removing all meaning from mathematics and reducing it to symbol manipulation. The charge is unfair. Hilbert's aim was to deal with the *foundations* of mathematics rather than with the daily practice of the mathematician. (Of course the same can be said of Russell and Whitehead's objective in connection with the logicist thesis.) And to show that mathematics is free of inconsistencies one first needed to formalize the subject. This was formalism in the service of informality.

As we know, Hilbert's grand design was laid to rest by Godel's incompleteness theorems of 1931. These showed the inherent limitations of the axiomatic method. The consistency of a large class of axiomatic systems, including those for arithmetic and set theory, cannot be established within the systems. Moreover, if consistent, these systems are incomplete (see [13, 42, 66a, 78] for details). In connection with the first result, Weyl remarked: "God exists since mathematics is consistent and the devil exists since we cannot prove the consistency." [43, p. 1206] The second result has elicited the comment that Gödel gave a formal demonstration of the inadequacy of formal demonstrations.

Chaitin notes that Gödel's work "demands the surprising and, for many, discomforting conclusion that there can be no definitive answer to the question 'What is a proof'?" [9, p. 51]. Just as in the nineteenth century, following the invention of non-Euclidean geometries, noncommutative algebras, and other developments, mathematics lost its claim to (absolute) truth, so in the twentieth century, following Godel's work, it lost its claim to certainty. In the nineteenth century truth in mathematics was replaced by validity (relative truth) and, in the twentieth century, certainty by faith. For a formal twentieth-century notion of truth in mathematics and its relation to proof, see [71]. In any case, although Godel's results are of fundamental philosophical consequence, they have not affected the daily work of most mathematicians. (See however [9] for a discussion of a connection between Godel's theorems and random numbers.)

7.8.4 Intuitionism

The intuitionists, headed by L.E.J. Brouwer, claimed that no formal analysis of axiomatic systems is necessary. In fact, mathematics should not be founded on systems of axioms. The mathematician's intuition, beginning with that of number, will guide him in avoiding contradictions. He must, however, pay special attention to definitions and methods of proof. These must be constructive and finitistic. In particular, the law of the excluded middle, completed infinities, the axiom of choice, and proof by contradiction are all outlawed. Hilbert protested that

> taking the principle of the excluded middle from the mathematician would be the same, say, as proscribing the telescope to the astronomer or to the boxer the use of his fists [42, p. 246].

Among the results unacceptable to the intuitionists is the law of trichotomy. Given any real number N, either $N > 0$ or $N = 0$ or $N < 0$. The following example substantiates that point [23, p. xx]:

Define a real number N as follows: $N = \sum_{n=2}^{\infty} a_n/10^n$, where

$$a_n = \begin{cases} 1, & \text{if } 2n \text{ is the first even integer that is not the sum of two primes,} \\ & n > 1, n \text{ even,} \\ -1, & \text{if } 2n \text{ is the first even integer that is not the sum of two primes,} \\ & n > 1, n \text{ odd,} \\ 0, & \text{otherwise.} \end{cases}$$

The definition of N is acceptable to both formalists and intuitionists (its digits can be calculated – at least in theory – to any degree of accuracy). But to the intuitionists, none of $N > 0$, $N < 0$, or $N = 0$ is meaningful since it is not known if Goldbach's conjecture is true or false. Thus the law of trichotomy fails.

A prominent feature of nineteenth-century mathematics was nonconstructive existence results. These were almost unknown before that time. Thus Gauss proved the fundamental theorem of algebra about the existence of roots of polynomial equations without showing how to find them. Cauchy and others proved the existence of solutions of differential equations without providing the solutions explicitly. Cauchy proved the existence of the integral of an arbitrary continuous function but often was unable to evaluate integrals of specific functions. He gave tests of convergence of series without indicating what they converge to. Late in the century Hilbert proved the existence of, but did not explicitly construct, a finite basis for any ideal in a polynomial ring. Dedekind constructed the real numbers by using completed infinities. Such examples abound. All were rejected by the intuitionists. (Weyl said of nonconstructive proofs that they inform the world that a treasure exists without disclosing its location [43, p. 1203].) On the other hand, the proofs of the intuitionists are certainly acceptable to the formalists. Many results in analysis, and more recently in algebra, have been reconstructed, thanks to the pioneering effort of Errett Bishop, using finitistic methods (see [4, 7, 8, 54, 60]). In fact, as early as 1924 Brouwer and Weyl gave constructive proofs yielding a root of a complex polynomial; however, it may take up to 10^{10} years to find it! Manin suggests that the mathematician "should at least be willing to admit that proof can have objectively different 'degrees of proofness'" [55, p. 17]. See [8, 42, 54, 60] for details.

The differences between the formalists and the intuitionists, and their nineteenth-century forerunners, were genuine. For the first time, mathematicians were seriously and irreconcilably divided over what constitutes a proof in mathematics. Moreover, this division seems to have had an impact on the work that at least some mathematicians chose to pursue, as the testimony of two of the most prominent practitioners of that epoch – Von Neumann and Weyl, respectively – indicates:

> In my own experience... there were very serious substantive discussions as to what the fundamental principles of mathematics are; as to whether a large chapter of mathematics is really logically binding or not.... It was not at all clear exactly what one means by absolute rigor, and specifically, whether one should limit oneself to use only those parts of mathematics which nobody questioned. Thus, remarkably enough, in a large fraction of mathematics there actually existed differences of opinion! [77, p. 480]

> Outwardly it does not seem to hamper our daily work, and yet I for one confess that it has had a considerable practical influence on my mathematical life. It directed my interests to fields I considered relatively 'safe,' and has been a constant drain on the enthusiasm and determination with which I pursued my research work [80, p. 13].

It is probably safe to say, however, that most mathematicians are untroubled, at least in their daily work, about the debates concerning the various philosophies of mathematics.

Davis and Hersh put the issue in perspective [13, p. 318]:

> If you do mathematics every day, it seems the most natural thing in the world. If you stop to think about what you are doing and what it means, it seems one of the most mysterious.

Weyl puts it more lyrically:

> The question of the ultimate foundations and the ultimate meaning of mathematics remains open; we do not know in what direction it will find its final solution or even whether a final objective answer can be expected at all. 'Mathematizing' may well be a creative activity of man, like language or music, of primary originality, whose historical decisions defy complete objective rationalization [42, p. 6].

Another point of philosophical contention is between Platonists, who believe that mathematics is discovered, and formalists, who claim it is invented (see [13, 42] for details). Davis and Hersh suggest that "the typical working mathematician is a Platonist on weekdays and a formalist on Sundays" [13, p. 321].

For elaboration of vairous points discussed in this section see [4, 7–9, 23, 24, 28, 38, 54, 77–80].

7.9 The Era of the Computer

While mathematics in the twentieth century's first two thirds, especially in the period 1930–1960, stressed the formulation of general methods and abstract theories – for example, abstract algebra, algebraic topology, the theory of distributions, homological algebra, and category theory – more attention has since been paid to the solution of specific problems – for example, the Kepler conjecture, the four-color problem, the Bieberbach conjecture, the proof of Fermat's Last Theorem, Mordell's conjecture, and the Poincaré conjecture. Clearly many counterexamples to this trend can be given; and, of course, the general theories were instrumental in the solution of these major problems.

The computer played a major role in this development. It has helped stimulate the growth of new mathematical fields – for example, algebraic coding theory, theory of automata, analysis of algorithms, optimization theory, and experimental mathematics – and has aided in the revival of older fields – for example, combinatorics and graph theory. More importantly from our perspective, it has assisted in the making, testing, and disproving of conjectures, and in the proving of theorems. "The intruder [the computer] has changed the ecosystem of mathematics, profoundly and permanently," asserted Lynn Steen [66, p. 34]. Neither the axiomatic method nor strict adherence to very rigorous mathematical proof are hallmarks of these developments. These changes have occasioned a rethinking of the meaning and role of proof in mathematics.

The catalyst has been Appel and Haken's 1976 computer-aided proof of the four-color theorem. The proof required the verification, by computer, of 1,482 distinct configurations. Some critics argued that this type of proof was a major departure from tradition. They advanced several reasons:

(a) The proof contained thousands of pages of computer programs *that were not published* and were thus not open to the traditional procedures of verification by the mathematical community. The proof was "not surveyable," in the words of Tymoczko, one of its forceful critics (see [75] and responses in [16, 69]), and was thus "*permanently and in principle* incomplete" [13, p. 380]. (There were also genuine concerns about the completenes of the search, but these were apparently laid to rest following the reworking of the proof in 1997 [23, p. 41].)

(b) Both computer hardware and computer software are subject to error. Hence also the tendency to feel that verification of the computer results by independent computer programs was not as reliable as the standard method of checking proofs. This introduces a measure of quasi-empiricism into the proof of the four-color theorem – the computer is an experimental tool.

(c) "Proof, in its best instances, increases understanding by revealing the heart of the matter," note Davis and Hersh [13, p. 151]. "A good proof is one which makes us wiser," echoes Yu. I. Manin [55, p. 18]. Thus, even if we believe that the proof of the four-color theorem is valid, we cannot *understand* the theorem unless we are (or can be) involved in the *entire* process of proof; and that is not possible in this case except for the very few.

The objections to the proof of the four-color theorem apply, *mutatis mutandis*, to the proofs of several other major theorems. One of them is the proof by Feit and Thompson in the 1960s of the solvability of all finite groups of odd order, another is the classification, carried out jointly by many mathematicians in the 1980s, of finite simple groups. The first proof takes up over 300 pages of an entire issue of the *Pacific Journal of Mathematics* and is based on much previous work. Chevalley once undertook to give a complete account of this proof in a seminar, but gave up after two years [10, p. 11].

The second proof consists of over 11,000 pages (!) of close mathematical reasoning scattered in many journals over many years. Daniel Gorenstein, one of the major contributors to the field, said of the proof [16, pp. 811–812]:

> It seems beyond human capacity to present a closely reasoned, several-hundred-page argument with absolute accuracy... how can one guarantee that the "sieve" has not let slip a configuration which leads to yet another simple group? Unfortunately, there are no guarantees – one must live with this reality. [A second-generation proof was completed c. 2004, but that proof too has gaps [2].]

Speaking of the Feit-Thompson theorem, and other results whose proofs are very long, Jean-Pierre Serre observed [10, p. 11]:

> What shall one do with such theorems, if one has to use them? Accept them on faith? Probably. But it is not a very comfortable situation.

He continued:

> I am also uneasy with some topics, mainly in differential topology, where the author draws a complicated picture (in two dimensions), and asks you to accept it as a proof of something taking place in five dimensions or more. Only the experts can "see" whether such a proof is correct or not – if you can call this a proof.

There are other examples of very long proofs – for example, the proofs of the two Burnside conjectures, c. 500 pages apiece [55, p. 17] (see also [15,46,57,65]). Some believe that long proofs are becoming the norm rather than the exception; the reason is that there are, in their view, relatively few interesting results with short proofs compared to the total number of interesting mathematical results [46]. On the other hand, Joel Spencer suggests that the mathematical counterpart of Einstein's credo that "God does not play dice with the universe" is that "short interesting theorems have short proofs" [65, p. 366]. But the four-color theorem, the Feit-Thompson theorem, the classification of the finite simple groups, the Kepler conjecture (now a theorem), and the Poincaré conjecture (also now a theorem) are – at present – illustrious counterexamples to this claim.

Largely as a result of these developments, a novel philosophy of mathematical proof seems to be emerging. It goes under various names – public proof, quasi-empiricist proof, proof as a social process. Its essence, according to its advocates, is that *proofs are not infallible*. Thus, mathematical theorems cannot be guaranteed absolute certainty. And this applies not only to the theorems requiring very long proofs or the assistance of a computer, but to many "run of the mill" theorems. This is so because proofs of theorems usually rely on the correctness of other theorems. And published proofs, it is argued, are usually read carefully only by the author (and perhaps by some referees), so mistakes are inevitable:

> Stanislaw Ulam estimates that mathematicians publish 200,000 theorems every year [written in 1979]. A number of these are subsequently contradicted or otherwise disallowed, others are thrown into doubt, and most are ignored. Only a tiny fraction come to be understood and believed by any sizable group of mathematicians [15, p. 272].

The truth of a theorem, then, has a certain probability, usually <1, attached to it. The probability increases as more mathematicians read, discuss, and use the theorem. In the final analysis, the acceptance of a theorem, that is, the acceptance of the validity of its proof, is a social process and is based on the confidence of the mathematical community in the social systems that it has established for purposes of validation [13, p. 390]:

> If a theorem has been published in a respected journal, if the name of the author is familiar, if the theorem has been quoted and used by other mathematicians, then it is considered established.

Imre Lakatos, in a brilliant polemic [50], also comes to the conclusion that mathematics is fallible, although his focus and arguments differ from those in the above analysis. Mathematical theorems, Lakatos claims, are not immutable – they are subject to constant examination and possible rejection through counterexamples. Proofs are not instruments of justification but tools of discovery, to be employed in the development of concepts and the refinement of conjectures. The interplay between conjecture, proof, counterexample, and refinement of conjecture is the lifeblood of mathematics. For instance, a counterexample may compel us to tighten a definition or to broaden a theorem. These ideas are masterfully illustrated with the example of the history of the Descartes-Euler formula $V - E + F = 2$ for a polyhedron. A proof is first presented, then counterexamples are introduced, the conjecture $V - E + F = 2$ is refined (that is, the notion of polyhedron is refined), and a new proof is given. The give-and-take of this historical-philosophical-pedagogical interplay encompasses about 200 years of historical analysis and continues for over 100 pages [50].

Examples of the interplay between theorem, proof, and counterexample abound. In ancient times, the Pythagorean theory of proportion applied only to commensurable magnitudes until the "counterexample" of the incommensurability of the side and diagonal of a square was discovered. A new concept of ratio was then introduced and the theory of proportion was revised [76]. In more recent times, Cauchy "proved," as we indicated earlier, that the sum of an infinite series of continuous functions is continuous. Following Abel's counterexample, the concept of uniform convergence was introduced and the above result and its proof were revised. See [40, Chap. 10] or [50, Appendix 1] for details.

There has been another interesting development in the evolution of proof: the notion of *probabilistic proof*. It has been shown that some results, even if theoretically decidable, have such long proofs that they can never be written down – either by humans or by computer. This is the case, for example, for almost all the familiar decidable results in logic (see [15, 56, 67]), as well as for tests of large numbers for primality. Michael Rabin proposed in 1976 to relax the notion of proof by allowing for probabilistic proofs [59]. For example, he found a quick way to determine, with a very small probability of error, say one in a billion, whether or not an arbitrarily chosen large number is a prime. Thus, he has shown that $2^{400}–593$ is a prime "for all practical purposes." (It has subsequently been shown that this number is indeed a prime [58, p. 102].) Such results can apparently be applied with impunity to cryptography, which is the main field of application of primality testing. It is noteworthy, moreover, that the proofs of such results use highly sophisticated abstract mathematics such as abelian varieties and Faltings' results dealing with the Mordell conjecture. See [45], which also contains an update of Rabin's work.

Another instance of a probabilistic proof comes from graph theory. If two graphs are nonisomorphic, it is very difficult to establish this rigorously, but easy to show it with very high probability.

Some have argued that there is no essential difference between such probabilistic proofs and the deterministic proofs of standard mathematical practice. Both are convincing arguments. Both are to be believed with a certain probability of error. In fact, many deterministic proofs, it is claimed, have a higher probability of error than probabilistic ones. The counterargument is that there is a fundamental *qualitative* difference between the two types of proof. Although both may be subject to error, an important philosophical distinction must be made. If probabilistic proofs were routinely admitted into the domain of mathematics, this would considerably strengthen the thesis of the quasi-empirical nature of mathematics and would entail a radical departure from the traditional view of the subject. See [23, 45, 58].

We conclude with two very recent and very interesting examples having to do with fallibility of proofs and computers. The first concerns "experimental mathematics." This is a new field, founded by Jonathan Borwein, David Bailey, and others (see [3, 23, pp. 33–59]), who define it as

> the methodology of doing mathematics that includes the use of computations for gaining insight and intuition, discovering new patterns and relationships, using graphical displays to sugest underlying mathematical principles, testing and especially falsifying conjetures, exploring a possible result to see if it is worth formal proof, suggesing approaches for formal proof, replacing lengthy hand derivations with computer-based derivations, confirming analytically derived results. [Borwein & Bailey, *Mathematics by Experiment*, 2nd ed., A K Peters, 2008, pp. 2–3.]

The methods of this field are thus for the most part akin to those of the scientist: experimenting, much of it done by the computer and its increasingly sophisticated tools, formulating hypetheses, and testing them by further experimentation. Not that proof is to be abandoned, but the focus is elswhere. As Borwein, who calls himsmelf a computer-assisted fallibilist, asserts [23, p. 34] and [3, p. 26]:

> In my view, it is now both necessary and possible to admit quasi-empirical inductive methods fully into mathematical argument. In doing so we will enrich mathematics.... Mathematics is primarily about *secure knowledge*, not proof .... Proofs are often out of reach – but understading, even certainty, is not.

As an illustration, Borwein gives the following example [23, p. 37]:

> Given an interesting identity buried in a long and compliated paper on an unfamiliar subject, which would give you more confidence in its correctness: staring at the proof, or confirming computationally that it is corect to 10,000 decimal places? Here is such a formula [which arose in quantum field theory]: $[24/7\sqrt{7}]\int_{\pi/3}^{\pi/2}\log|(\tan t+\sqrt{7})/(\tan t-\sqrt{7})|\mathrm{d}t = \sum_{n=0}^{\infty}[1/(7n+1)^2+1/(7n+2)^2+1/(7n+3)^2+1/(7n+4)^2+1/(7n+5)^2+1/(7n+6)^2]$.

See [3, 23, pp. 33–59] for further details.

The second example is Thomas Hales' proof in 2005/2006 of Kepler's conjecture about sphere packing. Hales posted the first version of a complete proof, found with the aid of his former student S. P. Ferguson and massive use of a computer, in 1998. (The proof had about 300 pages of text and relied on about 40,000 lines of custom computer code. According to Hales, it is one of the most complicated proofs ever produced.) In the same year, the *Annals of Mathematics*, arguably the most prestigious US research journal, solicited the paper for publication, and in early 1999 hosted a conference aimed at understanding the poof. A panel of 12 referees, headed by Gabor Fejes Toth, was assigned to verify the correctness of the proof!

After 4 years, Toth stated that he was 99% certain that the proof is correct. Robert MacPherson, then editor of the *Annals*, wrote to Hales [35] (unless indicated otherwise, all the quotations below come from [35]):

> The news from the referees is bad, from my perspective. They have not been able to verify the correctness of the proof, and will not be able to certify it in the future, because they have run out of energy to devote to the problem. This is not what I had hoped for.

He continues:

> Fejes Toth thinks that this situation will occur more and more often in mathematics. He says it is similar to the situation in experimental science – other scientists acting as referees can't certify the correctness of an experiment, they can only subject the paper to consistency checks. He thinks that the mathematical community will have to get used to this state of affairs.

And more:

> You may ask whether this degree of certification is enough checking for a mathematical paper, and whether it's not in fact comparable to the level of checking for most mathematical papers. Both the referees and the Editors think that it is not enough for complete certification as correct, for two reasons. First, rigor and certainty was what this particular problem was about in the first place. Second, there are not so many general principles and theoretical consistencies as there were, say, in the proof of Fermat ['s Last Theorem], so as to make you convinced that even if there is a small error, it won't affect the basic structure of the proof.

Nevertheless, an abridged version of the proof was published in 2005 in the *Annals* [34]. The complete proof appeared the following year in *Discrete & Computational Geometry*, and in revised form in 2009 in the same journal.

Hales observes that "this paper has brought about a change in the journal's policy on computer proof. It will no longer attempt to check the correctness of computer code." In fact, only the "human part" of the proof will be printed. The computer code and documentation will be maintained on the *Annals* website.

Finally, here is another interesting insight into proof related to Kepler's conjecture. Hales notes that "there is a way to proceed [with the proof of Kepler's conjecture] that more fully preserves the integrity of mathematics. This is the way of formal proof, [in which] all the intermediate logical steps are supplied, [and] no appeal is made to intuition." This is what Hales, with the assistance of many colleagues and computers, is attempting to do in the enormous Flyspeck Project [33,35]. Among other results, the Prime Number theorem, the Jordan Curve theorem, and the Four-Color theorem have already been "formally proved" ([33]; see the article by Wiedijk).

For amplification of the issues examined in this section, see [23,32,35,57,58,65, 66,69,74].

References

1. K. Appel and W. Haken, The four-color problem. In *Mathematics Today—Twelve Informal Essays*, ed. by L. A. Steen, Springer-Verlag, 1978, pp. 153–190.
2. M. Aschbacher, The status of the classification of finite simple groups, *Notices of the Amer. Math. Soc.* 51 (2004) 736–740.
3. D. Bailey, J. Borwein et al, *Experimental Mathematics in Action*, A K Peters, 2007.
4. E. Bishop, *Foundations of Constructive Analysis,* McGraw-Hill, 1967.
5. U. Bottazzini, *The Higher Calculus: A History of Real and Complex Analysis from Euler to Weierstrass,* Springer-Verlag, 1986.
6. N. Bourbaki, The architecture of mathematics, *Amer. Math. Monthly* 57 (1950) 221–232.
7. D. Bridges and F. Richman, *Varieties of Constructive Mathematics*, Cambridge Univ. Press, 1987.
8. A. Caldar, Constructive mathematics, *Scientific Amer*. 241 (October 1979) 146–171.
9. E. J. Chaitin, Randomness and mathematical proof, *Scientific Amer.* 232 (May 1975) 47–52.
10. C. T. Chong and Y. K. Leong, An interview with Jean-Pierre Serre, *Math. Intell.* 8:4 (1986) 8–13.

11. B. A. Cipra, Computer search solves an old math problem, *Science* 242 (December 16, 1988) 1507–1508.
12. P. J. Davis, Fidelity in mathematical discourse: Is one and one really two?, *Amer. Math. Monthly* 79 (1972) 252–262.
13. P. J. Davis and R. Hersh, *The Mathematical Experience,* Birkhäuser, 1981. (Revised Study Edition, with E. Marchisotto, published in 1995.)
14. R. Dedekind, *Essays on the Theory of Numbers*, Dover, 1963 (orig. 1901).
15. R. A. De Millo, R. J. Lipton, and A. J. Perlis, Social processes and proofs of theorems and programs, *Communications of the ACM* 22 (1979) 271–280.
16. M. Detlefsen and M. Luker, The four-color theorem and mathematical proof, *Jour. of Phil.* 77 (1980) 803–820.
17. C. H. Edwards, *The Historical Development of the Calculus,* Springer-Verlag, 1979.
18. H. M. Edwards, Kronecker's algorithmic mathematics, *Math. Intell.* 31:2 (2009) 11–14.
19. S. Engelsman, *Families of Curves and the Origins of Partial Differentiation,* North-Holland, 1984.
20. H. Eves, *Great Moments in Mathematics*, 2 vols. (before 1650, and after 1650, resp.), Math. Assoc. of America, 1983.
21. S. Feferman, What does logic have to tell us about mathematical proofs?, *Math. Intell.* 2:1 (1979) 20–24.
22. C. G. Fraser, The calculus as algebraic analysis: some observations on mathematical analysis in the 18th century, *Arch. Hist. Exact Sc.* 39 (1989) 317–335.
23. B. Gold and A. Simons (eds), *Proof & Other Dilemmas: Mathematics and Philosophy*, Math Assoc. of America, 2008.
24. N. Goodman, Mathematics as an objective science, *Amer. Math. Monthly* 86 (1979) 540–551.
25. J. V. Grabiner, Who gave you the epsilon: Cauchy and the origins of rigorous calculus, *Amer. Math. Monthly* 90 (1983) 185–194.
26. J. V. Grabiner, The changing concept of change: The derivative from Fermat to Weierstrass, *Math. Mag.* 56 (1983) 195–206.
27. J. V. Grabiner, Changing attitudes toward mathematical rigor: Lagrange and analysis in the 18th and 19th centuries. In *Epistemological and Social Problems of the Sciences in the Early 19th Century*, ed. by H. Jahnke and M. Otte, D. Reidel, 1981, pp. 311–330.
28. J. V. Grabiner, *The Origins of Cauchy's Rigorous Calculus*, M.I.T. Press, 1981.
29. J. V. Grabiner, Is mathematical truth time-dependent? Amer. *Math. Monthly* 81 (1974) 354–365.
30. J. Gray, Review of *Uber die Enstehung von David Hilberts 'Grundlagen der Geometrie'*, *Hist. Math.* 15 (1988) 181–183.
31. I. Hacking, Proof and eternal truths: Descartes and Leibniz. In *Descartes: Philosophy, Mathematics and Physics, ed.* by S. Gaukroger, Barnes and Noble Books, 1980, pp. 169–180.
32. W. Haken, An attempt to understand the four-color problem, Jour. *Graph Theory* 1 (1977) 193–206.
33. T. Hales (ed.), Formal proof, *Notices of the Amer. Math. Soc.* 55 (2008) 1370–1380. (This is part of a Special Issue on Formal Proof, Vol. 55, no. 11, 2008, with articles by Hales, Gonthier, Harrison, and Wiedijk.)
34. T. Hales, A proof of the Kepler comjecture, *Annals of Mathematics* 162 (2005) 1065–1185.
35. T. Hales, The Flyspeck Project, http://code.google.com/p/flyspeck/
36. G. Hanna, *Rigorous Proof in Mathematics Education,* Ontario Institute for Studies in Education Press, 1983.
37. V. Hendricks et al (eds), *Proof Theory: Historical and Philosophical Significance*, Kluwer, 2000.
38. R. Hersh, *What is Mathematics, Really?*, Oxford Univ. Press, 1997.
39. V. J. Katz, *A History of Mathematics: An Introduction*, 3rd. ed., Addison-Wesley, 2009.
40. P. Kitcher, *The Nature of Mathematical Knowledge*, Oxford Univ. Press, 1983.
41. I. Kleiner, *A History of Abstract Algebra, Birkhäuser*, 2007.
42. M. Kline, *Mathematics: The Loss of Certainty*, Oxford Univ. Press, 1980.

43. M. Kline, *Mathematical Thought from Ancient to Modem Times*, Oxford Univ. Press, 1972.
44. W. R. Knorr, On the early history of axiomatics: The interaction of mathematics and philosophy in Greek antiquity. In *Theory Change, Ancient Axiomatics and Galileo's Methodology*, ed. by J. Hintikka et al, D. Reidel, 1980, pp. 145–186.
45. G. Kolata, Prime tests and keeping proofs secret, *Science* 233 (Aug. 1986) 938–939.
46. G. Kolata, Mathematical proofs: The genesis of reasonable doubt, *Science* 192 (June 1976) 989–990.
47. I. Lakatos, A renaissance of empiricism in the recent philosophy of mathematics? In *New Directions in the Philosophy of Mathematics*, ed. by T. Tymoczko, Birkhäuser, 1986, pp. 29–48.
48. I. Lakatos, What does a mathematical proof prove? In New *Directions in the Philosophy of Mathematics*, ed. by T. Tymoczko, Birkhäuser, 1986, pp. 153–162.
49. I. Lakatos, Cauchy and the continuum: The significance of non-standard analysis for the history and philosophy of mathematics, *Math. Intell.* 1:3 (1978) 151–161.
50. I. Lakatos, *Proofs and Refutations*, Cambridge Univ. Press, 1976.
51. D. Laugwitz, Infinitely small quantities in Cauchy's textbooks, *Hist. Math.* 14 (1987) 258–274.
52. G. E. R. Lloyd, *Magic, Reason and Experience: Studies in the Origin and Development of Greek Science*, Cambridge Univ. Press, 1979.
53. P. Mancosu, On mathematical explanation. In *The Growth of Mathematical Knowledge*, ed. by E. Grosholz and H. Breger, Kluwer, 2000, pp. 103–119.
54. M. Mandelkern, Constructive mathematics, *Math. Mag.*58 (1985) 272–280.
55. Yu. I. Manin, How convincing is a proof? *Math. Intell.* 2:1 (1979) 17–18.
56. A. R. Meyer, The inherent computational complexity of theories of ordered sets. In *Proc. Int. Congr. of Mathematicians*, Vancouver, 1974, pp. 477–482.
57. F. H. Norwood, Long proofs, *Amer. Math. Monthly* 89 (1982) 110–112.
58. C. Pomerance, Recent developments in primality testing, *Math. Intell.* 3:3 (1981) 97–105.
59. M. O. Rabin, Probabilistic algorithms. In *Algorithms and Complexity: New Directions and Recent Results,* ed. by J. F. Traub, Academic Press, 1976, pp. 21–40.
60. F. Richman, Existence proofs, *Amer. Math. Monthly* 106 (1999) 303–308.
61. E. Robson, Mathematics in Ancient Iraq: A Social History, Princeton Univ. Press, 2008.
62. G. – C. Rota, The phenomenology of mathematical proof, *Synthese* 111 (1997) 183–196.
63. G. F. Simmons, *Differential Equations,* McGraw-Hill, 1972.
64. S. Smale, Algebra and complexity theory, *Bull. Amer. Math. Soc.* 4 (1981) 1–36.
65. J. Spencer, Short theorems with long proofs, *Amer. Math. Monthly* 90 (1983) 365–366.
66. L. A. Steen, Living with a new mathematical species, *Math. Intell.* 8:2 (1986) 33–40.
66a. J. Stillwell, *Roads to Infinity: The Mathematics of Truth and Proof*, A K Peters, 2010.
67. L. J. Stockmeyer and A. K. Chandra, Intrinsically difficult problems, *Scientific Amer.* 240 (May 1979) 140–159.
68. D. J. Struik, A *Source Book in Mathematics, 1200–1800*, Harvard Univ. Press, 1969.
69. E. R. Swart, The philosophical implications of the four-color problem, *Amer. Math. Monthly* 87 (1980) 697–707.
70. A. Szabo, *The Beginnings of Greek Mathematics*, D. Reidel, 1978.
71. A. Tarski, Truth and proof, *Scientific Amer.* 220 (June 1969) 63–77.
72. *The Two-Year College Mathematics Journal* 12:2 (March 1981). This issue features the concept of proof. It includes articles by Appel and Haken, Galda, Renz, and Tymoczko.
73. T. Tymoczko, Making room for mathematicians in the philosophy of mathematics, *Math. Intell.* 8:3 (1986) 44–50.
74. T. Tymoczko, Computers, proofs and mathematicians: a philosophical investigation of the four-color proof, *Math. Mag.*53 (1980) 131–138.
75. T. Tymoczko, The four-color problem and its philosophical significance, *Jour. of Phil.* 76:2 (1979) 57–83.
76. B. L. Van der Waerden, *Science Awakening*, Noordhoff, 1954.

77. J. Von Neumann, The role of mathematics in the sciences and in society. In *Collected Works*, Vol. 6, ed. by A. H. Taub, Macmillan, 1963, pp. 477–490.
78. J. Von Neumann, The mathematician. In *The World of Mathematics,* Vol. 4, ed. by J. R. Newman, Simon and Schuster, 1956, pp. 2053–2063.
79. H. Weyl, Axiomatic versus constructive procedures in mathematics, *Math. Intell.* 7:4 (1985) 10–17.
80. H. Weyl, Mathematics and logic, *Amer. Math. Monthly* 53 (1946) 2–13.
81. R. Wilder, *Evolution of Mathematical Concepts,* John Wiley and Sons, 1968.
82. R. L. Wilder, The role of the axiomatic method, *Amer. Math. Monthly* 74 (1967) 115–127.

Chapter 8
Paradoxes: What Are They Good For?

8.1 Introduction

A paradox has been described as a truth standing on its head to attract attention. Undoubtedly, paradoxes captivate. They also cajole, provoke, amuse, exasperate, and seduce. More importantly, they arouse curiosity, they stimulate, and they motivate.

In this chapter we present examples of paradoxes from the history of mathematics which have inspired the clarification of basic concepts and the introduction of major results. Our examples will deal with numbers, logarithms, functions, continuity, tangents, infinite series, sets, curves, and decomposition of geometric objects. This is but a small sample. For further examples see [6, 9, 16, 18, 24, 38, 44, 53, 63].

We will use the term "paradox" in a broad sense to mean an inconsistency, a counterexample to widely held notions, a misconception, a true statement that seems to be false, or a false statement that seems to be true. It is in these various senses that paradoxes have played an important role in the evolution of mathematics. Indeed, as E. T. Bell and P. J. Davis, respectively, put it:

> The mistakes and unresolved difficulties of the past in mathematics have always been the opportunities of its future [4, p. 283].

> One of the endlessly alluring aspects of mathematics is that its thorniest paradoxes have a way of blooming into beautiful theories [15, p. 55].

Paradoxes can also serve a useful role in the classroom. The temporary confusion and insecurity which they may engender in students can be put to good use. Conflict and predicament are useful pedagogical devices, provided, of course, that they are dealt with. They may foster a positive attitude to "getting stuck," provide the opportunity to participate in debate and controversy over mathematical issues, and promote the realization that mathematics often develops in this very way. Teachers may gain a better appreciation of students' difficulties in coming to grips with concepts and results with which some of the greatest mathematicians of all

I. Kleiner, *Excursions in the History of Mathematics*,
DOI 10.1007/978-0-8176-8268-2_8,

time grappled. Such concepts and results, while challenging at the time, became commonplaces in subsequent generations. In the words of Kasner and Newman [29, p. 193]:

> The testament of science is so continually in a flux that the heresy of yesterday is the gospel of today and the fundamentalism of tomorrow.

8.2 Numbers

The evolution of the concept of number has been beset by paradoxes almost every step of the way. In the words of Davis [14, p. 305]:

> It is paradoxical that while mathematics has the reputation of being the one subject that brooks no contradictions, in reality it has a long history of successful living with contradictions. This is best seen in the extensions of the notion of number that have been made over a period of 2500 years. From limited sets of integers, to fractions, negative numbers, irrational numbers, complex numbers, transfinite numbers, each extension, in its way, overcame a contradictory set of demands.

The first sentence in the above quotation may be thought of as a "metaparadox" – a nontechnical, paradoxical statement about technical, paradoxical phenomena. We will point out a variety of such metaparadoxes; they are interesting in their own right as issues for philosophical discussion or contemplation. But now to some paradoxes dealing with the evolution of various number systems.

8.2.1 *Incommensurables*

The Pythagoreans of the sixth century BC believed that every line segment can be measured by a positive integer or the ratio of two such integers. This was to them not merely a very plausible fact but an article of faith, an aspect of their philosophy. Moreover, the idea formed the basis of the Pythagorean theory of proportion [55]. It must therefore have been a great shock – a paradox – when they discovered that the diagonal of a unit square cannot be measured by a whole number or by a ratio of whole numbers; or, as the Greeks put it, that the diagonal and side of a square are *incommensurable.* Their proof of this result is essentially the one we use today to show that $\sqrt{2}$ is irrational. The paradox was arrived at by using the Pythagorean theorem. Thus the

Metaparadox: The Pythagorean theorem was the undoing of the Pythagorean philosophy and the Pythagorean theory of proportion.

The discovery of the incommensurability of the diagonal and side of a square had far-reaching consequences for Greek mathematics. On the positive side, it inspired Eudoxus to found a sophisticated theory of proportion which applied to both commensurable and incommensurable magnitudes. This, in turn, motivated

Dedekind more than two millennia later to define the real numbers via Dedekind cuts. On the debit side, the discovery turned the direction of Greek mathematics, at least in its very productive, classical period, from a harmonious collaboration of number and geometry to an almost exclusive concern with geometry.

8.2.2 *Negative Numbers*

The introduction of negative numbers into mathematics and their subsequent use occasioned considerable consternation and difficulties. A major conceptual framework that had to be abandoned was the prohibition of subtracting a greater from a smaller number. As Wallis in the seventeenth century put it [42, p. 438]: "[How can] any magnitude… be less than nothing, or any number fewer than none?"

Among other paradoxes having to do with negative numbers are the following two:

(a) Wallis "proved" that negative numbers are greater than infinity. He argued that since (for positive a) $a/0 = \infty$, a/a negative number $> \infty$; this is so because decreasing the denominator increases the fraction.
(b) In a letter to Leibniz, Arnauld, a seventeenth-century mathematician and philosopher, objected to the equality $1/-1 = -1/1$ on the grounds that the ratio of a greater to a smaller quantity cannot equal the ratio of a smaller to a greater. Leibniz agreed this was a difficulty, but argued for the tolerance of negative numbers because they are useful and, in general, lead to consistent results. See [10, pp. 39–40].

Justification of inexplicable notions on the grounds that they yield useful results has occurred frequently in the evolution of mathematics. This brings up the following

Metaparadox: How can inexplicable, little understood, things be so useful? Of course, out of confusion emerged, in time, clarity and understanding.

8.2.3 *Complex Numbers*

The solution by radicals of cubic equations was one of the great achievements of sixteenth-century mathematics. Cardan's solution of the cubic $x^3 = ax + b$ can be expressed, using modern notation, by the formula

$$x = \sqrt[3]{\frac{b}{2} + \sqrt{\left(\frac{b}{2}\right)^2 - \left(\frac{a}{3}\right)^3}} + \sqrt[3]{\frac{b}{2} - \sqrt{\left(\frac{b}{2}\right)^2 - \left(\frac{a}{3}\right)^3}}.$$

Bombelli applied it to the equation $x^3 = 15x + 4$ to obtain $x = \sqrt[3]{2 + \sqrt{-121}} + \sqrt[3]{2 - \sqrt{-121}}$. Cardan had earlier denied the applicability of his formula to such equations since it introduced square roots of negative numbers, which he rejected. But Bombelli noted (by inspection) that $x = 4$ is a solution of $x^3 = 15x + 4$. (The other two roots, $-2 \pm \sqrt{3}$, are also real.) Here was a paradox: The roots of $x^3 = 15x + 4$ are real, yet the formula yielding the roots involved complex, and at the time meaningless, numbers. "The whole matter seemed to rest on sophistry rather than on truth," noted Bombelli [43, p. 19]. And he set himself the task of resolving that sophistry by devising rules for manipulating expressions of the form $a + b\sqrt{-1}$, thereby showing that (one of the values of) $\sqrt[3]{2 + \sqrt{-121}} + \sqrt[3]{2 - \sqrt{-121}}$ is indeed 4. It was the birth of complex numbers. Birth, however, did not entail legitimacy. It took another two and a half centuries before complex numbers were accepted as bona fide mathematical entities. See Chap. 12.

8.3 Logarithms

The issue of the meaning of logarithms of negative and complex numbers arose in the early eighteenth century in connection with integration. In analogy with the real case, Johann Bernoulli integrated $1/(x^2 + a^2)$ as follows:

$$\begin{aligned}\int \mathrm{d}x/(x^2 + a^2) &= \int \mathrm{d}x/(x + a\mathrm{i})(x - a\mathrm{i})\\ &= -1/2a\mathrm{i} \int [1/(x + a\mathrm{i}) - 1/(x - a\mathrm{i})]\,\mathrm{d}x\\ &= -1/2a\mathrm{i}\,[\log(x + a\mathrm{i}) - \log(x - a\mathrm{i})]\\ &= -(1/2a\mathrm{i})\log[(x + a\mathrm{i})/(x - a\mathrm{i})].\end{aligned}$$

In an exchange of letters, begun in 1712 and lasting sixteen months, Bernoulli and Leibniz argued about the meaning of $\log[(x + a\mathrm{i})/(x - a\mathrm{i})]$, and, in particular, about the meaning of $\log(-1)$. Bernoulli asserted that $\log(-1)$ is real while Leibniz claimed it is imaginary, each advancing various arguments to support his view. (To Leibniz "imaginary" meant "not real," but not necessarily complex; he did not exclude other kinds of "imaginaries.") For example, Bernoulli reasoned that since $\mathrm{d}x/x = \mathrm{d}(-x)/-x$, $\int \mathrm{d}x/x = \int \mathrm{d}(-x)/-x$, hence $\log x = \log(-x)$. In particular, $\log(-1) = \log 1 = 0$.

Among Leibniz' arguments were the following:

(a) Since the range of $\log a$, for $a > 0$, comprises all real numbers, it follows that $\log a$, for $a < 0$, must be imaginary, because the real numbers have already been "spoken for."
(b) If $\log(-1)$ were real, then $\log \mathrm{i}$ would also be real, since $\log \mathrm{i} = \log(-1)^{1/2} = (1/2)\log(-1)$. But this is clearly absurd, alleges Leibniz.

Fig. 8.1 Gottfried Wilhelm Leibniz (1646–1716)

(c) Putting $x = -2$ in the expansion $\log(1+x) = x - x^2/2 + x^3/3 - \ldots$ yields $\log(-1) = -2 - 4/2 - 8/3 - \ldots$. Since the series on the right diverges, it cannot be real, hence it must be imaginary.

The above are indeed interesting examples of the art – not to say "science" – of symbolic manipulation practiced by some of the greatest mathematicians of the seventeenth and eighteenth centuries. The resulting paradoxes had "for a long time... tormented me," noted Euler [37, p. 72]. He resolved them in a 1745 paper. We quote from its interesting introduction [34, p. 4]:

> Since logarithms are clearly part of pure mathematics it may well be surprising to learn that they have been until now the subject of an embarrassing controversy in which whatever side is taken contradictions appear that seem completely impossible to resolve. Meanwhile if truth is to be universal there can be no doubt that these contradictions..., however unresolved they seem, can only be apparent.... I will bring out fully all the contradictions involved so that it may be seen how difficult it is to discover truth and to guard against inconsistency even when two great men are working on the problem.

The crux of Euler's solution was the Euler–Cotes formula $e^{i\theta} = \cos\theta + i\sin\theta$. It implies that $e^{i(\pi+2n\pi)} = \cos(\pi+2n\pi) + i\sin(\pi+2n\pi) = \cos\pi + i\sin\pi = -1$, so that $\log(-1) = i(\pi+2n\pi)$, where $n = 0, \pm1, \pm2, \ldots$. Thus $\log(-1)$ is *multivalued*, in fact, infinite-valued, and all its values are complex. Both Bernoulli and Leibniz were wrong, the former "more so" than the latter.

This solution, however, did not satisfy Euler's contemporaries, in particular d'Alembert, who persisted in subscribing to Bernoulli's solution until 1761, even though in 1749 Euler gave a "simpler" solution. See [33, p. 178] as well as [8, 10].

8.4 Functions

The concept of function originated in the early eighteenth century. Newton and Leibniz invented the calculus in the latter part of the seventeenth century. Here, then, is a

Metaparadox: Calculus without functions.

Indeed, the calculus of Newton and Leibniz was a calculus of curves, given by equations, rather than a calculus of functions. See Chap. 4.

A function was viewed at different times as a formula, a curve, or an arbitrary correspondence. Paradoxes turned up to dethrone one or another of these views of functionality. Even the very meaning of a formula, as well as its scope (i.e., the functions that it represents), changed over time, and were often subjects of considerable controversy. For example:

8.4.1 The Eighteenth Century

To Euler and his contemporaries of the mid-eighteenth century a function meant a formula, where the latter concept, though not rigorously defined, was broadly construed to allow, among other things, infinite sums and products in its formation. There were several implicit assumptions:

(a) The function had to be given by a *single* expression. For example,

$$f(x) = \begin{cases} x, & x > 0, \\ -x, & x \leq 0, \end{cases}$$

was not considered a function.

(b) The independent variable had to range over *all* real numbers, except possibly for isolated points, as in $f(x) = 1/x$. For instance, $f(x) = x,\ 0 \leq x \leq 1$, was not considered a function.

(c) Two functions which agreed on an interval were assumed to agree everywhere on the line.

The significance of these assumptions was the fact that the algorithms of calculus applied at that time only to such functions. See Chap. 5.

8.4.2 Nineteenth-Century Views

Many of the eighteenth-century conceptions about functions were overturned by Fourier's work on heat conduction in the early decades of the nineteenth century. As a result of this work Fourier claimed to have shown that *any* function defined on some interval can be represented on that interval as an infinite series of sines and cosines – a *Fourier series.* Given *our* conception of function, this result is, of course, incorrect in the generality which Fourier claimed for it, although his contemporaries would have been hard put to find an exception. Fourier's result implied that if, for example,

$$f(x) = \begin{cases} -1, & -\pi < x < 0, \\ 0, & x = 0 \\ 1, & 0 < x < \pi, \end{cases}$$

then

$$f(x) = (4/\pi)(\sin x/1 + \sin 3x/3 + \sin x5x/5 + \ldots) \text{ for all } x \in (-\pi, \pi).$$

Several fundamental departures concerning functions resulted from Fourier's work:

1. It became legitimate, and important, to consider functions whose domain is an *interval* rather than the entire real line.
2. Two functions could agree on an interval but differ outside the interval.
3. A function given by two or more distinct expressions could equal a function given by a single expression.

In an 1829 paper on Fourier series Dirichlet introduced the so-called Dirichlet function

$$D(x) = \begin{cases} 1, & \text{if } x \text{ is rational} \\ 0, & \text{if } x \text{ is irrational} \end{cases}$$

This function was neither a formula nor a curve. It was a new type of function, described by a correspondence. It was the first of many functions which came to be called "pathological" – but not for very long [58].

At the end of the nineteenth century, Baire extended the notion of formula. To him it meant an expression obtained from a variable and constants by a possibly countable iteration of additions, multiplications, and the taking of limits. He called such a function *analytically representable* and showed that the Dirichlet function is of this type: $D(x) = \lim_{m\to\infty} \lim_{n\to\infty} \cos(m!\pi x)^n$. Thus, the "pathological" Dirichlet function became a "tame," analytically representable function.

Is analytic representability a universal mode of representability of functions? That is, are there functions which are not analytically representable (in Baire's sense)? Yes and no. If you are a formalist, you can show by a counting argument that the set of analytically representable functions has cardinality c, while the set of all functions (clearly) has cardinality 2^c. Thus, there are uncountably many functions which are not analytically representable. But, no one has given a *constructive* example of even one. See Chap. 5.

8.5 Continuity

8.5.1 Euler and Cauchy

Although the concept of continuity is nowadays fundamental in mathematics, its modern definition was not formulated until the nineteenth century, about 150years after the invention of the calculus by Newton and Leibniz. In the eighteenth century, Euler did define a notion of continuity in response to the famous vibrating-string controversy (see Chap. 5). To him a continuous function was one given by a single expression (formula), while a function given by several expressions was considered *discontinuous*. For example, the function

$$f(x) = \begin{cases} x, & x > 0 \\ -x, & x \leq 0 \end{cases}$$

was discontinuous, while the function comprising the two branches of a hyperbola was considered continuous, since it is given by the single expression $g(x) = 1/x$ [36, p. 301]. (In the second half of the eighteenth century Euler extended the notion of function, so that expressions such as $f(x)$ were now considered functions. See Sect. 8.4.1 and Chap. 5.)

The work on Fourier series showed the untenability of the eighteenth-century notion of continuity. For example, the function

$$f(x) = \begin{cases} -1, & -\pi < x < 0 \\ 0, & x = 0 \\ 1, & 0 < x < \pi \end{cases}$$

could (as we have seen) be represented by a single expression, namely its Fourier series, hence it was and was not continuous in the eighteenth-century sense of that concept.

In an 1821 work Cauchy initiated a reappraisal and reformulation of the foundations of eighteenth-century calculus. In this work he defined continuity essentially as we understand the concept today, although he used the then-prevailing language of infinitesimals rather than the now-accepted $\varepsilon - \delta$ formulation given by Weierstrass in the 1850s (see Chap. 4). The shift in point of view from Euler's to Cauchy's conceptions was fundamental. In the former case, continuity was a global property while in the latter it was local. But the concept proved to be very subtle, and was not completely understood even by Cauchy and his contemporaries of the early to mid-nineteenth century. For example:

Cauchy "proved" that an infinite sum (a convergent series) of continuous functions is a continuous function [8, p. 110], but of course this is incorrect. A counterexample was given by Abel in the 1820s – it is essentially the series $\sum_0^\infty \sin(2n+1)x/2n+1$ that we encountered earlier, which is discontinuous at $x = k\pi, k = 0, \pm 1, \pm 2, \ldots$. The error in Cauchy's proof resulted from his failure to distinguish between convergence and uniform convergence of a series of functions. In fact, "the realization of the central role of the concept of uniform convergence in analysis came about slowly in the last [nineteenth] century" [48, p. 97]. (Lakatos claims that Cauchy's proof is not erroneous if it is reinterpreted in terms of Robinson's infinitesimals [32].)

8.5.2 *Continuity and Differentiability*

Euler's continuous functions were, in practice, differentiable, except possibly at isolated points. So were Cauchy's. In fact, Cauchy and his contemporaries believed, and some of them "proved," that continuity implies differentiability (except at obvious, isolated points) [57]. It was therefore astonishing when Weierstrass gave an example in the 1860s of a continuous function which is *nowhere* differentiable, namely $f(x) = \sum_1^\infty b^n \cos(a^n \pi x)$, a an odd integer, b a real number in (0, 1), and $ab > 1 + 3\pi/2$. This and similar examples showed for the first time that the notion of continuity is considerably broader than that of differentiability, and thus established continuity as an important concept of investigation in its own right. The examples also showed the limitations of intuitive geometric reasoning in analysis, and thus the need for careful, analytic formulations of basic notions.

In a modern development of a different kind, Schwartz and Sobolev showed in the 1940s that every continuous function is, indeed, "differentiable." But, the derivative is now a "generalized function" (a "distribution"). For example, if

$$f(x) = \begin{cases} 1, & \text{if } x > 0 \\ 1/2, & \text{if } x = 0 \\ 0 & \text{if } x < 0 \end{cases}$$

then

$$f'(x) = \begin{cases} 0, & x \neq 0 \\ \infty, & x = 0 \end{cases},$$

which is the Dirac delta "function" $\delta(x)$. As this example shows, there are even discontinuous functions which are differentiable (in the Schwartz/Sobolev sense) – a shocking realization it would have been for mathematicians of the second half of the nineteenth century. See Chap. 5 for details.

8.6 Aspects of Calculus Other than Continuity

8.6.1 Tangents

Calculus was invented, independently, by Newton and Leibniz in the last third of the seventeenth century. But many of its important ideas were foreshadowed in early seventeenth-century work of prominent mathematicians, notably Fermat. In the late 1630s he devised a method for dealing with problems of tangents and of maxima and minima. The following example illustrates Fermat's approach [17, p. 122]:

Suppose we wish to find the tangent to the parabola $y = x^2$ at some point (x, x^2). Let $x+e$ be a nearby point on the x-axis and let s denote the *subtangent* of the curve at the point (x, x^2) (see Fig. 8.2). Similarity of triangles yields $x^2/s = k/(s+e)$. Fermat notes that k is "approximately equal" (he calls it "adequal") to $(x+e)^2$; writing this as $k \approx (x+e)^2$ we get $x^2/s \approx (x+e)^2/s+e$.

Solving for s we have

$$s \approx ex^2/[(x+e)^2-x^2] = ex^2/[x^2+2ex+e^2-x^2] = ex^2/e(2x+e) = x^2/(2x+e),$$

hence $x^2/s \approx 2x+e$. Note that x^2/s is the slope of the tangent to the parabola at (x, x^2). Fermat now deletes e and claims that the slope of the tangent is $2x$.

Fermat's method was severely criticized by some of his contemporaries, in particular by Descartes. They objected to his introduction and subsequent suppression of the mysterious e. Dividing by e meant regarding it as not zero. Discarding e implied treating it as zero. This is inadmissible, they rightly claimed. In a somewhat different context, but with equal justification, Bishop Berkeley in the eighteenth

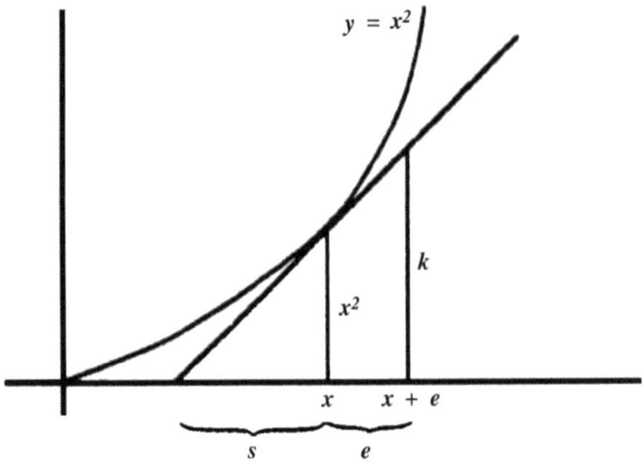

Fig. 8.2 Fermat's "algebraic" method of tangents

century would refer to such e's as "the ghosts of departed quantities," arguing that "by virtue of a twofold mistake... [one] arrive[d], though not at a science yet at the truth" [30, p. 428].

The justification of seventeenth- and eighteenth-century algorithms of calculus was that they yielded correct results – another important example of the utility of inexplicable procedures (see Sect. 8.2). The end seemed to have justified the means. *Rigorous* justification of the calculus – of one kind – came with the 1821 introduction of limits by Cauchy, and – of another kind – with the 1960 introduction of infinitesimals by Robinson.

Metaparadox: How can calculus be founded on two distinct, and in some ways incompatible, theories: limits, based on the real numbers, and infinitesimals, based on the hyperreal numbers? As Steen put it: "The epistemological foundation of mathematical analysis is far from settled" [52, p. 92].

8.6.2 *Infinite Series*

Power series were a potent tool in seventeenth- and especially eighteenth-century calculus. They were manipulated as polynomials, with little if any attention paid to questions of convergence. In fact, Euler and others consciously used *divergent* series to great advantage. The results thus obtained were impressive and important, but errors and paradoxes became unavoidable. Here are two:

(a) There is undoubtedly a touch of the metaphysical in the mathematical infinite. The following example, due to Euler, confirms it [30, p. 447]:

Letting $x = -1$ in $(1+x)^{-2} = 1 - 2x + 3x^2 - 4x^3 + \ldots$, he gets

$$\infty = 1 + 2 + 3 + 4 + \ldots \tag{8.1}$$

Letting $x = 2$ in $(1-x)^{-1} = 1 + x + x^2 + x^3 + \ldots$, one has

$$-1 = 1 + 2 + 4 + 8 + \ldots \tag{8.2}$$

Since each term on the right side of (8.2) is greater than or equal to the corresponding term on the right side of (8.1), $-1 > \infty$. But clearly $\infty > 1$. Hence, $-1 > \infty > 1$. Euler infers that ∞ must be a sort of limit between the positive and negative numbers, and in this respect resembles 0 [30, p. 447].

(b) Occasionally, seventeenth-and eighteenth-century mathematicians reveled in the art of series-manipulation – if for no better reason (it would seem) than to demonstrate their prowess. For example, putting $x = 1$ in $\log(1 + x) =$

$x - x^2/2 + x^3/3 - \ldots$ yields $\log 2 = 1 - 1/2 + 1/3 - 1/4 + \ldots$. So far so good. But now, the argument went, the right side equals

$$(1 + 1/3+1/5+\ldots)+(1/2+1/4+1/6+\ldots)-2(1/2+1/4$$
$$+1/6 + \ldots) = (1 + 1/2 + 1/3 + 1/4 + 1/5 + \ldots)$$
$$-(1 + 1/2 + 1/3 + 1/4 + 1/5 + \ldots) = 0,$$

hence $\log 2 = 0$. It was only in the mid-nineteenth century that Riemann resolved this paradox by proving that the sum of a conditionally convergent series can assume, upon rearrangement, *any* value. "The discovery of this apparent paradox contributed essentially to a re-examination and rigorous founding... of the theory of infinite series," notes Remmert [48, p. 30].

8.7 Sets

During the last three decades of the nineteenth century Cantor developed many important set-theoretic ideas using an intuitive ("naive") notion of set. Eventually this proved inadequate and led to paradoxes. Perhaps the best known is Russell's classic paradox of 1902: Let $R = \{x : x \notin x\}$. Then $R \in R$ if and only if $R \notin R$. This paradox had a profound effect on a number of mathematicians [41]. It devastated the logician Frege, who had just completed a two-volume treatise on the foundations of arithmetic which relied on set-theoretic notions. Learning of Russell's paradox, he lamented [30, p. 1192]:

> A scientist can hardly meet with anything more undesirable than to have the foundation give way just as the work is finished. I was put in this position by a letter from Mr. Bertrand Russell when the work was nearly through the press.

On the other hand, the paradoxes of set theory had positive effects. In particular, they provoked mathematicians to give precise meaning to the notion of set by devising various axiomatizations of set theory, such as the Zermelo–Fraenkel axioms, the Russell and Whitehead theory of types, the Gödel–Bernays system. Although such axiom systems avoided the known paradoxes, they did not guarantee that new ones would not emerge. As Poincaré put it picturesquely [30, p. 1186]:

> We have put a fence around the herd to protect it from the wolves but we do not know whether some wolves were not already within the fence.

Here are two metaparadoxes resulting from Cantor's work in set theory:

Metaparadox 1: Infinity comes in different sizes, in fact in *infinitely many* different sizes.

The second metaparadox comes from juxtaposing the following two quotations by Poincaré and Hilbert, respectively [30], p. 1003]:

Metaparadox 2: (a) "Later generations will regard *Mengenlehre* [set theory] as a disease from which one has recovered."

(b) "No one shall expel us from the paradise which Cantor created for us."

8.8 Curves

The notion of curve is fundamental in geometry. To Euclid it meant "breadthless length." The collection of curves known to his contemporaries was small – the conic sections, the conchoid, the cissoid, the spiral, the quadratrix, and a very few others. The situation changed dramatically with the invention of analytic geometry in the seventeenth century. Now any equation in two variables came to represent a (plane) curve, although, as historian Bos observes [7, p. 296]:

> Seventeenth-century mathematicians did not have a uniform definition of the concept of curve (nor apparently did they feel the need for such a definition).

The study of curves was pursued vigorously for the next three centuries, attracting some of the best mathematicians, who attacked it by geometric, analytic, algebraic, arithmetic, and topological means.

"Pathological" functions introduced in the second half of the nineteenth century raised questions about the nature of curves. For example, in what sense does a continuous nowhere-differentiable function represent a curve? Jordan responded in 1887 with what came to be the first formal definition of a curve (other than perhaps Euclid's). To him a curve was the path of a continuously moving point. More precisely, it was $\{(f(t), g(t)) | f, g\text{: } [0, 1] \to R \text{ are continuous functions}\}$. It was in this context that he stated and proved – incorrectly, as it later turned out – the celebrated Jordan-curve theorem.

In 1890 Peano gave his famous and astounding example of a "space-filling curve" – that is, he exhibited a continuous mapping of the unit interval onto a square, including its interior. But according to Jordan's definition, that made the square into a curve – a not very desirable state of affairs. "How was it possible that intuition could so deceive us?" wondered Poincaré [56, p. 123]. Jordan's definition was too broad and had to be modified.

But Jordan's definition also turned out to be too narrow. For we would surely want the graph of $y = \sin 1/x$ and its limit points on the y-axis, that is, $\{(x, \sin 1/x)\colon x \in (-\infty, 0) \cup (0, \infty)\} \cup \{(0, y) : -1 \leq y \leq 1\}$, to be called a curve, but it is not the image of a continuously moving point. (This is intuitively clear, although to prove it we need topological notions. See [26, p. 1968].

Metaparadox: How can a definition be both too broad and too narrow?

A satisfactory resolution of the dilemma was achieved by Menger and Urysohn only in the 1920s. First one had to clarify the notion of dimension [39]. (That notion, too, was challenged by the paradoxical Peano curve which implied that a square is one-dimensional since it is the continuous image of the unit interval. Cantor's proof of the one–one correspondence between an interval and a square also put into question the intuitive notion of dimension.) When this was done, a curve was defined as a one-dimensional continuum [62]. (A continuum is a closed, connected set of points.) The definition proved adequate until the 1970s when Mandelbrot introduced curves – his fractals – whose dimensions are fractions. See [22].

8.9 Decomposition of Geometric Objects

8.9.1 Doubling the Cube

In 1924 Banach and Tarski proved that a pea and the sun are equidecomposable. That is, a pea may be cut up into finitely many pieces (it was shown in the 1940s that five pieces suffice; in fact, no number less than five will do) which can be rearranged to yield the sun (in volume if not in substance). This is the celebrated *Banach–Tarski paradox* [59]. Moreover, Banach and Tarski have shown that *any* two bounded sets in Euclidean space R^n are equidecomposable if they contain interior points and if $n > 2$ [5, p. 351]. If one allows for *denumerable* decompositions, this result holds also for $n = 2$ [5, p. 351].

Of course, the pieces into which the pea is cut in the Banach–Tarski decomposition are *not measurable;* that is, they do not have a volume. They are not the kinds of pieces that can be obtained using scissors or other cutting devices. They are obtained using the axiom of choice.

Metaparadox: How can simple assumptions – for example, the axiom of choice, have such formidable consequences – for example, the Banach–Tarski paradox?

Of course, the axiom of choice may not be such a simple assumption after all [41]. But it would have been very helpful to the Delians of Greek antiquity [59, p. v]:

> Delians: "How can we be rid of the plague?"
> Delphic Oracle: "Construct a cubic altar double the size of the existing Altar."
> Banach and Tarski: "Can we use the axiom of choice?"

8.9.2 Squaring the Circle

"At long last, the circle has been squared." This is no hoax. It is the title of an article which appeared in the reputable *Notices of the American Mathematical Society* [23]. In 1988 the Hungarian mathematician Laczkovich showed that a circle can be decomposed into finitely many pieces which can be reassembled to give a square of equal area. But the pieces are not measurable (none has an area) and the decomposition is secured using the axiom of choice [23].

8.10 Conclusion

We have presented a variety of mathematical paradoxes from different historical periods. They resulted from, among other things, debates and controversies among mathematicians, counterexamples to what were thought to be immutable notions, failures to see the need for tightening (broadening) a concept or broadening (tightening) a result, and the application of a "principle of continuity" which suggested the transferability of procedures from a given case to what appeared to be like cases (see Chap. 9). We saw that such paradoxical phenomena have had a very substantial impact on the development of mathematics through the refinement and reshaping of concepts, the broadening of existing theories, and the rise of new ones. Moreover, this process is ongoing.

We have also suggested roles for paradoxes in the teaching and learning of mathematics. They can generate curiosity, increase motivation, create an effective environment for debate, encourage the examination of underlying assumptions, and show that faulty logic and erroneous arguments are not uncommon features of the mathematical enterprise.

References

1. E. Akin, A spiteful computer: a determinism paradox, *Math. Intell.* 14:2 (1992) 45-47.
2. R. Arnot and K. Small, The economics of traffic congestion, *Amer. Scientist* 82 (Sept. 1994) 446-455.
3. T. Bass, Road to ruin, *Discover* 13:5 (May 1993) 56-61.
4. E. T. Bell, *The Development of Mathematics*, 2nd ed., McGraw-Hill, 1945.
5. L. M. Blumenthal, A paradox, a paradox, a most ingenious paradox, *Amer. Math. Monthly* 47 (1940) 346-353.
6. B. Bolzano, *Paradoxes of the Infinite*, Routledge and Kegan Paaul, 1950 (orig. 1848).
7. H. J. M. Bos, On the representation of curves in Descartes' *Géométrie, Arch. Hist. Exact Sc.* 24 (1981) 295-338.
8. U. Bottazzini, *The Higher Calculus: A History of Real and Complex Analysis from Euler to Weierstrass,* Springer-Verlag, 1986.
9. V. M. Bradis et al, *Lapses in Mathematical Reasoning*, The Macmillan Co., 1963.
10. F. Cajori, History of exponential and logarithmic concepts, *Amer. Math. Monthly* 20 (1913) 5-14, 35-47, 75-84, 107-117, 148-151, 173-182, 205-210.
11. M. T. Carroll et al, The wallet paradox revisited, *Math. Mag.* 74 (2001) 378-383.
12. J. Case, Paradoxes involving conflicts of interest, *Amer. Math. Monthly* 107 (2000) 33-43.
13. T. Chow, The surprise examination or unexpected hanging paradox, *Amer. Math. Monthly* 105 (1998) 41-51.
14. P. J. Davis, *The Mathematics of Matrices,* Blaisdell, 1965.
15. P. J. Davis, Number, *Scientific Amer.* 211 (Sept. 1964) 51-59.
16. A. De Morgan, *A Budget of Paradoxes*, Open Court, 1915 (orig. 1872).
17. C. H. Edwards, *The Historical Development of the Calculus,* Springer-Verlag, 1979.
18. N. Faletta, *The Paradoxicon*, Doubleday & Co., 1983.
19. P. Fjelstad, The repeating integer paradox, *Coll. Math. Jour.* 26 (1995) 11-15.
20. J. Fleron, Gabriel's wedding cake, *Coll. Math. Jour.* 30 (1999) 35-38.
21. D. Gale, Paradoxes and a pair of boxes, *Math. Intellligencer* 13:2 (1991) 31-33.
22. M. Gardner, Mathematical games, in which 'monster' curves force redefinition of the word 'curve,' *Scientific Amer.* 235 (Dec. 1976) 124-133.
23. R. J. Gardner and S. Wagon, At long last the circle has been squared, *Notices Amer. Math. Soc.* 36 (1989) 1338-1343.
24. R. Gethner, Can you paint a can of paint? *Coll. Math. Jour.* 36 (2005) 400-402.
25. R. K. Guy, The strong law of small numbers, *Amer. Math. Monthly* 95 (1988) 697-712.
26. H. Hahn, The crisis in intuition. In *The World of Mathematics,* ed. by J. R. Newman, Simon & Schuster, 1956, Vol. 3, pp. 1956-1976.

27. R. W. Hamming, The tennis ball paradox, *Math. Mag.* 62 (1989) 268-269.
28. P. J. Hilton et al, *Mathematical Vistas*, Springer, 2002.
29. E. Kasner and J. R. Newman, *Mathematics and the Imagination,* Simon & Schuster, 1967.
30. M. Kline, *Mathematical Thought from Ancient to Modem Times*, Oxford Univ. Press, 1972.
31. J. Kocik, Proof without words: Simpson's paradox, *Math. Mag.* 74 (2001) 399.
32. I. Lakatos, *Proofs and Refutations*, Cambridge Univ. Press, 1976.
33. D. Laugwitz, *Controversies about numbers and functions*. In *The Growth of Mathematical Knowledge*, E. Grosholz and H. Berger (eds), Kluwer, 2000, pp. 177-198.
34. Leapfrogs, *Imaginary Logarithms,* E. G. M. Mann & Son (England), 1978.
35. E. Linzer, The two envelope paradox, *Amer. Math. Monthly* 10 (1994) 417-419.
36. J. Lützen, Euler's vision of a generalized partial differential calculus for a generalized kind of function, *Math. Mag.* 56 (1983) 299-306.
37. P. Marchi, The controversy between Leibniz and Bernoulli on the nature of the logarithms of negative numbers, In *Akten das II Inter. Leibniz-Kongress* (Hanover, 1972), Bnd II, 1974, pp. 67-75.
38. E. A. Maxwell, *Fallacies in Mathematics*, Cambridge Univ. Press, 1961.
39. K. Menger, What is dimension?, *Amer. Math. Monthly* 50 (1943) 2-7.
40. K. G. Merryfield et al, The wallet paradox, *Amer. Math. Monthly* 104 (1997) 647-649.
41. G. H. Moore, *Zermelo's Axiom of Choice: Its Origins, Development, and Influence,* Springer-Verlag, 1982.
42. E. Nagel, 'Impossible numbers': a chapter in the history of modern logic, *Stud. in the Hist. of Ideas* 3 (1935) 429-474.
43. P. J. Nahin, *An Imaginary Tale: The Story of* $\sqrt{-1}$, Princeton Univ. Press, 1998.
44. R. Nelson, Pictures, probability, and paradox, *Coll. Math. Jour.* 10 (1979) 182-190.
45. E. P. Northrop, *Riddles in Mathematics: A Book of Paradoxes*, Van Nostrand, 1944.
46. Z. Pogoda and M. Sokolowski, Does mathematics distinguish certain dimenisons of space? *Amer. Math. Monthly* 104 (1997) 860-869.
47. A. Rapoport, Escape from paradox, *Sc. Amer.* 217 (July 1967) 50-56.
48. R. Remmert, *Theory of Complex Functions,* Springer-Verlag, 1991.
49. D. Saari, a chaotic exploration of aggregation paradoxes, *SIAM Review* 37:1 (March 1995) 37-52.
50. R. P. Savage, The paradox of nontransitive dice, *Amer. Math. Monthly* 101 (1994) 429-436.
51. P. Scholten and A. Simoson, The falling ladder paradox, *Coll. Math. Jour.* 27 (1996) 49-54.
52. L. A. Steen, New models of the real-number line, *Scientific Amer.* 225 (Aug. 1971) 92-99.
53. G. Szekely and D. Richards, The St. Petersburg paradox and the crash of high-tech stocks in 2000, *Amer. Statistician* 58:3 (2004) 225-231.
54. H. Thurston, Can a graph be both continuous and discontinuous? *Amer. Math. Monthly* 96 (1989) 814-815.
55. B. L. Van der Waerden, *Science Awakening I*, Scholar's Bookshelf, 1988 (orig. 1954).
56. N. Ya. Vilenkin, *Stories About Sets,* Academic Press, 1968.
57. K. Volkert, Zur Differentzierbarkeit stetiger Funktionen—Ampere's Beweis und seine Folgen, *Arch. Hist. Exact Sc.* 40 (1989) 37-112.
58. K. Volkert, Die Geschichte der pathologischen Funktionen—Ein Beitrag zur Enstehung der mathematischen Methodologie, *Arch. Hist. Exact Sc.* 37 (1987) 193- 232.
59. S. Wagon, *The Banach-Tarski Paradox*, Cambridge Univ. Press, 1985.
60. J. S. Walker, An elementary resolution of the liar paradox, *Coll. Math. Jour.* 35 (2004) 105-111.
61. J. R. Weeks The twin paradox in a closed universe, *Amer. Math. Monthly* 108 (2001) 585-590.
62. G. T. Whyburn, What is a curve?, *Amer. Math. Monthly* 49 (1942) 493-497.
63. F. Zames, Surface area and the cylinder area paradox, *Coll. Math. Jour.* 8 (1977) 207-211.

Chapter 9
Principle of Continuity: Sixteenth–Nineteenth Centuries

9.1 Introduction

The Principle of Continuity was a very broad law, used widely and importantly – though often not explicitly formulated – throughout the seventeenth, eighteenth, and nineteenth centuries. In general terms, the Principle of Continuity says that what holds in a given case continues to hold in what appear to be like cases. Specifically, it maintains that

1. What is true for positive numbers is true for negative numbers.
2. What is true for real numbers is true for complex numbers.
3. What is true up to the limit is true at the limit.
4. What is true for finite quantities is true for infinitely small and infinitely large quantities.
5. What is true for polynomials is true for power series.
6. What is true for a given figure is true for a figure obtained from it by continuous motion.
7. What is true for ordinary integers is true for (say) Gaussian integers.

Each of these assumptions was used by mathematicians at one time or another, as we shall see. No doubt they realized that not *all* properties holding in a given case carry over to what appear to be like cases; they chose the properties that suited their purposes. And these purported analogies, even when not fully borne out, were often starting points for fruitful theories.

André Weil, in his essay "From metaphysics to mathematics," gives poetic expression to some of the above thoughts [34, p. 408]:

> Mathematicians of the eighteenth century were accustomed to speak of "the metaphysics of the calculus," or "the metaphysics of the theory of equations." They understood by this a vague set of analogies, difficult to grasp and difficult to formulate, which nonetheless seemed to them to play an important role at a given moment in mathematical research and discovery ... .
> All mathematicians know that nothing is more fertile than these obscure analogies, these troubled reflections of one theory in another, these furtive caresses, these inexplicable

DOI 10.1007/978-0-8176-8268-2_9,

> misunderstandings; also nothing gives more pleasure to the investigator. A day comes when ... the metaphysics has become mathematics, ready to form the material whose cold beauty will no longer know how to move us.

We begin our story with Kepler, although the Principle of Continuity, in one form or another, was used earlier, as we shall see. In the early seventeenth century Kepler enunciated a version of the Principle in connection with his study of conics. All conics, he claimed, are of the same species. For example, a parabola may be regarded as a limiting case of an ellipse or a hyperbola, in which one of the foci has gone to infinity. And "a straight line goes over into a parabola through infinite hyperbolas, and through infinite ellipses into a circle" [32, p. 744]. (Desargues and Pascal thought along similar lines.) See also [23].

It was Leibniz, however, who made the Principle of Continuity – he called it *lex continui* – into an all-embracing law. (He owed some of his ideas to Descartes.) It appears throughout his work – in mathematics, philosophy, and science. Here are several ways in which he expressed it [15, pp. 291–294]:

1. Nature makes no leaps ... We pass from the small to the great, and the reverse, through the medium.
2. When the essential determinations of one being approximate those of another, all the properties of the former should also gradually approximate those of the latter.
3. Since we can move from polygons to a circle by a continuous change and without making a leap, it is also necessary not to make a leap in passing from the properties of polygons to those of a circle, otherwise the law of continuity would be violated.

Leibniz' rationale for this encompassing Principle was that "the sovereign wisdom, the source of all things, acts as a perfect geometrician.... [And geometry is] but the science of the continuous" [15, p. 292].

In this chapter we will focus on examples from several areas of mathematics – analysis, algebra, geometry, and number theory – to illustrate the Principle of Continuity "in action," in several of its guises. We will also highlight in each case the transition from the metaphysics to the mathematics, from vague analogies to fruitful theories.

9.2 Analysis

The seventeenth century saw the rise of calculus, one of the great intellectual achievements of all time. The subject was founded independently by Newton and Leibniz during the last third of that century, although practically all of the prominent mathematicians of Europe around 1650 could solve many of the problems in which elementary calculus is now used. On the other hand, it took another two centuries to provide the subject with rigorous foundations. The immediate task of Newton and Leibniz – the "basic problem" – was this:

Basic problem: To devise general methods for discovering and deriving results in analysis.

It is in response to "basic problems" that Principles of Continuity were usually devised.

9.2.1 Leibniz and Robinson

Central to Leibniz' approach in dealing with this problem, as with many others, was the notion of "differential," the difference between two infinitesimally close "adjacent" points (see Sect. 4.3.4). He computed with differentials as if they were real numbers, although he at times had to make "adjustments." Here is an example:

Leibniz searched for some time to find the rules for differentiating products and quotients. When he found them, the "proofs" were easy. Here is his discovery/derivation of the product rule: $\mathrm{d}(xy) = (x + \mathrm{d}x)(y + \mathrm{d}y) - xy = xy + x\mathrm{d}y + y\mathrm{d}x + (\mathrm{d}x)(\mathrm{d}y) - xy = x\mathrm{d}y + y\mathrm{d}x$. Leibniz omits $(\mathrm{d}x)(\mathrm{d}y)$, noting that it is "infinitely small in comparison with the rest" [11, p. 255].

The $\mathrm{d}x$ and $\mathrm{d}y$ are the differentials of the variables x and y, respectively. The notions of derivative and of function – used nowadays to formulate the product rule – were introduced only in the following century (though Newton's "fluxion" is a derivative with respect to time; see Sect. 4.4.3). Note that Leibniz has here both discovered and derived the product rule. Discovery and derivation ("proof") often went hand-in-hand. Of course Leibniz' demonstration would not be acceptable to us, but standards of rigor have changed, and in any case Leibniz' contemporaries were, for the most part, not looking for rigorous proof. (But see [22] for an example of rigorous proofs given by Leibniz. The article was first published in 1993, so its contents might not have been known to Leibniz' contemporaries.) They were satisfied with what Polya would call "plausible reasoning" [29] and what Weil would describe as "metaphysics."

The metaphysics (1670s–): What holds for the real numbers also holds for the "hyperreal" numbers (essentially the reals and the infinitesimals/differentials), *with some exceptions* (in this case, ignoring higher differentials).

Basic problem: To determine which concepts and results of the calculus are transferable from the reals to the hyperreals. Put another way, to give precise meaning to the exceptions.

It took about 300 years to fix the problem, to turn the metaphysics into mathematics. The fixing was done by Robinson.

Robinson and Keisler, respectively, explain the long delay:

> What was lacking at the time [of Leibniz] was a formal language which would make it possible to give a precise expression of, and delimitation to, the laws which were supposed to apply equally to the finite numbers and to the extended system including infinitely small and infinitely large numbers [31, p. 266].

> The reason Robinson's work was not done sooner is that the Transfer Principle for the hyperreal numbers is a type of axiom that was not familiar in mathematics until recently [19, p. 904].

The "formal language" was model theory, and the "Transfer Principle" was a law that decreed the conditions under which transferability of concepts and results between the reals and hyperreals was permissible.

The mathematics (1960): Robinson's nonstandard analysis.

Robinson saw nonstandard analysis as a vindication of Leibniz' (and Euler's) use of infinitesimals (differentials) [31, p. 2]. He put it as follows [31, p. 269]:

> Leibniz's theory of infinitely small and infinitely large numbers, ... in spite of its inconsistencies, ... may be regarded as a genuine precursor of the theory in the present book.

He argued, moreover, that the history of the calculus had to be rewritten in light of nonstandard analysis [26, pp. 260–261]. Bos, in a spirited rejoinder, objected to these views [4, pp. 81–86].

As far as the Principle of Continuity goes, we do not claim that the Leibnizian calculus marched inexorably toward its natural resolution in nonstandard analysis, only that Robinson's work provided a rigorous justification of Leibniz' use of differentials (see Sect. 4.4.7). The same comment applies, mutatis mutandis, to our other examples. All that we claim in each case of transition from the metaphysics to the mathematics is that the latter was a suitable rigorous formulation of the former, not that it was the inevitable consequence.

9.2.2 *Euler and Cauchy*

Already in the seventeenth century, but especially in the eighteenth, power series became a fundamental tool in analysis. They were usually treated like polynomials, with little concern for convergence (but see [22]). The operative (and philosophical) principle, even if not explicitly stated in general form, was that the rules applicable to polynomials could also be applied to power series. Newton, Euler, and Lagrange, among others, subscribed to this view.

An excellent example of Euler's use of these ideas is his discovery/derivation of the formula $1 + \left(\frac{1}{2}\right)^2 + \left(\frac{1}{3}\right)^2 + \left(\frac{1}{4}\right)^2 + \ldots = \frac{\pi^2}{6}$. This is how he argues:

The roots of $\sin x$ are $0,\ \pm\pi, \pm 2\pi, 3\pi \ldots$ These, then, are also the roots of the "infinite polynomial" $x - x^3/3! + x^5/5! - \ldots$, which is the power-series expansion of $\sin x$. Dividing by x, hence eliminating the root $x = 0$, implies that the roots of $1 - x^2/3! + x^4/5! - \ldots$ are $\pm\pi,\ \pm 2\pi,\ \pm 3\pi,\ \ldots$.

Now, the infinite polynomial obtained by expansion of the infinite product $[1 - x^2/\pi^2][1 - x^2/(2\pi)^2][1 - x^2/(3\pi)^2]\ldots$ has precisely the same roots and the same constant term as $1 - x^2/3! + x^4/5! - \ldots$, hence the two infinite polynomials are identical (cf. the case of "ordinary" polynomials):

$$1 - x^2/3! + x^4/5! - \ldots = [1 - x^2/\pi^2][1 - x^2/(2\pi)^2][1 - x^2/(3\pi)^2]\ldots.$$

Comparing the coefficients of x^2 on both sides yields $-1/3! = -[1/\pi^2 + 1/(2\pi)^2 + 1/(3\pi)^2 + \ldots]$. Simplifying we get

$$1 + 1/2^2 + 1/3^2 + \ldots = \pi^2/6.$$

What a tour de force! One stands in awe of Euler's wizardry. The result was quite a coup for him: Neither Leibniz nor Jakob Bernoulli was able to find the sum of the series $1 + 1/2^2 + 1/3^2 + 1/4^2 + \ldots$. Note that, as in the previous example, discovery and demonstration went hand-in-hand, although even some of Euler's contemporaries objected to his demonstration.

The metaphysics: What holds for polynomials continues to hold for power series.

Basic problem: Justification of "algebraic analysis" (a term coined by Lagrange). That is, how do we justify analytic procedures by using formal algebraic manipulations?

What made seventeenth- and especially eighteenth-century mathematicians put their trust in the power of symbols? First and foremost, the use of such formal methods led to important results. Moreover, the methods were often applied to problems, the reasonableness of whose solutions "guaranteed" the correctness of the results and, by implication, the correctness of the methods. In an interesting article on eighteenth-century analysis, Fraser puts the issue thus [12, p. 331]:

> The 18th-century faith in formalism, which seems to us today rather puzzling, was reinforced in practice by the success of analytical [algebraic] methods. At base it rested on what was essentially a philosophical conviction.

Those attitudes gradually began to change. Two very important "practical" problems – the vibrating-string problem and the heat-conduction problem, of the eighteenth and early nineteenth centuries, respectively – raised questions about central issues in calculus that could no longer be addressed by algebraic analysis. They necessitated, in particular, the clarification of the concepts of function, convergence, continuity, and the integral (see Chaps. 4, 5). This Cauchy proceeded to do. Thus,

The mathematics (1820s): Cauchy provided rigorous foundations for analysis by *eliminating* algebra as a foundational basis for it. He put it thus [14, p. 6]:

> As for my methods, I have sought to give them all the rigor which exists in [Euclidean] geometry, so as never to refer to reasons drawn from the generalness of algebra. Reasons of this [latter] type, though often enough admitted, especially in passing from convergent series to divergent series, and from real quantities to imaginary expressions, can be considered only ... as inductions, sometimes appropriate to suggest truth, but as having little accord with the much-praised exactness of the mathematical sciences.... Most [algebraic] formulas hold true only under certain conditions, and for certain values of the quantities they contain. By determining these conditions and these values, and by fixing precisely the sense of all the notations I use, *I make all uncertainty disappear* [Cauchy's italics].

Cauchy accomplished the task by selecting a few fundamental concepts, namely limit, continuity, convergence, derivative, and integral, establishing the limit concept as the one on which to base all the others, and deriving by fairly modern and rigorous means the major results of calculus. That this sounds commonplace to us today is in large part a tribute to Cauchy's program – a grand design, brilliantly executed.

Cauchy's new proposals for the rigorization of calculus gave rise to their own problems and enticed a new generation of mathematicians to tackle them. Cauchy, too, was not immune to occasional metaphysical reasoning. For example, he believed that every continuous function is differentiable, except possibly at isolated points, and he "proved" the following:

Theorem (1821). *An infinite sum (a convergent series) of continuous functions is a continuous function.*

The metaphysics: Continuity of functions carries over from finite to infinite sums.

Cauchy's proof of the above theorem relied on infinitesimals; this masked the distinction between pointwise and uniform convergence of a series of functions. For an analysis of where Cauchy went wrong see [6]. Laugwitz [27] argues that with an appropriate interpretation of Cauchy's use of infinitesimals his proof can be made rigorous.

In 1826 Abel gave a counterexample to the above theorem. He put it delicately [6, p. 113]:

> But it seems to me that this [Cauchy's] theorem admits exceptions. For example, the series $\sin x - (1/2)\sin 2x + (1/3)\sin 3x - \ldots$ is discontinuous for every value $(2m+1)\pi$ of x, m being a whole number. There are, as we know, many series of this kind. [Note: $\sin x - (1/2)\sin 2x + (1/3)\sin 3x - \ldots = x/2$ for $x \in (-\pi, \pi)$, but if $x = \pi$, $\pi/2 \neq \sin\pi - (1/2)\sin 2\pi + \ldots = 0$.]

We should keep in mind that the concept of continuity is very subtle and was not very well understood in Cauchy's time. Moreover, "the fact that a statement has been refuted does not mean that it will be clear where the incriminating point lies" [6, p. 202]. And the fact that there are different ways to consider convergence of series of functions emerged only gradually over the next several decades.

The mathematics (late 1840s): Seidel and Weierstrass introduced, independently, *uniform convergence* [6]. It is, of course, a *uniformly* convergent series of continuous functions that is continuous.

9.3 Algebra

For about three millennia, until the early nineteenth century, "algebra" meant solving polynomial equations, mainly of degree four or less. This is now known as *classical algebra*. By the early decades of the twentieth century, algebra had evolved into the study of axiomatic systems, known collectively as *abstract algebra*. The transition occurred in the nineteenth century. In the first example, we focus on one aspect of this transition: English contributions to algebra in the first half of that century.

9.3.1 British Symbolical Algebra

The study of the solution of polynomial equations inevitably leads to the study of the nature and properties of various number systems, for of course the solutions of the equations are numbers. Thus the study of number systems constituted an important aspect of classical algebra.

The negative (and complex) numbers, although used frequently in the eighteenth century, were often viewed with misgivings and were little understood. For example, Newton described negative numbers as quantities "less than nothing," and Leibniz said that a complex number is "an amphibian between being and nonbeing." Although rules for the *manipulation* of negative numbers, such as $(-1)(-1) = 1$, had been known since antiquity, no mathematical *justification* for these rules had been given in the past.

During the late eighteenth and early nineteenth centuries, mathematicians began to ask *why* such rules as $(-1)(-1) = 1$ should hold. Members of the Analytical Society at Cambridge University made important advances on this question. In the early nineteenth century, mathematics at Cambridge was part of liberal arts studies, and was viewed as a paradigm of absolute truths employed for the logical training of young minds. It was therefore important, these mathematicians felt, to base algebra, and in particular the laws of operation with negative numbers, on firm foundations [30].

Basic problem: To justify the laws of operation with negative numbers.

The most comprehensive work on this topic was Peacock's *Treatise of Algebra* of 1830 (improved edition, 1845). (Peacock and other members of the Analytical Society were building on the ideas of seventeenth-century continental mathematicians [24].) His main idea was to distinguish between "arithmetical algebra" and "symbolical algebra." The former referred to laws and operations on symbols that stood only for *positive* numbers and thus, in Peacock's view, needed no justification. For example, $a - (b - c) = a - b + c$ is a law of arithmetical algebra when $b > c$ and $a > b - c$. It becomes a law of symbolical algebra if no restrictions are placed on $a - c$. In fact, *no interpretation of the symbols is called for*. Thus *symbolical algebra* was the subject, newly founded by Peacock (and others), of operations with symbols that need not refer to specific objects but that obey the laws of arithmetical algebra. (Cf. Newton's designation of algebra as "universal arithmetic.")

Peacock justified his identification of the laws of symbolical algebra with those of arithmetical algebra by means of his Principle of Permanence of Equivalent Forms – in effect, a Principle of Continuity. This said that [30, p. 38]:

> Whatever form is Algebraically equivalent to another, when expressed in general symbols, must be true whatever those symbols denote. Conversely, if we discover an equivalent form in Arithmetical Algebra or any other subordinate science, when the symbols are general in form though specific in their nature [that is, referring to positive numbers], the same must be an equivalent form, when the symbols are general in their nature [that is, not referring to specific objects] as well as in their form.

In short, *the laws of arithmetic shall also be the laws of algebra.* What these laws were was not made explicit at the time. The laws were clarified in the second half of the nineteenth century, when they turned into axioms for rings and fields [21, Chaps. 3, 4].

It is noteworthy that what *we* do in introducing an algebraic system is not very different from what Peacock did: we too *decree* what the laws of operation of the system shall be. These decrees we call axioms. Of course our decrees are not arbitrary, but neither were Peacock's.

The metaphysics: What holds for positive numbers continues to hold for negative numbers.

Peacock's Principle of Permanence turned out to be very useful. For example, it enabled him to prove the following

Theorem (1845). $(-a)(-b) = ab$.

Proof. Since $(a - b)(c - d) = ac + bd - ad - bc(^{**})$ is a law of arithmetical algebra whenever $a > b$ and $c > d$, it becomes, by the Principle of Permanence, a law of symbolical algebra, which holds without restriction on a, b, c, d. Letting $a = 0$ and $c = 0$ in (**) yields $(-b)(-d) = bd$.

Peacock's work, and that of others, signaled a fundamental shift in the essence of algebra from a focus on the *meaning* of symbols to a stress on their *laws of operation.*

Witness Peacock's description of symbolical algebra [30, p. 36]:

> In symbolical algebra, the rules determine the meaning of the operations ... we might call them arbitrary assumptions, in as much as they are arbitrarily imposed upon a science of symbols and their combinations, which might be adapted to any other assumed system of consistent rules.

This was a very sophisticated idea, well ahead of its time. In fact, however, Peacock paid only lip service to the arbitrary nature of the laws. In practice, they remained the laws of arithmetic. In the next several decades, English mathematicians put into practice what Peacock had preached by introducing algebras with properties which differed in various ways from those of arithmetic (see [21, Sect. 3.1.1]). In the words of Bourbaki [7, p. 52]:

> The algebraists of the English school bring out first, between 1830 and 1850, the abstract notion of law of composition, and enlarge immediately the field of Algebra by applying this notion to a host of new mathematical objects: the algebra of Logic with Boole, vectors, quaternions and general hypercomplex systems with Hamilton, matrices and non-associative laws with Cayley.

Thus, whatever its limitations, symbolical algebra provided a positive climate for subsequent developments in algebra. Laws of operation on symbols began to take on a life of their own, becoming objects of study in their own right rather than a language to represent relationships among numbers.

The mathematics: Advent of abstract (axiomatic) thinking in algebra. See [21].

9.3.2 *Cubic Equations and Complex Numbers*

Here is a sixteenth-century application of the Principle of Continuity. For centuries mathematicians adhered to the following view of square roots of negative numbers: since the squares of positive as well as of negative numbers are positive, square roots of negative numbers do not – in fact, cannot – exist. All this changed in the sixteenth century, following work on the solution of equations by several Italian mathematicians.

A solution by radicals of the cubic was first published by Cardano in his *Ars Magna* (*The Great Art*, referring to algebra) of 1545. What came to be known as Cardano's formula for the solution of the cubic $x^3 = ax + b$ is given by

$$\mathrm{x} = \sqrt[3]{b/2 + \sqrt{(b/2)^2 - (a/3)^3}} + \sqrt[3]{b/2 - \sqrt{(b/2)^2 - (a/3)^3}}.$$

Square roots of negative numbers arise "naturally" when Cardano's formula is used to solve cubic equations. For example, application of the formula to the equation $x^3 = 9x + 2$ gives $x = \sqrt[3]{1 + \sqrt{-26}} + \sqrt[3]{1 - \sqrt{-26}}$.

What was one to make of this solution? Since Cardano was suspicious of negative numbers, he certainly had no taste for their square roots, so he regarded his formula as inapplicable to equations such as $x^3 = 9x + 2$. He concluded that such expressions are "as refined as [they are] useless" [18, p. 404]. Judged by past experience, these were not unreasonable sentiments.

The crucial breakthrough was achieved by Bombelli. In his important book *Algebra* of 1572 he applied Cardano's formula to the equation $x^3 = 15x + 4$ and obtained $x = \sqrt[3]{2 + \sqrt{-121}} + \sqrt[3]{2 - \sqrt{-121}}$. But he could not dismiss this solution, for he noted (by inspection) that $x = 4$ is also a root of the equation. (Its other two roots, $-2 \pm \sqrt{3}$, are also real.) This gave rise to a paradox: while all three roots of the cubic $x^3 = 15x + 4$ are real, the formula used to obtain them involved square roots of negative numbers – meaningless at the time.

Basic problem: How was one to resolve this paradox?

Bombelli adopted the rules for real quantities to manipulate "meaningless" expressions of the form $a + \sqrt{-b}$ $(b > 0)$, and thus managed to show that $\sqrt[3]{2 + \sqrt{-121}} = 2 + \sqrt{-1}$ and $\sqrt[3]{2 - \sqrt{-121}} = 2 - \sqrt{-1}$, hence that $x = \sqrt[3]{2 + \sqrt{-121}} + \sqrt[3]{2 - \sqrt{-121}} = (2 + \sqrt{-1}) + (2 - \sqrt{-1}) = 4$ [28, p. 18].

The metaphysics: What holds for real numbers continues to hold for complex numbers.

Bombelli had made the bold assumption that square roots of negative numbers could be manipulated in a meaningful way to yield significant results. In his own words [28, p. 19]:

> It was a wild thought in the judgment of many; and I too was for a long time of the same opinion. The whole matter seemed to rest on sophistry rather than on truth. Yet I sought so long, until I actually proved this to be the case.

Fig. 9.1 Rafael Bombelli (1526–1572)

This signified the birth of complex numbers. But birth did not entail legitimacy. For the next two centuries complex numbers were shrouded in mystery, little understood, and often ignored. Only after their geometric representation in 1831 by Gauss as points in the plane were they accepted as bona fide elements of the number system. (The earlier works of Argand and Wessel on this topic were not well-known among mathematicians.) See Chap. 12.

The mathematics: Complex numbers are admitted as legitimate mathematical entities.

9.4 Geometry

9.4.1 Projective Geometry

For several millennia, until the early nineteenth century, "geometry" meant Euclidean geometry. The nineteenth century witnessed an explosive growth in the subject, both in scope and in depth. New geometries emerged: projective geometry (Desargues' 1639 work on this subject came to light only in 1845), hyperbolic geometry, elliptic geometry, Riemannian geometry, and algebraic geometry. Poncelet founded (synthetic) projective geometry in the early 1820s as an independent subject, but lamented its lack of general principles. For example, the proof of each result had to be handled differently. Thus, the

Basic problem: To develop tools for the emerging subject of projective geometry.

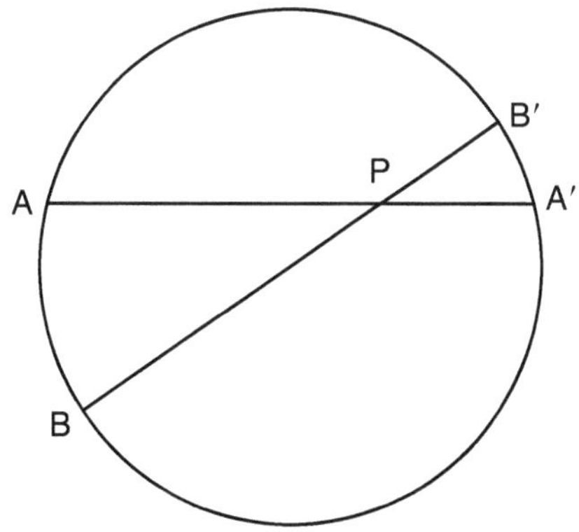

Fig. 9.2 Theorem in geometry about equality of products of segments of intersecting chords in a circle

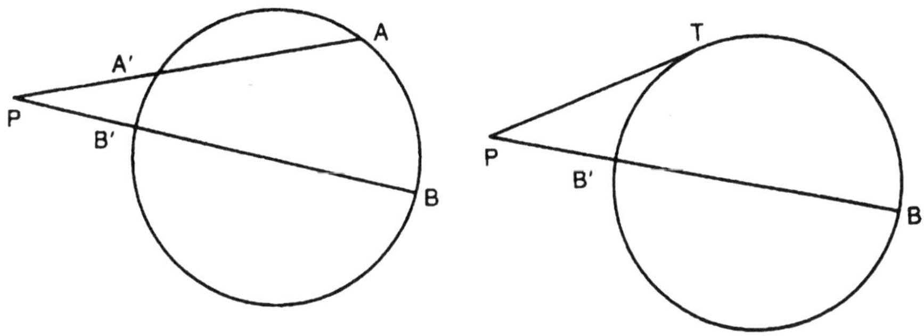

Fig. 9.3 Two results arrived at by Poncelet using his Principle of Continuity

This Poncelet began to do by introducing a Principle of Continuity in his 1822 book *Traité des propriétés projectives des figures*.

The metaphysics: Poncelet's Principle of Continuity [8, p. 136]:

> A property known of a figure in sufficient generality also holds for all other figures obtainable from it by continuous variation of position.

As an elementary illustration of his Principle, Poncelet cited the well-known, and easily established, theorem about the equality of the products of the segments of intersecting chords in a circle: $\mathrm{PB} \times \mathrm{PB}' = \mathrm{PA} \times \mathrm{PA}'$ (Fig. 9.2). The Principle of Continuity then implies that $\mathrm{PB} \times \mathrm{PB}' = \mathrm{PA} \times \mathrm{PA}'$ and $\mathrm{PB} \times \mathrm{PB}' = (\mathrm{PT})^2$ (Fig. 9.3).

A much more substantial result that Poncelet proved using his Principle of Continuity was the so-called

Closure theorem: Let C and D be two conics. Let P_1 be a point of C and L_1 a tangent to D through P_1. Let $P_1, L_1, P_2, L_2, P_3, L_3, \ldots$ be a "Poncelet transverse" between C and D, that is, P_i is on C, L_i is tangent to D and P_i is the intersection of $L_{\mathrm{i}-1}$ and L_i. We say that the Poncelet transverse closes after n steps if $P_{n+1} = P_1$. The closure theorem says that if a transverse, starting at P_1 on C, closes after n steps, then a Poncelet transverse from *any* point on C will close after n steps (Fig. 9.4).

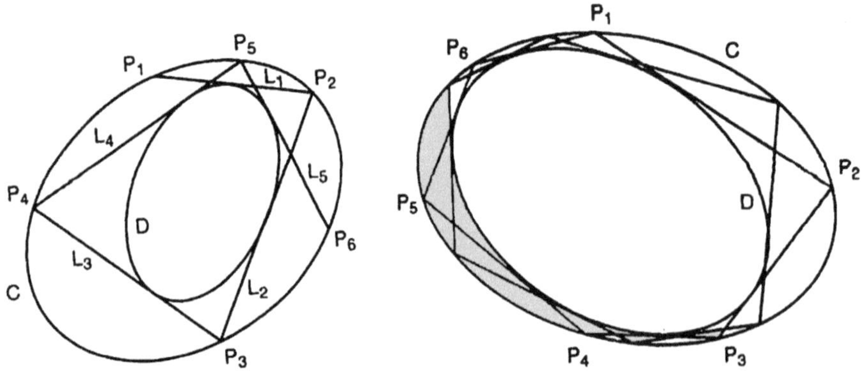

Fig. 9.4 The Closure Theorem proved by Poncelet with the aid of his Principle of Continuity in geometry

Thus, if there is one inscribed n-gon between C and D, then there are infinitely many such n-gons. (Poncelet's formulation of this result is somewhat different [5].)

Bos et al. give three different proofs of the Closure Theorem: Poncelet's, in 1822, using the Principle of Continuity, Jacobi's, in 1828, using elliptic functions, and Griffiths', in 1976, using elliptic curves [5].

The Principle of Continuity was criticized, by, among others, Cauchy, for being vague, but it was a powerful tool, used by Poncelet to great effect to establish projective geometry as a central discipline. In fact, it was he who coined the term "Principle of Continuity." But there arose a

Basic problem: What *is* projective geometry?

Two major issues emerged: the relationship of projective to Euclidean geometry and the validity of the principle of duality. For Poncelet, the major problem of projective geometry was the determination of all properties of geometric figures that do not change under projections. In his development of the subject he used notions from Euclidean geometry (length and angle). Thus to him projective geometry was a subgeometry of Euclidean geometry. Other geometers began to believe that projective geometry is more basic than Euclidean geometry. In 1859 Cayley showed that, in fact, Euclidean geometry is a subgeometry of projective geometry. See [16].

Poncelet and others formulated the principle of duality in projective geometry. Although it appeared to be a working principle, its validity was in question. A vigorous debate raged in the early decades of the nineteenth century about the relative merits of the synthetic versus the analytic approaches to geometry. The principle of duality seems to have been a test case for the two schools of thought. Poncelet, as we noted, developed projective geometry synthetically. Gergonne and Plücker were fervent proponents of the analytic approach. Both introduced homogeneous coordinates; this made the principle of duality *analytically* transparent. In 1882 Pasch supplied an axiomatic treatment of projective geometry, which made that principle *synthetically* transparent. See [16, 18].

The mathematics: Clarification of the nature of projective geometry.

9.4.2 *What Is Geometry?*

In the second half of the nineteenth century the question about the nature of projective geometry was incorporated in a broader

Basic problem: What is geometry?

There were good reasons to pose this question. The nineteenth century was a golden age in geometry. New geometries arose, as we have noted. Geometric methods competed for supremacy: the metric versus the projective, the synthetic versus the analytic. And important new ideas entered the subject: elements at infinity (points and lines), use of complex numbers (e.g., complex projective space), the principle of duality, use of calculus, extension of geometry to n dimensions, Grassmann's calculus of extension (this involved important geometric ideas), invariants (e.g., the Cayley–Sylvester invariant theory of forms), and groups (e.g., groups of the regular solids). An important development was Klein's proof that not only Euclidean, but also non-Euclidean geometries, both hyperbolic and elliptic, are subgeometries of projective geometry. For a time it was said that "projective geometry is all geometry." A broad look at the subject of geometry was in order.

In a lecture in 1872 at the University of Erlangen, entitled A Comparative Review of Recent Researches in Geometry, Klein classified the various geometries using the unifying notions of group and invariance. He defined a geometry of a set S and a group G of permutations of S as the totality of properties of the subsets of S that are invariant under the permutations of G [20]. This conception of geometry, although not all-encompassing (e.g., it excluded Riemannian geometry, of which Klein seems to have been unaware in 1872), had considerable influence on the development of the subject [3].

The mathematics: Klein's definition of geometry: the so-called *Erlangen Program.*

Under Klein's view of geometry, projective geometry, say of the plane, is the totality of properties of the projective plane left invariant under collineations (those transformations that take lines into lines). His ideas also made transparent the relationship of projective geometry to several other geometries [16, 18]. As for Poncelet's Principle of Continuity, its "mathematical content is today reduced to the identity theorem for analytic functions and the fundamental theorem of algebra" [8, p. 136].

9.5 Number Theory

The study of number theory goes back several millennia. Its two main contributors in ancient Greece were Euclid (c. 300 BC) and Diophantus (c. 250 AD). Their works differ fundamentally, both in method and in content. Euclid's comprises Books VII–IX of the *Elements* and is in the theorem-proof style. Here Euclid introduced some of the subject's main concepts, such as divisibility, prime and composite integers, greatest common divisor and Least common multiple, and established some of its main results, among them the Euclidean algorithm, the infinitude of primes, results on perfect numbers, and what some historians consider to be a version of the Fundamental Theorem of Arithmetic. (Much of the number-theoretic work in the *Elements* is due to earlier mathematicians.).

Diophantus' work appeared in his book *Arithmetica* – a collection of about 200 problems, each giving rise to one or more Diophantine equations, many of degree two or three. These are equations in two or more variables, with integer coefficients, for which the solutions sought are integers or rational numbers. Diophantus found rational solutions for these equations, often by ingenious methods. Their study has since Diophantus become a central topic in number theory. See [2, 33].

Basic problem: To develop tools for solving Diophantine equations.

We consider two celebrated examples.

9.5.1 The Bachet Equation

The Bachet equation, $x^2 + k = y^3$ (k an integer), is an important type of Diophantine equation. (It is an example of an elliptic curve.) The special case $x^2 + 2 = y^3$, which we focus on here, appears already in the *Arithmetica* (Problem VI.17). Fermat gave its positive solution, $x = 5$, $y = 3$, but did not publish a proof of the fact that this is the *only* such solution. It was left for Euler, over 100 years later, to do that.

Euler introduced a fundamental new idea to solve $x^2 + 2 = y^3$. He factored its left-hand side, which yielded the equation $(x+\sqrt{2}i)(x-\sqrt{2}i) = y^3$. This was now an equation in a domain D of "complex integers," where $D=\{a+b\sqrt{2}i : a, b \in Z\}$. Here was the first use of complex numbers – "foreign objects" – in number theory.

Euler now proceeded as follows: If *a, b, and c* are integers such that $ab = c^3$, and $(a, b) = 1$, then $a = u^3$ and $b = v^3$, with u and v integers. This is a well-known and easily established result in number theory. (It holds with the exponent three replaced by any integer, and for any number of factors a, b, ...) Euler carried it over – *without acknowledgment* – to the domain D. Since $(x + \sqrt{2}i)(x - \sqrt{2}i) = y^3$, and $(x + \sqrt{2}i, x - \sqrt{2}i) = 1$ (Euler claimed, *without substantiation*, that $(m, n) = 1$ in Z implies $(m + n\sqrt{2}i,\ m - n\sqrt{2}i) = 1$ in D), it follows that $x + \sqrt{2}i = (a+b\sqrt{2}i)^3 = (a^3-6ab^2)+(3a^2b-2b^3)\sqrt{2}i$ for some integers a and b. Equating real and imaginary parts we get $x = a^3-6ab^2$ and $1 = 3a^2b-2b^3 = b(3a^2-2b^2)$. Since a and b are integers, we must have $a = \pm 1$, $b = 1$, hence $x = \pm 5$, $y = 3$. These, then, are the only solutions of $x^2 + 2 = y^3$. See [33] and Sect. 2.6.

Now to our second example.

9.5.2 *Fermat's Last Theorem*

Fermat's Last Theorem asserts the unsolvability in nonzero integers of the equation $x^p + y^p = z^p$, p an odd prime. In 1847 Lamé claimed before the Paris Academy to have proved the theorem. He argued as follows:

Assume that the equation $x^p + y^p = z^p$ *has* nonzero integer solutions (we can assume that $z > 0$). Factor its left-hand side to obtain $(x + y)(x + yw)(x + w^2)\ldots(x + yw^{p-1}) = z^p$ (**), where w is a primitive p-th root of 1 (that is, w is a root of $x^p = 1$, $w \neq 1$). This is now an equation in the domain $D_p = \{a_0 + a_1 w + \ldots + a_{p-1} w^{p-1} : a_i \in Z\}$ of so-called *cyclotomic integers.*

Lamé claimed, not unlike Euler, that since the product on the left-hand-side of (**) is a p-th power, each factor must be a p-th power. (By multiplication by an appropriate constant he was able to make the factors relatively prime in pairs.) He then showed that there are nonzero integers u, v, *and* w such that $u^p + v^p = w^p$, with $0 < w < z$. Continuing this process ad infinitum leads to a contradiction. So Fermat's Last Theorem is proved.

Both Euler's and Lamé's proofs were essentially correct, on the assumption – which they both implicitly made – that the domains under consideration (D and D_p) possess unique factorization.

The metaphysics: The unique factorization property, which holds for the domain of ordinary integers, continues to hold for various domains of "complex integers."

Of course, this is not always the case. While unique factorization holds in D, and in D_p for $p < 23$, it fails in D_p for *all* $p \geq 23$. So Euler's proof was essentially correct, while Lamé's failed for $p \geq 23$. But it was a driving force behind important developments. Mathematicians began to address questions such as: For which "integer domains" (such as D and D_p) does unique factorization hold? What *is* an "integer domain"? When unique factorization fails, can it be restored in some way?

The mathematics: The study of unique factorization in various domains. This led in the second half of the nineteenth century to the introduction of fundamental algebraic concepts, such as ring, ideal, and field, and to the rise, in the hands of Dedekind and Kronecker, of *algebraic number theory*. See [21].

9.6 Conclusion

Underlying the use of the Principle of Continuity is the tension between rule and context. In the final analysis, context is of course all-important, but the rule took centre-stage in the mathematical breakthroughs we have discussed. Even the cases in which the Principle of Continuity was inapplicable – the cautionary tales, if you will – were often starting points for fruitful developments (cf. Lamé's "proof" of Fermat's Last Theorem).

The interplay between rule and context, between computation and conceptualization, between algorithm and proof, is central in mathematics – both in research and in teaching. Whitehead and Freudenthal, respectively, give expression to some of these thoughts:

> It is a profoundly erroneous truism, repeated by all copybooks and by eminent people when they are making speeches, that we should cultivate the habit of thinking of what we are doing. The precise opposite is the case. Civilization advances by extending the number of important operations which we can perform without thinking about them. Operations of thought are like cavalry charges in battle – they are strictly limited in number, they require fresh horses, and must only be made at decisive moments [35, pp. 41–42].

> I have observed, not only with other people but also with myself . . . that sources of insight can be clogged by automatisms. One finally masters an activity so perfectly that the question of how and why is not even asked any more, cannot be asked any more, and is not even understood any more as a meaningful and relevant question [13, p. 469].

The Principle of Continuity is of course not a universal law. In particular, there are many important instances in which progress was made by disregarding it, bucking what appeared to be immutable laws. Here are three examples:

1. Ignoring the commutative law of multiplication – a "sine qua non" for number systems – in attempts to extend the multiplication of complex numbers, first to triples, and when that failed, to quadruples, enabled Hamilton in the 1840s to invent/discover quaternions [17].
2. Ignoring the law that the whole is greater than any of its parts – one of Euclid's "common notions" – overcame a major obstacle in Cantor's introduction of infinite cardinals and ordinals in the 1870s [10].
3. Ignoring the received wisdom that a function must be given by a formula or a curve – the seventeenth- and eighteenth-century view of functions – enabled the introduction of "pathological" functions, for example, everywhere continuous and nowhere differentiable functions, and the rise of mathematical analysis. See Chap. 5.

The Principle of Continuity can be thought of as an argument by analogy. We have only scratched the surface of this vast topic. See, for example, Polya's *Mathematics and Plausible Reasoning*, which is addressed to students and teachers [29]. In this chapter we have considered a rather restricted notion of analogy, in which mathematical arguments, objects, or theories are carried over from given cases to what appear to be like cases, for example, from positive to negative numbers, real to complex numbers, polynomial to power series, and ordinary integers to "complex integers." And – most important – in the examples we have given mathematicians *assumed* that the analogies were valid.

The power of analogy in mathematics often stems from seeing similarities between theories not readily visible to the "naked eye." And, of course, nowadays we would have to *prove* that the analogies held. The following is an important example of analogy – a Principle of Continuity, if you will – in this broader sense.

In the 1850s Riemann introduced the fundamental notion of a Riemann surface to study algebraic functions. But his methods were nonrigorous. Dedekind and Weber, in an important 1882 paper, set themselves the task of "justify[ing] the theory of algebraic functions of a single variable ... from a simple as well as rigorous and completely general viewpoint" [26, p. 154]. To accomplish this, they carried over to algebraic functions the ideas that Dedekind had introduced in the 1870s for algebraic numbers. This was a singular achievement, pointing to what was to become an important analogy between algebraic number theory and algebraic geometry. See [21, Chaps. 3, 4].

For further remarks on analogy in mathematics see [9, Chap. 4], [23, 25, 34].

We conclude with a quotation from Atiyah's 1975 Bakerian Lecture on Global Geometry [1, p. 717]:

> Mathematics can, I think, be viewed as the *science of analogy*, and the widespread applicability of mathematics in the natural sciences, which has intrigued all mathematicians of a philosophical bent, arises from the fundamental role which comparisons play in the mental process we refer to as "understanding."

References

1. M. F. Atiyah, Bakerian lecture, 1975: global geometry, *Amer. Math. Monthly* 111 (2004) 716–723.
2. I. G. Bashmakova, *Diophantus and Diophantine Equations*, Math. Assoc. of Amer., 1997.
3. G. Birkhoff and M. K. Bennett, Felix Klein and his "Erlanger Programm." In *History and Philosophy of Modern Mathematics*, ed. by W. Aspray and P. Kitcher, Univ. of Minnesota Press, 1988, pp. 145–176.
4. H. J. M. Bos, Differentials, higher-order differentials and the derivative in the Leibnizian calculus, *Arch. Hist. Exact Sc.* 14 (1974) 1–90.
5. H. J. M. Bos, C. Kers, F. Oort, and D. W. Raven, Poncelet's closure theorem, *Expositiones Math.* 5 (1987) 289–364.
6. U. Bottazzini, *The Higher Calculus: A History of Real and Complex Analysis from Euler to Weierstrass*, Springer-Verlag, 1986.
7. N. Bourbaki, *Elements of the History of Mathematics*, Springer-Verlag, 1991.
8. E. Brieskorn and H. Knörrer, *Plane Algebraic Curves*, Birkhäuser, 1986.
9. D. Corfield, *Towards a Philosophy of Real Mathematics*, Cambridge Univ. Press, 2003.
10. J. W. Dauben, *Georg Cantor: His Mathematics and Philosophy of the Infinite*, Harvard Univ. Press, 1979.
11. C. H. Edwards, *The Historical Development of the Calculus*, Springer-Verlag, 1979.
12. C. G. Fraser, The calculus as algebraic analysis: some observations on mathematical analysis in the 18th century, *Arch. Hist. Exact Sci.* 39 (1989) 317–335.
13. H. Freudenthal, *The Didactic Phenomenology of Mathematical Structures*, Reidel, 1983.
14. J. V. Grabiner, *The Origins of Cauchy's Rigorous Calculus*, MIT Press, 1981.
15. H. Grant, Leibniz and the spell of the continuous, *College Math. Jour.* 25 (1994) 291–294.
16. J. Gray, *The Worlds out of Nothing: A Course in the History of Geometry in the 19th Century*, Springer, 2007.
17. T. L. Hankins, *Sir William Rowan Hamilton*, Johns Hopkins Univ. Press, 1980.
18. V. J. Katz, A History of Mathematics: An Introduction, 3rd ed., Addison-Wesley, 2009.
19. J. Keisler, *Elementary Calculus: An Infinitesimal Approach*, 2nd ed., Prindle, Weber & Schmidt, 1986.
20. F. Klein, A comparative review of recent researches in geometry, *New York Math. Soc. Bull.* 2 (1893) 215–249.

21. I. Kleiner, *A History of Abstract Algebra*, Birkhäuser, 2007.
22. E. Knobloch, Leibniz's rigorous foundation of infinitesimal geometry by means of Riemannian sums, *Synthese* 133 (2002) 59–73.
23. E. Knobloch, Analogy and the growth of mathematical knowledge. In *The Growth of Mathematical Knowledge*, ed. by E. Grosholz and H. Breger, Kluwer Acad. Publ., Amsterdam, 2000, pp. 295–314.
24. E. Koppelman, The calculus of operations and the rise of abstract algebra, *Arch. Hist. Exact Sc.* 8 (1971/72) 155–242.
25. M. H. Krieger, Some of what mathematicians do, *Notices of the Amer. Math. Soc.* 51 (2004) 1226–1230.
26. D. Laugwitz, *Bernhard Riemann, 1826–1866*, Birkhäuser, 1999.
27. D. Laugwitz, On the historical development of infinitesimal mathematics I, II, *Amer. Math. Monthly* 104 (1997) 447–455, 660–669.
28. P. J. Nahin, *An Imaginary Tale: The Story of* $\sqrt{-1}$, Princeton Univ. Press, 1998.
29. G. Polya, *Mathematics and Plausible Reasoning*, 2 Vols., Princeton Univ. Press, 1954.
30. H. M. Pycior, George Peacock and the British origins of symbolical algebra, *Hist. Math.* 8 (1981) 23–45.
31. A. Robinson, *Non-Standard Analysis*, North-Holland, 1966.
32. B. A. Rosenfeld, The analytic principle of continuity, *Amer. Math. Monthly* 112 (2005) 743–748.
33. A. Weil, *Number Theory: An Approach through History*, Birkhäuser, 1984.
34. A. Weil, De la métaphysique aux mathématiques, *Collected Papers*, Vol. 2, Springer-Verlag, 1980, pp.408–412. (Translation into English in J. Gray, Open Univ. Course in History of Math., Unit 12, p. 30.)
35. A. N. Whitehead, *An Introduction to Mathematics*, Oxford Univ. Press, 1948.

Chapter 10
Proof: A Many-Splendored Thing

10.1 Introduction

In a 1993 article, Jaffe and Quinn, concerned that "today in certain areas there is again a trend toward basing mathematics on intuitive reasoning without proof" [38, p. 1], have suggested a framework for dealing with the issue which includes attaching labels to "speculative and intuitive" work. The article has engendered a fascinating debate within the mathematical community about the nature and function of proof in mathematics and, inevitably, about the nature of the mathematical enterprise, for it is sometimes (often?) difficult to isolate "proving" from the general fabric of doing mathematics [5, 7, 9, 29, 36, 38, 45, 72, 81].

A historical perspective on the subject would be useful. The current debate is a renewal of a 2,000-year-old tradition! From antiquity onwards we note a divergence of views among mathematicians on how best to do mathematics, what methods to use for attacking problems and establishing results. Some employed formal, rigorous proofs, others intuitive, heuristic ones. Adherents of the synthetic method battled supporters of the analytic method. Idealists confronted empiricists and formalists opposed intuitionists, to use current terms rather loosely. There were in general no established criteria of what constitutes an acceptable proof, and perceptions of the role of proof varied considerably. "Indeed," observed Euler [12, p. 143],

> It cannot be denied that mathematics contains the sort of speculations which put eminent geometers in great disagreement. Not only applied mathematics but also pure abstract mathematics itself, strange as it may be, has supplied remarkable sources of dissent.

Of course the tensions among "eminent geometers" have, in general, been healthy for mathematics, though perhaps less so for the protagonists.

This chapter consists of historical examples illustrating the pluralistic nature of mathematical practice, bearing in mind the debate initiated by the Jaffe and Quinn article, to which we return at the end. For further examples, including some of the issues raised in this chapter, see [21, 22, 32, 35, 37, 41, 52, 65, 66].

I. Kleiner, *Excursions in the History of Mathematics*,

DOI 10.1007/978-0-8176-8268-2_10,

10.2 Heuristics vs. Rigor

10.2.1 Ancient Mathematics

There is general agreement among historians of mathematics that the Babylonians of c. 1600 BC knew the Pythagorean theorem and knew how to solve quadratic equations, yet we have no evidence that they had proofs of these results. Mathematics without proof?

Proof as deduction from explicitly stated postulates was conceived in ancient Greece, although when, how, and why this came about is open to conjecture. But there were apparently two distinct approaches to the subject in the crucial, formative period of c. 450–350 BC, what Struik calls Democritean materialism and Platonic idealism [71, p. 47]. Thus Democritus, a founder of atomism, is mentioned by Archimedes as having shown (c. 430 BC) that the volume of a pyramid is one-third the area of the base times the height, presumably using the nonrigorous method of indivisibles employed subsequently by Archimedes in *The Method* and, centuries later, by Cavalieri and others. On the other hand, Hippocrates, at about the same time, obtained rigorously the quadrature of lunes. This is the first result known to have been obtained deductively.

Euclid was the paragon of rigor for over 2,000 years, and yet, as we now recognize, the proof of his very first proposition in Book I of the *Elements*, which gives the construction of an equilateral triangle, is faulty: While Euclid implicitly assumes that two circles intersect, this requires an axiom of continuity, supplied two millennia later by Hilbert. Relativity of rigor!

10.2.2 Calculus

We make a great leap forward and come to the "modern" period – the seventeenth century. This saw the introduction of (modern) number theory, probability, analytic and projective geometry, but above all – calculus. Among its early, illustrious practitioners was Fermat, whose methods of maxima and minima, tangents, and quadratures were conceptually the most advanced prior to the works of Newton and Leibniz. The following example illustrates Fermat's approach to tangents.

Suppose we want to find the tangent to the parabola $y = x^2$ at some point (x, x^2) – a far from trivial problem for early-seventeenth-century mathematicians. Fermat lets $x + e$ be another point on the x-axis and denotes by s the subtangent of the curve at the point (x, x^2) (see Fig. 10.1). Similarity of triangles yields $x^2/s = k/(s + e)$. Fermat notes that k is "approximately equal" to $(x + e)^2$; writing this as $k \approx (x + e)^2$ we get $x^2/s \approx (x + e)^2/(s + e)$. Solving for s we have $s \approx ex^2/[(x+e)^2 - x^2] = ex^2/(x^2 + 2ex + e^2 - x^2) = ex^2/e(2x+e) = x^2/(2x+e)$, hence $x^2/s \approx 2x + e$. Note that x^2/s is the slope of the tangent to the parabola at (x, x^2). Fermat now "deletes" e and claims that the slope of the tangent is $2x$.

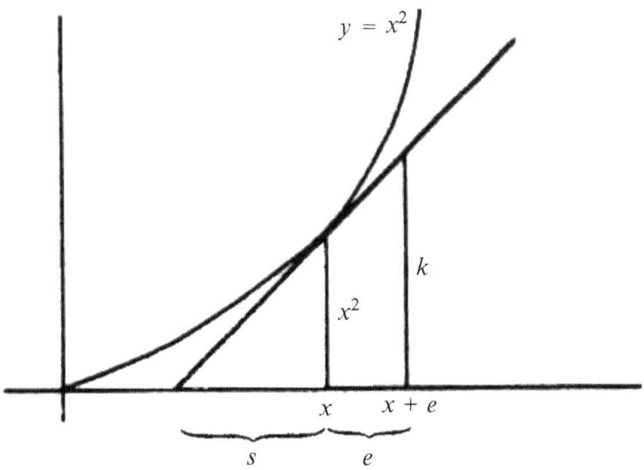

Fig. 10.1 Fermat's method for finding the tangent to $y = x^2$ at an arbitrary point

Fermat's method was severely criticized by some of his contemporaries, notably Descartes. They objected to his introduction and subsequent deletion of the mysterious e. Dividing by e meant regarding it as not zero. Discarding e implied treating it as zero. This is inadmissible, they rightly claimed. In a somewhat different context, but with equal justification, Bishop Berkeley in the eighteenth century would refer to such e's as "the ghosts of departed quantities," arguing that "by virtue of a twofold mistake... [one] arrive[d], though not at a science yet at the truth" [42, p. 428]. Of course Fermat's e embodied a crucial idea – the giving of a "small" increment to a variable.

Newton's and Leibnitz' methods were, logically, no better – as both were well aware. Newton affirmed of his fluxions that they "rather briefly explained than narrowly demonstrated" [27, p. 201], and Leibniz said of his differentials that "it will be sufficient simply to make use of them as a tool that has advantages for the purpose of calculation, just as the algebraists retain imaginary roots with great profit" [27, p. 265]. (In Leibniz' time complex numbers had no greater logical legitimacy than fluxions or differentials.) Euler used infinitely small and infinitely large numbers, as well as divergent series, "with abandon" and with brilliance, achieving remarkable results, for example showing that the sum of the squares of the reciprocals of the positive integers is $\pi^2/6$. Polya wrote of the method Euler used to show this [60, p. 21]:

> In strict logic... [it] was not justified. Yet it was justified by analogy, by analogy of the most successful achievements of a rising science that he called... the 'Analysis of the Infinite.'

Unsuccessful attempts *were* made in the eighteenth century to provide rigorous foundations for the calculus, notably by Maclaurin via geometry and by Lagrange via algebra. D'Alembert, following a definition of limit (not sufficiently precise from our point of view), observed prophetically that "the theory of limits is the true metaphysics of the calculus," but it was left to Cauchy and Weierstrass a century later to work out the "details." In the final analysis, seventeenth- and eighteenth-century mathematicians were confident of the calculus because it yielded correct results – of this they had no doubt. See Chap. 4.

10.2.3 Riemann and Weierstrass

Rigorous mathematics as a going concern is a rather rare phenomenon in the 4,000-year-old history of mathematics. It was consciously practiced for around 200 years in ancient Greece and again for about the same time beginning in the nineteenth century. Both periods were preceded by two millennia (give or take several centuries) of often vigorous but rarely rigorous mathematical activity which produced "crises," for example, the Pythagoreans' discovery of the incommensurability of the diagonal and side of a square, and Fourier's "proof" that *any* function is representable in a Fourier series. These created a need for examination of underlying foundations. Moreover, both rigorous periods sustained competing traditions in which intuition and heuristic reasoning were the animating forces of mathematical practice. Two shining exemplars of the respective traditions in the second period of rigor are Riemann and Weierstrass.

Both mathematicians (as well as Cauchy) founded complex function theory, around the mid-nineteenth century, but they had fundamentally different approaches to the subject. Riemann's global, geometric conception was based on the notion of a Riemann surface and on the Dirichlet Principle, while Weierstrass' local, algebraic theory was grounded in power series and analytic continuation. Klein compares their respective approaches to mathematics [15, p. 291]:

> Riemann is a man of brilliant intuition. When his interest is awakened he begins fresh, without letting himself be diverted by the tradition and without acknowledging the requirements of systematization. Weierstrass is first of all a logician; he proceeds slowly, systematically, step-by-step. When he works, he strives for the definitive form.

Indeed, Riemann had a penetrating geometric and physical intuition. But his two major "tools" – the Riemann surface and the Dirichlet Principle – were mathematically not well grounded. The former was defined rigorously only in 1913 (by Weyl) while the latter was severely criticized by Weierstrass, who produced a counterexample. Following this, Weierstrass' approach to complex analysis became dominant [58, p. 98]:

> Only with the works of Klein and the rehabilitation of the Dirichlet Principle by Hilbert could the Riemannian theory again gradually recover from the blow delivered it by Weierstrass.

As for Riemann's longer-term impact, Ahlfors wrote in 1953: "Riemann's writings are full of almost cryptic messages to the future. . . . The spirit of Riemann will move future generations as it has moved us" [2, pp. 3, 13]. See also [51].

Fig. 10.2 Bernhard Riemann (1826–1866)

10.3 Analysis vs. Synthesis

The terms "analysis" and "synthesis" have changed meaning over the centuries, to the point where what they meant at one time was the opposite of what they had meant at an earlier time. Among the pairs of meanings of analysis–synthesis were, respectively, discovery–demonstration, algebra–geometry, manipulation–rigor, discrete–continuous, symbolic–visual, and rigorous–intuitive [16]. Several examples will illustrate some of these meanings.

10.3.1 Ancient Greece

In ancient Greece, analysis was associated with discovery and synthesis with proof; they went hand in hand, two sides of the same coin. In the analysis stage one assumed what was required to prove or to construct in order to find necessary conditions for the desired result or construction. In the synthesis stage one would try to reverse the steps, to show that the necessary conditions were also sufficient. The only extant Greek work on the subject is Pappus' *On the Domain of Analysis* (fourth century AD), perhaps because the Greeks were reluctant to put their methods of discovery on public display (at least this is what they were accused of by some seventeenth-century mathematicians). Archimedes' *The Method* is a counterexample. See [40].

10.3.2 Leibniz and Newton

Viète called his groundbreaking work in algebra, in which he introduced symbols for parameters and unknowns, *The Analytic Art*, and Descartes believed that in his analytic geometry he was rediscovering the ancient method of analysis. But both Newton and Leibniz spoke out vigorously against the "analytic method" in geometry. Here is Leibniz's statement [63, p. 171]:

> I am still not satisfied with algebra because it does not give the shortest method or the most beautiful construction in geometry. This is why I believe that, so far as geometry is concerned, we need still another analysis which is distinctly geometric or linear and which will express *situation* [situs] directly as algebra expresses *magnitude* directly. And I believe that I have found the way and that we can represent figures and even machines and movements by characters, as algebra represents numbers or magnitudes.

It was Newton who was largely responsible for the extension of the term "analysis" to calculus. His first exposition of the subject, based on infinite series, was entitled "On Analysis of Equations with Infinitely Many Terms." Although he applied this method to the solution of geometric problems – tangents, areas, volumes, he was disdainful of the algebra (coordinate geometry) of Descartes, calling it "The analysis of the bunglers" [16, p. 456]. He was very emphatic about the desirability of solving (classical) geometric problems synthetically rather than analytically [77, p. 380]:

> Men of recent times, eager to add to the discoveries of the ancients, have united the arithmetic of variables with geometry. Benefiting from that, progress has been broad and far-reaching if your eye is on the profuseness of output, but the advance is less of a blessing if you look at the complexity of its conclusions. For these computations... often express in an intolerably roundabout way quantities which in geometry are designated by the drawing of a single line.

It is perhaps ironic that the two creators of the calculus – analysis par excellence – rejected the use of analysis in *geometry*.

10.3.3 Eighteenth and Nineteenth Centuries

In the second half of the eighteenth century mathematicians began to debate the relative theoretical merits of the analytic vs. the synthetic method. Some contended that the analytic method offered more power, generality, and economy of thought. Others countered that calculation (analysis) replaces thinking while pure geometry stimulates it. Analysis conceals the sources of discovery and is particularly unsuitable for solving problems in geometry. Indeed, geometry and intuition are inseparable, and the analytic method removes all intuition from geometry. Boyer relates the vivid, although perhaps not unbiased, description of Malfatti, a foremost geometer of the period (c. 1780), of the practitioners of the two methods [16, p. 458]:

> Malfatti compared the synthesist to a traveller who is in no hurry to reach his goal but stops frequently to enjoy everything beautiful which presents itself; the analyst, who shuts himself up in his carriage and is carried along mechanically, may reach his goal more quickly, but he will have seen less.

The early nineteenth century witnessed a great resurgence of the synthetic method, largely dormant in much of the eighteenth century, initially through the works in descriptive geometry of Monge, one of the founders of the Ecole Polytechnique. The battle of the synthesists vs. the analysts was joined in full vigor in connection with the rise of projective geometry in the early decades of that century. "Perhaps no other controversy in mathematics," notes Boyer, "has exceeded in bitterness that between analysts and synthesists during the second and third decades of the nineteenth century" [16, p. 459]. The fundamental Principle of Duality in projective geometry seems to have been a test case for the two schools of thought. The question was: On what grounds does one justify that principle? Why does it work? In this instance the analysts had the upper hand – the use of homogeneous coordinates made the Principle of Duality transparent. The Principle became synthetically transparent at the end of the nineteenth century when projective geometry was supplied with an axiomatic basis.

10.4 Pure vs. Applied

10.4.1 Introduction

Pure mathematics, in the sense of mathematics for its own sake, goes back (at least) to the Babylonians of c. 1600 BC, who solved quadratic equations and found Pythagorean triples "for fun" (so it seems). Applied mathematics may be older. In any case, since time immemorial there has been an intimate relationship, admittedly with discontinuities, between mathematics and the physical world, which has enriched our understanding of both. In ancient Greece that relationship took significant steps toward its flowering in the seventeenth century. It has since been sustained and strengthened.

In the eighteenth century in particular there was no clear distinction between analysis and several of its fields of application, such as mechanics, optics, hydrostatics, elasticity, and astronomy. Moreover, the very men who fashioned the infinitesimal concepts and methods – the Bernoullis, Euler, d'Alembert, Lagrange et al – also formulated and derived the laws governing the motions of fluids and of rigid bodies, the bending of beams, and the vibrations of elastic bodies. And they saw no distinction between the two types of activities. Now to some examples.

The calculus of variations began in 1696 when Johann Bernoulli posed (and solved) the *Brachistochrone Problem*, which asked for the path of quickest descent of a particle falling under gravity from a point A to a point B not directly below it. The eighteenth century's vibrating-string problem and the nineteenth century's heat-conduction problem gave rise to (among other important matters) the study of partial differential equations. In fact, Truesdell asserts that the theory of partial differential equations is the great gift of continuum mechanics to analysis [73]. Klein claimed that much of Riemann's work was characterized by his striving to put "in mathematical form a unified formulation of the laws which lie at the basis of all natural phenomena" [14, p. 4]. And Poincaré pioneered the qualitative theory of differential equations to deal with the three-body problem.

In none of these cases would the mathematics stand up to logical scrutiny. We focus on two pivotal examples.

10.4.2 The Vibrating-String Problem

Vibrating strings and the musical tones they emit have been studied since antiquity. In the mid-eighteenth century d'Alembert made a crucial advance by providing a model for the vibrating string in the form of the equation $\partial^2 y/\partial t^2 = a^2(\partial^2 y/\partial x^2)$– the *wave equation*, where $y(x,t)$ describes the shape of the string with fixed ends, released from some initial position. D'Alembert proceeded to solve this equation – the first historical example of the solution of a partial differential equation.

Euler wrote a paper on the same problem in 1748, in which he agreed with d'Alembert's solution but differed from him on its *interpretation*. Euler contended that d'Alembert's solution was not the "most general" as the latter had claimed, since it did not allow for the "initial position" of the string to have the shape of (say) an inverted V, nor of a curve drawn free-hand. Basing his argument on physical and intuitive considerations, Euler asserted that such initial shapes were permissible.

Although the Euler-d'Alembert's debate centered on the allowable initial shapes of vibrating strings, it was driven by their differing approaches to mathematics as a whole. According to Langer [50, p. 17],

> Euler's temperament was an imaginative one. He looked for guidance in large measure to practical considerations and physical intuition, and combined with a phenomenal ingenuity, an almost naïve faith in the infallibility of mathematical formulas and the results of manipulations upon them. D'Alembert was a more critical mind, much less susceptible to conviction by formalisms.

D'Alembert's "critical mind" dictated that the initial shape of the vibrating string must be a twice-differentiable function since it satisfied the wave equation $\partial^2 y/\partial t^2 = a^2(\partial^2 y/\partial x^2)$, hence he pronounced Euler's ideas to be "against all rules of analysis."

Daniel Bernoulli entered the picture in 1753 by giving yet another solution of the vibrating-string problem. Bernoulli, who was essentially a physicist, based his argument on the physics of the problem and the known facts about musical vibrations, discovered earlier by Rameau and others. It was generally recognized at the time that musical sounds, and in particular vibrations of a musical "string," are composed of fundamental frequencies and their harmonic overtones.

This physical evidence, and some "loose" mathematical reasoning, convinced Bernoulli that the solution to the vibrating-string problem must be given by $y(x,t) = \Sigma_1^{\infty} b_n \sin(n\pi x/l)\cos(n\pi at/l)$.

Although Bernoulli's ideas were eventually vindicated, both d'Alembert and Euler declared them absurd. Bernoulli countered that d'Alembert's and Euler's solutions constitute "beautiful mathematics, but what has it to do with vibrating strings?" [62, p. 78]. Ravetz characterized the essence of the debate as one between d'Alembert's mathematical world, Bernoulli's physical world, and Euler's "no-man's land" between the two [62, p. 81]. Beyond the specifics of the debate, however, its broader significance was to bring "the whole of eighteenth-century analysis... under inspection: the theory of functions, the role of algebra, the real-line continuum and the convergence of series..." [31, p. 2].

10.4.3 The Heat-Conduction Problem

Fourier startled the mathematical community of the early nineteenth century with his work on heat conduction, resulting in what came to be known as Fourier series. The major mathematical result of his 1822 classic *Analytic Theory of Heat* was that *any* function $f(x)$ over $(-l,l)$ is representable over this interval by a series of sines and cosines, $f(x) = a_0 + \sum_1^{\infty}[a_n \cos(n\pi x/l) + b_n \sin(n\pi x/l)]$, where the coefficients a_n and b_n are given by $a_n = 1/l \int f(t)\cos(n\pi t/l)dt$ and $b_n = 1/l \int f(t)\sin(n\pi t/l)dt$.

Fourier's proof of this theorem was loose; in fact, it was formalism in the spirit of the eighteenth century – "a play upon symbols in accordance with accepted rules but without much or any regard for content or significance," according to Langer [50, p. 33].

Nevertheless, Langer continues with considerable insight,

> it was, no doubt, partially because of his very disregard for rigor that he was able to take conceptual steps which were inherently impossible to men of more critical genius [50, p. 33].

Fourier's result is of course false in the stated generality. Both Euler and Lagrange knew that *some* functions have Fourier series representations, as above. But the "principle of continuity" of eighteenth and early nineteenth-century mathematics (see Chap. 9) suggested that this could not be true for *all* functions: Since the sine and cosine are continuous, the same must be true of a sum of such terms (no distinctions were made between finite and infinite sums). But to refute Fourier's claim one needed – but lacked – clear notions of continuity, convergence, and the integral. These were soon to be conceived by Cauchy, Abel, Dirichlet, and others. And of course Fourier's result, properly modified, remains one of the profound insights of analysis.

10.5 Legitimate vs. Illegitimate

To outsiders, mathematicians appear rational and dispassionate, at least in dealing with their subject. But of course *we* know better. Such expressions about a colleague's work as "illegitimate," "against all rules," "outside of mathematics," "the work of the devil," are not unheard of. Here is the evidence.

10.5.1 The Quaternions

Hamilton's invention (discovery) of the quaternions was conceptually a revolutionary development, for it liberated algebra from the canons of arithmetic. But like all revolutions it was not received with universal approval. John Graves, Hamilton's mathematician friend, had this to say when Hamilton informed him of his discovery [34, p. 300]:

> There is still something in the system [of quaternions] which gravels me. I have not yet any clear view as to the extent to which we are at liberty arbitrarily to create imaginaries and to endow them with supernatural properties.

Although Graves and others soon came around to Hamilton's point of view by accepting operations which did not satisfy all the laws of arithmetic, this was not the end of the quaternion saga. A 50-year war ensued which pitted quaternionists against vector-analysts. Among the latter were Oliver Heaviside and Lord Kelvin, neither of whom had kind words for the quaternions: "A positive evil of no inconsiderable magnitude," asserted Heaviside; "an unmixed evil to those who touched them in any way," echoed Kelvin [48, pp. 9, 10].

10.5.2 Functions

The notion of function has a long and tortuous history. It was first viewed as a formula – a so-called "analytic expression" (although what that meant was not clear), then as a curve, a formula again, an arbitrary correspondence, a formula, a set of ordered pairs ... (see Chap. 5). Dirichlet's function $D(x) = c$ for x rational and d otherwise, given in a paper on Fourier series in 1829, was a watershed in the evolution of the function concept. In an 1870 work on functions, Hankel laments the lack of clarity of this concept. He says that some define it according to Euler, others according to Dirichlet, still others by saying that a function is a law, and some do not define it at all, "but everyone draws from this concept conclusions that are not contained in it" [15, p. 198]. Hankel then proposes to set the record straight by distinguishing between "legitimate" and "illegitimate" functions, consigning Dirichlet's function $D(x)$ to the illegitimate category [15, p. 198].

Dirichlet's function was just the tip of the iceberg in a collection of "unacceptable" functions introduced in the second half of the nineteenth century, designated (by some) as "pathological" [75].

10.5.3 Continuity

Continuity proved to be a very subtle concept in the first half of the nineteenth century. For example, Cauchy "proved" that a convergent series of continuous functions is continuous (but see [46]). Everyone believed, and some "proved," that a continuous function is differentiable, except possibly at isolated points [74]. It was thus "shocking" when Riemann and Weierstrass gave examples of continuous, nowhere differentiable functions. As expected, not everyone took kindly to these examples: "I turn away with freight and horror from this lamentable evil of functions without derivatives," pronounced Hermite in 1893 [42, p. 973]. Poincaré was not as accommodating [42, p. 973]:

> Logic sometimes makes monsters. For half a century we have seen a mass of bizarre functions which appear to be forced to resemble as little as possible honest functions which serve some purpose. More of continuity, or less of continuity, more derivatives, and so forth. Indeed, from the point of view of logic, these strange functions are the most general; on the other hand those which one meets without searching for them, and which follow simple laws, appear as a particular case which does not amount to more than a small corner.
>
> In former times when one invented a new function it was for a practical purpose; today one invents them purposely to show up defects in the reasoning of our fathers and one will deduce from them only that.
>
> If logic were the sole guide of the teacher, it would be necessary to begin with the most general functions, that is to say with the most bizarre. It is the beginner that would have to be set grappling with this teratological museum.

10.5.4 Definitions in Mathematics

In 1904 Zermelo introduced the Axiom of Choice and used it to prove that every set can be well ordered. In 1905, in a spirited exchange of letters, Baire, Borel, Hadamard, and Lebesgue discussed Zermelo's proof but also ranged more widely, dealing with such questions as: Does the definition of a mathematical object, say a number or function, ensure its existence? And does a (nonconstructive) proof of existence of a mathematical object legitimize its acceptance and use without further consideration? Baire, Borel, and Lebesgue lined up against Hadamard on the major issues (and against Zermelo's proof), although the first three differed on various matters of substance [56]. In particular, the use of uncountably many choices in definitions and existence proofs was classified by Borel as "outside mathematics" [56, p. 93]. Hadamard, who did not subscribe to this view, made a telling response [56, pp. 96–97]:

> From the infinitesimal calculus to the present, it seems to me, the essential progress in mathematics has resulted from successively annexing notions which, for the Greeks or the Renaissance geometers or the predecessors of Riemann, were "outside mathematics" because it was impossible to define them.

10.5.5 Abstraction

Closer to home, we have the now-famous 1964 review by Mordell of Lang's book, *Diophantine Geometry*, and a 1995 rejoinder by Lang, in the light of a 1964 letter from Siegel to Mordell which comments on Lang's book [49, 57]. Siegel speaks briefly, and Lang extensively, about number theory and its relation to algebraic geometry. Siegel lashes out at what he sees as uncalled-for abstraction perpetrated in the subject by the likes of Lang. Here are several choice phrases: "I was disgusted with the way in which my own contributions to the subject had been disfigured and made unintelligible;" and "I see a pig broken [sic] into a beautiful garden and rooting up all the flowers and trees;" and again, "I am afraid that mathematics will perish before the end of this century if the present trend for senseless abstraction – as I call it: a theory of the empty set – cannot be blocked up" [49, p. 339]. Mathematical passions!

It is remarkable how differently three foremost number theorists saw the subject in the 1960s, and how unaccepting some are of others' approaches. In fact, all five of the above episodes illustrate how resistant mathematicians can be to new points of view.

10.6 Idealists vs. Empiricists

These are terms introduced (in mathematics) by du Bois-Reymond in 1887 and again by Lebesgue in 1905 [56, p. 100]. They signified roughly what we would now designate as formalists and intuitionists, respectively. This was prior to the formalist-intuitionist controversy of the first decades of the twentieth century. That controversy had strong roots in the nineteenth century, some of which we spell out below. But we first want to observe that both formalist and intuitionist "tendencies" are already present in the mathematics of ancient Greece – in the forms, respectively, of the axiomatic method and of constructions. The latter, it has been argued, are in fact existence proofs [43].

10.6.1 Ideals

The foremost preintuitionist was undoubtedly Kronecker, and the foremost pre-formalist (prelogicist?), perhaps Dedekind. Their respective "philosophies" of mathematics – constructive vs. conceptual – were reflected in their works [28]. For example, they developed distinct, but essentially equivalent, theories of algebraic numbers, Kronecker's based on divisors, Dedekind's on ideals. Dedekind defined an ideal as a collection of algebraic integers satisfying certain closure properties. The definition was nonconstructive and used the completed infinite. Both ideas were anathema to Kronecker, who opted for the constructive divisors. But "for me," said Dedekind [28, p. 10],

> the notion of a 'form with variables' [in terms of which divisors were defined] contains something far more abstract than that of an ideal, which seems to be thoroughly *concrete* as an assemblage of completely determined *numbers* having the two basic properties [of closure]...

10.6.2 "Pathological" Functions

We spoke earlier of the introduction in the nineteenth century of continuous, nowhere differentiable functions. Among other "monstrous" functions which made their appearance around the same time were integrable functions which are discontinuous everywhere, continuous functions that are not piecewise monotonic, and nonintegrable functions that are limits of integrable functions. All were deemed artificial by Hermite [42, p. 1035]:

> I believe that the numbers and functions of analysis are not the arbitrary product of our mind; I believe that they exist outside of us with the same character of necessity as the objects of objective reality; and we find or discover them and study them as do the physicists, chemists and zoologists.

10.6.3 Invariants

One of the prime examples of the computational vs. the conceptual in the nineteenth century was invariant theory, an important and vigorous field of research in the second half of that century. First, specific invariants were found for certain forms (for example for binary quartic forms), but soon the problem shifted to finding a (finite) *complete system* of invariants – a basis – for a given form. Gordan found – that is, *constructed* – such a basis for *binary* forms, which was a major accomplishment. Hilbert astonished the mathematical community in 1890 by showing that forms of any degree, in any number of variables, have a basis. This came to be known as the *Hilbert Basis Theorem*. But Hilbert's proof was noncomputational – it did not exhibit a basis. Gordan protested: "This is not mathematics, it is theology" [42, p. 930]. The theology of the 1890s, however, became the mathematical gospel of the 1920s. But not of the 1990s, according to Rota [64, p. 95]:

> We witness today a return to the concrete mathematics of the nineteenth century after a long period of abstraction; algorithms and techniques which were once made fun of

(as Hilbert did of Gordan's invariant theory) are now revalued after a century of interruption. Contemporary mathematics, with its lack of a unifying trend, its historical discontinuities and its lapses into the past, is a further step on the way to the end of the embarrassing Victorian heritage: the idea of progress and the myth of definitiveness.

10.6.4 Weyl and Von Neumann

The formalist and intuitionist philosophies as expounded in the early twentieth century by Hilbert and Brouwer, respectively, are well known [78]. Their effect on the thoughts and works of two of the foremost mathematicians of the twentieth century are perhaps less familiar:

> Outwardly it does not seem to hamper our daily work, and yet I for one confess that it has had a considerable practical influence on my mathematical life. It directed my interests to fields I considered relatively "safe," and has been a constant drain on the enthusiasm and determination with which I pursued my research work (Weyl [79, p. 13]).

> In my own experience... there were very serious substantive discussions as to what the fundamental principles of mathematics are; as to whether a large chapter of mathematics is really logically binding or not... It was not at all clear exactly what one means by absolute rigor, and specifically, whether one should limit oneself to use only those parts of mathematics which nobody questioned. Thus, remarkably enough, in a large part of mathematics there actually existed differences of opinion! (Von Neumann [76, p. 480]).

Speaking about "differences of opinion," here are three views of the Continuum Hypothesis. To the formalists it is independent (of the standard axioms of set theory), to the platonists it is true or false (in *some* system of set theory yet to be determined), and to the intuitionists it is meaningless (no matter what set theory is under consideration). Democracy in mathematics!

10.7 Short vs. Long Proofs

The age of innocence is gone. Ideally, a proof should be simple, elegant, transparent, and relatively short – sufficiently short that a mathematician familiar with the subject should be able to convince herself or himself of its validity in reasonable time. But some proofs in the latter part of the twentieth century and the first decade of the twenty-first have been anything but that. The star counterexample is the proof of the so-called Enormous Theorem, which gives the classification of finite simple groups [30]. The proof is about 15,000 pages long, is scattered over some 500 articles, almost all published between the late 1940s and early 1980s, and was a combined effort of over 100 mathematicians from six countries [30]. Moreover, "gaps remain in the proof, even a decade after [the theorem] was 'established,' not to mention that some details depend on computer verification" [29, pp. 672–673]. Yet the classification result is accepted as a theorem by those in the know, hence by almost everyone else, and important consequences have been drawn from it. For those reluctant to use it, and some *are* (see [19]), help is on the way: A second-generation short proof (3,000–5,000 pages) is expected "well into the next century" [19, p. 794] (written in 1995). A third-generation program is now under way [8] (written in 2004).

Speaking more generally of theorems with long proofs, Serre wonders: "What shall one do with such theorems if one has to use them? Accept them on faith? Probably. But this is not a very comfortable situation" [18, p. 11].

There are other examples of theorems with very long proofs, for example the Feit-Thompson Theorem, the Four-Color Theorem, Fermat's Last Theorem, the Kepler sphere-packing theorem, and the Poincaré conjecture (now a theorem). Some believe that long proofs are becoming the norm rather than the exception. The reason is that there are, in their view, relatively few interesting results with short proofs compared to the total number of interesting mathematical results [44]. On the other hand, Spencer suggests that the mathematical counterpart of Einstein's credo that "God does not play dice with the universe" is that "short interesting theorems have short proofs" [69, p. 366]. There are currently several very significant counterexamples to this view.

10.8 Humans vs. Machines

Another fundamental departure of the latter twentieth century concerning proof is computer-assisted proofs. The unease which mathematicians feel about such proofs stems from the following:

(a) Control over the subject must be shared with a foreign agent – a machine. In particular, how can a referee check the entire proof?
(b) Mathematics seems to resemble an experimental science. For example, to help increase one's confidence in a computer-assisted proof one "repeats the experiment" by letting another machine run the program.
(c) Computer-assisted proofs are, in general, not "surveyable:" The computer programs used in such proofs are (usually) not published and are thus not open to the traditional procedures of verification.
(d) Both computer hardware and computer software are subject to error. "Computer programs always seem to have bugs," observes Thurston [72, p. 169]. "To certify reliability of large systems is all but impossible," notes Babai [11, p. 2].
(e) While the acceptability of a traditional mathematical proof is a social process, that of a computer proof is not, observe De Millo et al [23, p. 34]:

> Being unreadable and – literally – unspeakable, verifications [of computer proofs] cannot be internalized, transformed, generalized, used, connected to other disciplines, and eventually incorporated into a community consciousness. They cannot acquire credibility gradually, as a mathematical theorem does; one either believes them blindly, as a pure act of faith, or not at all.

(f) The function of a proof is – or should be – to enlighten the reader, in addition to validating the result which it purports to prove. "A good proof is one which makes us wiser," observes Manin [53, p. 18]. But surely a computer proof does not.

It seems, then, that computer-assisted proofs are a major deviation from the traditional "hand-made" proofs. But on closer scrutiny, they do resemble, in some ways, at least the *long* human proofs: One can exercise little control over both types of proof, both are unsurveyable, and both are subject to error. In fact, "mathematics is much less formally complete and precise than computer programs" [72, p. 170], and "the probability of human error is considerably higher than that of machine error" [6, p. 179]. In any case, computer-assisted proofs are clearly here to stay.

10.9 Deterministic vs. Probabilistic Proofs

An interesting idea concerning the *concept* as well as the practice of proof is that of probabilistic proof. Its use in number theory, mainly for primality testing, was pioneered by Rabin, Solovay, and others in the 1970s [10, 61]. The idea of such proofs is to show *rigorously* that a given result is true with very high probability – in theory, as high as desired.

In 1990 Babai and others introduced, in connection with work in complexity theory, the notion of "transparent" or "holographic" proof. The idea is to test the validity of extremely long formal proofs by transforming them into "transparent" proofs which are easy to "spot check." By making a number of such checks one can assure oneself with very high probability that the original proof is correct [10, 20].

It seems, then, that when proofs are too long for humans we seek the aid of the computer – and when they are too long for the computer, we come back to humans for help.

Some have argued that there is no essential difference between such probabilistic proofs and the deterministic proofs of standard mathematical practice [44]. Both are convincing arguments. Both are to be believed with a certain probability of error. In fact many deterministic proofs, it is claimed, have a higher probability of error than probabilistic ones. (For example, can one be assured that the probability of error in the proof of the Enormous Theorem is less than $1/2^{500}$, as one can in proofs testing for primality?) Most mathematicians, however, would probably argue that *qualitatively* and *conceptually* the two types of proof are very different.

10.10 Theorems vs. Proofs

Perhaps a better heading would be "Theorems, independent of proofs." It is remarkable that theorems have a permanence denied to proofs. Proofs have often been questioned, while the results obtained by them were not. If a proof is wrong, we usually modify the proof, and sometimes refine the concepts or conditions that went into the making of the theorem. Theorems are accepted not so much because of the rigor in their proofs as because of the reasonableness of the results, their importance, the light they shed on other work, the prestige of the author, and the endorsement of other mathematicians.

The proofs of many of the theorems in Euclid's *Elements* are wrong from a modern perspective: There are not enough axioms, hence the arguments are logically unjustified. But none of the results in the *Elements* has ever been disqualified.

Most of the proofs (justifications) of results in the calculus of the seventeenth and eighteenth centuries, including those of Newton and Leibniz, are incorrect, yet hardly a theorem has been rejected. Gauss' first proof of the Fundamental Theorem of Algebra – his Ph.D. thesis – had an "immense gap," filled only in 1920 [68, p. 4]. Euler's theorem for polyhedra, $V - E + F = 2$, stands, but when he first published it in 1750 he stated that "he had no satisfactory proof but was convinced of its general validity by a wealth of examples" [33, p. 122]. He later gave a proof, but it was inadequate. In fact, "no one was capable of giving a *general* definition of what should be understood by a 'polyhedron'... [until] Poincaré's work in 1895" [25, p. 237]. See also [46].

The examples could be multiplied. But the following two statements capture splendidly the essence of our remarks:

> I have had my results for a long time, but I do not yet know how to arrive at them (Gauss [46, p. 9]).

> If I only had the theorems! Then I should find the proofs easily enough (Riemann [46, p. 9]).

10.11 The Recent Debate

In a 1993 article in the *Bulletin of the American Mathematical Society*, Jaffe and Quinn discuss what they see as the "start of the bifurcation of mathematics into theoretical [read: nonrigorous, speculative, intuitive, heuristic] and rigorous communities," claiming that "a new connection with physics is providing a good deal of the driving force toward speculation in mathematics" [38]. (All subsequent quotations from Jaffe and Quinn refer to this article.) They allow that "at times speculations have energized development in mathematics," but assert that "at other times they have inhibited it." The danger, in their view, is that "the failure to distinguish carefully between the two can cause damage both to the community of mathematics and to the mathematics literature." To meet the danger they describe "practices and guidelines which, if carefully implemented, should give a positive context for speculation in mathematics." Their main objective is to ensure "truth in advertising," and to that end they propose, among other things, that "theoretical [nonrigorous] work should be explicitly acknowledged as theoretical and incomplete."

Jaffe and Quinn "turn to the history of mathematics for examples illustrating [both] the benefits and hazards of nonrigorous work." Among the hazards, the "cautionary tales," they mention eighteenth-century calculus and the "Italian school" of algebraic geometry (early twentieth century). Concerning the former, they assert that "in the eighteenth century casual reasoning led to a plague of problems in analysis concerning issues like convergence of series and uniform convergence of functions."

Here is *our* reading of this period in history: Newton and Leibniz invented calculus in the late seventeenth century. Their eighteenth-century successors – Euler, Lagrange, and others – were well aware of its logical shortcomings, including issues of convergence, but nevertheless forged ahead, bravely and brilliantly, to exploit the marvellous subject bequeathed to them. They achieved extraordinary results, proved "speculatively." The absence of rigorous grounding in calculus caused them hardly any problems. The watershed came with Fourier's work in the early nineteenth century. In this case the lack of clarity in fundamental concepts – function, convergence, integral – cried out for elucidation, which was brought off by Cauchy, Weierstrass, and others. Fourier's nonrigorous work was thus a godsend.

As for the admittedly speculative Italian school of algebraic geometry, Jaffe and Quinn state that it "did not avoid major damage: it collapsed after a generation of brilliant speculation." But it was the attempts to tame this "brilliant [geometric] speculation" that inspired the splendid rigorous algebraic works of Zariski, Van der Waerden, Weil, and others. Zariski speaks with gratitude of "the geometric theories which we were fortunate to inherit from the Italian school [80, p. 77]," and Dieudonné points to "the new life in algebraic geometry brought by the arrival on the scene of ... Castelnuovo... [and] Enriques [members of the Italian school]" [26, p. 288]. The Italian school of algebraic geometry was breaking fundamentally new ground – the study of algebraic *surfaces* (rather than curves), and the major issues therein were *what* to prove rather than *how* to prove it.

Hilbert observed that every mathematical discipline goes through three periods of development: the *naïve*, the *formal*, and the *critical*. What we were witnessing in the case of eighteenth-century calculus was the formal period (formal manipulation with not much attention paid to rigorous justification), and in the case of the Italian school of algebraic geometry perhaps the naïve and formal periods. Atiyah, responding to the Jaffe and Quinn piece, puts the issue thus [9, p. 178]:

> If mathematics is to rejuvenate itself and break exciting new ground it will have to allow for the exploration of new ideas and techniques which, in their creative phase, are likely to be as dubious as in some of the great eras of the past. Perhaps we now have high standards of proof to aim at but, in the early stages of new developments, we must be prepared to act in more buccaneering style.

Are the mathematical areas of concern to Jaffe and Quinn not now in their "creative phase," which will likely be followed by a "critical" (rigorous) stage, if history is any guide?

But why not *label* speculative mathematics as such? The matter has been addressed in the various responses to Jaffe and Quinn, both as to the desirability and the feasibility of such a procedure [9], but perhaps (again) history can illuminate what we regard as a serious difficulty with labeling.

Incidentally, sporadic attempts at "labeling" have occurred in the past: Recall the labels "illegitimate," "outside of mathematics," "pathological," "against all rules," etc. It is ironic that while Jaffe and Quinn suggest attaching labels to nonrigorous mathematics, most of the above labels referred to rigorous mathematics which was deemed *too abstract*. Of course, for much of its history mathematics thrived without any need for labels, often (we believe) in circumstances similar to those of the contemporary scene.

Jaffe and Quinn are concerned that in the absence of labels, "failure to distinguish between the two types of activity [speculative and rigorous work] can lead students to emulate the more glamorous and less disciplined aspects and to end up unable to do more than manipulative jargon." But there will be high costs to be paid in the *presence* of labels. Heuristic work will inevitably be relegated to a second-class activity, and this will discourage young (and perhaps not so young) mathematicians from pursuing it, to the great loss of mathematics. (Jaffe and Quinn grant that "speculation, if properly undertaken [that is, if labeled], can be profoundly beneficial.") The following example from history has, we think, relevance.

Mathematics in ancient Greece was not labeled, but in the time of its greatest masters – Eudoxus, Euclid, Archimedes, and Apollonius – there was an implicit if not explicit understanding that the proper way to do mathematics was to do it rigorously. This was the result, Struik conjectures, of "the victory of Platonic idealism [rigor] over Democritean materialism [heuristics] in the realm of mathematical philosophy" [71, p. 47].

It is thus not unreasonable to speculate that it was because the Greeks could not – or would not – accommodate irrational numbers *rigorously* in the numerical realm, their mathematics turned from a fruitful collaboration of number and geometry to an almost exclusive concern with the latter. This was surely to the detriment of further vigorous progress in mathematics. We can also surmise what turn mathematics might have taken had Archimedes' *The Method* (apparently an isolated singularity in classical Greek mathematics) not been lost, or had similar heuristic work been pursued by others. Wallis in the seventeenth century was annoyed at the ancient Greek mathematicians for what he believed was the deliberate concealment of their methods of discovery. This forced Wallis and his contemporaries to reinvent them – 2,000 years after Archimedes' time!

Another historical example of implicit labeling, this time *rigorous* work having been stifled, concerns Darboux. Much of his research was in differential geometry, but he wrote three papers "which are informed by the conviction that French analysis was not practiced with sufficient rigor" [4, p. 179]. However, "the chilly reception afforded Darboux's three papers by the Paris mathematical community influenced Darboux's decision to publish no further articles on foundational matters" [4, p. 180].

In the first half of the twentieth century mathematics was rigorous, abstract, and "pure." Some – for example Courant, von Neumann, Weyl – warned of sterility setting in if connections with science were not made. In the last several decades important and beneficial connections *have been* made, but perhaps at the price of looser standards of rigor. The Jaffe and Quinn article has been useful in raising some of the resulting issues. To us the contemporary debate is not at all unlike those between (for example) Euler and d'Alembert, Riemann and Weierstrass, Kronecker and Dedekind, Gordan and Hilbert, or Borel and Hadamard. "We *are* a community," as Jaffe and Quinn note, but we do not, *and need not*, observe uniform "community norms and standards for behavior." Diversity is a sign of the health and vigor of mathematics.

References

1. A. Abdulle and G. Wanner, 200 years of least squares method, *Elemente der Mathematik* 57:2 (2002) 45–60.
2. L. V. Ahlfors, Development of the theory of conformal mapping and Riemann surfaces through a century, *Annals of Math. Studies* 30 (1953) 3–13.
3. D. Alexander, Dirty math, *MAA Focus* (March 2002) 6–7.
4. D. S. Alexander, Gaston Darboux and the history of complex dynamics, *Hist. Math.* 22 (1995) 179–185.
5. G. E. Andrews, The death of proof? Semi-rigorous mathematics? You've got to be kidding!, *Math. Intelligencer* 16:4 (1994) 16–18.
6. K. Appel and W. Haken, The four-color problem. In *Mathematics Today*, ed. by L. Steen, Springer-Verlag, 1978, pp.151–180.
7. V. I. Arnold, Will mathematics survive? Report on the Zurich Congress, *Math. Intelligencer* 17:3 (1995) 6–10.
8. M. Aschbacher, The status of the classification of the finite simple groups, *Notices of the Amer. Math. Soc.* 51 (2004) 736–740.
9. M. Atiyah et al., Responses to 'Theoretical mathematics: Toward a cultural synthesis of mathematics and theoretical physics,' by A. Jaffe and F. Quinn, *Bulletin of the Amer. Math. Soc.* 30 (1994) 178–211.
10. L. Babai, Probably true theorems, cry wolf?, *Notices of the Amer. Math. Soc.* 41 (1994) 453–454.
11. L. Babai, Transparent proofs, *MAA Focus* 12:3 (1992) 1–2.
12. E. J. Barbeau and P. J. Leah, Euler's 1760 paper on divergent series, *Hist. Math.* 3 (1976) 141–160.
13. B. E. Blank, *Book Review*: *The Universal Computer: The Road from Leibniz to Turing*, by M. Davis, Norton, 2000. In *Notices of the Amer. Math. Soc.* 48 (2001) 498–501.
14. U. Bottazzini and R. Tazzioli, *Naturalphilosophie* and its role in Riemann's mathematics, *Rev. d'Hist. Math.* 1 (1995) 3–38.
15. U. Bottazzini, *The Higher Calculus: A History of Real and Complex Analysis from Euler to Weierstrass*, Springer-Verlag, 1986.
16. C. B. Boyer, Analysis: Notes on the evolution of a subject and a name, *Math. Teacher* 47 (1954) 450–462.
17. G. J. Chaitin, Thoughts on the Riemann Hypothesis, *Math. Intelligencer* 26:1 (2004) 4–7.
18. C. T. Chong and Y. K. Leong, An interview with Jean-Pierre Serre, *Math. Intelligencer* 8:4 (1986) 8–13.
19. B. Cipra, At math meetings, Enormous Theorem eclipses Fermat, *Science* 267 (10 Febr. 1995) 794–795.
20. B. Cipra, New computer insights from 'transparent' proofs. In *What's Happening in the Mathematical Sciences*, vol. 1, Amer. Math. Society, 1993, pp. 7–11.
21. J. Dauben, Conceptual revolutions and the history of mathematics: two studies in the growth of knowledge. In *Revolutions in Mathematics*, ed. by D. Gillies, Oxford Univ. Press, 1992, pp. 49–82.

22. J. W. Dawson, *The Life and Work of Kurt Gödel*, A K Peters, 1997. (See especially the essay titled Logical Dilemmas.)
23. R. A. De Millo, R. J. Lipton, and A. J. Perlis, Social processes and proofs of theorems and programs, *Math. Intelligencer* 3:1 (1980) 31–40.
24. J. Dhombres and J. – B. Robert, Book Review: *Joseph Fourier, 1768–1830: Créateur de la Physique-Mathématique,* Berlin, 1998. In *Math. Intelligencer* 25:4 (2003) 77–80.
25. J. Dieudonné, *Mathematics—the Music of Reason*, Springer-Verlag, 1992.
26. J. Dieudonné, The beginnings of Italian algebraic geometry. In *Studies in the History of Mathematics*, ed. by E. R. Phillips, Math. Assoc. of Amer., 1987, pp. 278–299.
27. C. H. Edwards, *The Historical Development of the Calculus*, Springer-Verlag, 1979.
28. H. M. Edwards, Mathematical ideas, ideals, and ideology, *Math. Intelligencer* 14:2 (1992) 6–19.
29. D. Epstein and S. Levy, Experimentation and proof in mathematics, *Notices of the Amer. Math. Soc.* 42 (1995) 670–674.
30. D. Gorenstein, The Enormous Theorem, *Scientific Amer.* (Dec. 1985) 104–115.
31. I. Grattan-Guinness, *The Development of the Foundations of Mathematical Analysis from Euler to Riemann*, MIT Press, 1970.
32. J. Gray, Weierstrass, Luzin, and intuition, *Amer. Math. Monthly* 108 (2001) 865–870.
33. B. Grünbaum and G. C. Shephard, A new look at Euler's theorem for polyhedra, *Amer. Math. Monthly* 101 (1994) 109–128.
34. T. L. Hankins, *Sir William Rowan Hamilton*, The Johns Hopkins Univ. Press, 1980.
35. H. Hellman, *Great Feuds in Mathematics*, Wiley, 2006.
36. J. Horgan, The death of proof, *Scientific Amer.* 269 (Oct. 1993) 92–103.
37. A. Jackson, Comme appelé du néant—as if summoned from the void: the life of Alexandre Grothendieck, *Notices of the Amer. Math. Soc.* 51 (2004) 1038–1055.
38. A. Jaffe and F. Quinn, 'Theoretical mathematics': Toward a cultural synthesis of mathematics and theoretical physics, *Bulletin of the Amer. Math. Soc.* 29 (1993) 1–13.
39. D. M. Jesseph, *Squaring the Circle. The War between Hobbes and Wallis*, Univ. of Chicago Press, 1999.
40. V. J. Katz, *A History of Mathematics: An Introduction*, 3rd ed., Addison-Wesley, 2009.
41. G. Kampis, L. Kvasz and M. Stöltzner (eds.), *Appraising Lakatos: Mathematics, Methodology and the Man,* Kluwer, 2002.
42. M. Kline, *Mathematical Thought from Ancient to Modern Times*, Oxford Univ. Press, 1972.
43. W. Knorr, Construction as existence proof in ancient geometry, *Ancient Philosophy* 3 (1983) 125–148.
44. G. Kolata, Mathematical proofs: the genesis of reasonable doubt, *Science* 192 (June 1976) 989–990.
45. S. G. Kranz, The immortality of proof, *Notices of the Amer. Math. Soc.* 41 (1994) 10–13.
46. I. Lakatos, *Proofs and Refutations*, Cambridge Univ. Press, 1976.
47. C. W. H. Lam, How reliable is a computer-based proof?, *Math. Intelligencer* 12:1 (1990) 8–12.
48. J. Lambek, If Hamilton had prevailed: quaternions in physics, *Math. Intelligencer* 17:4 (1995) 7–15.
49. S. Lang, Mordell's review, Siegel's letter to Mordell, diophantine geometry, and 20th century mathematics, *Notices of the Amer. Math. Soc.* 42 (1995) 339–350.
50. R. E. Langer, Fourier series: the genesis and evolution of a theory, *Amer. Math. Monthly* 54**S** (1947) 1–86.
51. D. Laugwitz, *Bernhard Riemann, 1826–1866*, Birkhäuser, 1999.
52. Y. I. Manin, Truth, rigour, and common sense. In *Truth in Mathematics*, ed. by H. G. Dales and G. Oliveri, Oxford Univ. Press, 1998, pp. 147–159.
53. Y. I. Manin, How convincing is a proof?, *Math. Intelligencer* 2:1 (1979) 17–18.
54. A. F. Monna, Hermann Hankel, *Niew. Arch. voor Wisk.* 21 (1973) 64–87.
55. G. H. Moore, At cross-purposes: The Gödel-Zermelo letters, *Bulletin Can. Soc. Hist. Phil. Math.* 27 (Nov. 2000) 6–8.

56. G. H. Moore, *Zermelo's Axiom of Choice: Its Origins, Development, and Influence*, Springer-Verlag, 1982.
57. L. J. Mordell, Book review: *Diophantine geometry*, by S. Lang, *Bulletin of the Amer. Math. Soc.* 70 (1964) 491–498.
58. E. Neuenschwander, Studies in the history of complex function theory II: Interactions among the French School, Riemann, and Weierstrass, *Bulletin of the Amer. Math. Soc.* 5 (1981) 87–105.
59. J. Paradis, J. Pla, and P. Viader, Fermat and the quadrature of the folium of Descartes, *Amer. Math. Monthly* 111 (2004) 216–229.
60. G. Polya, *Induction and Analogy in Mathematics*, Princeton Univ. Press, 1953.
61. M. O. Rabin, Probabilistic algorithms. In *Algorithms and Complexity: New Directions and Recent Results*, ed. by J. F. Traub, Academic Press, 1976, pp. 21–40.
62. J. R. Ravetz, Vibrating strings and arbitrary functions. In *The Logic of Personal Knowledge: Essays Presented to M. Polanyi on his Seventieth Birthday*, The Free Press, 1961, pp. 71–88.
63. B. A. Rosenfeld, *A History of Non-Euclidean Geometry*, Springer-Verlag, 1988. (Translated from the Russian by A. Shenitzer.)
64. G.-C. Rota, The concept of mathematical truth. In *Essays on Humanistic Mathematics*, ed. by A. M. White, Math. Assoc. of Amer., 1993, pp. 91–96.
65. D. E. Rowe, On projecting the future and assessing the past—the 1946 Princeton bicentennial conference, *Math. Intelligencer* 25:4 (2003) 8–15.
66. D. E. Rowe, Episodes in the Berlin-Göttingen rivalry, 1870–1930, *Math. Intelligencer* 22:1 (2000) 60–69.
67. A. Shenitzer, The Cinderella career of projective geometry, *Math. Intelligencer* 13:2 (1991) 50–55.
68. S. Smale, Algebra and complexity theory, *Bulletin of the Amer. Math. Soc.* 4 (1981) 1–36.
69. J. Spencer, Short theorems with long proofs, *Amer. Math. Monthly* 90 (1983) 365–366.
70. M. Stöltzner, What Lakatos could teach the mathematical physicists. In *Appraising Lakatos: Mathematics, Methodology and the Man*, ed. by G. Kampis, L. Kvasz and M. Stöltzner, Kluwer, 2002, pp. 157–187.
71. D. J. Struik, *A Concise History of Mathematics*, 4th ed., Dover, 1987.
72. W. P. Thurston, On proof and progress in mathematics, *Bulletin of the Amer. Math. Soc.* 30 (1994) 161–177.
73. C. Truesdell, The rational mechanics of flexible or elastic bodies, 1638–1788. In L. Euler, *Opera*, Ser. 2, Vol. 11, 1960, Section 2.
74. K. Volkert, Zur Differentzierbarkeit stetiger Funktionen—Ampère's Beweis und seine Folgen, *Arch. Hist. Exact Sc.* 40 (1989) 37–112.
75. K. Volkert, Die Geschichte der pathologischen Funktionen—Ein Beitrag zur Enstehung der mathematischen Methologie, *Arch. Hist. Exact Sc.* 37 (1987) 193–232.
76. J. Von Neumann, The role of mathematics in the sciences and society, *Collected Works*, Vol. 4, ed. by A.H. Taub, Macmillan, 1963, pp. 477–490.
77. R. Westfall, *Never at Rest*, Cambridge Univ. Press, 1980.
78. H. Weyl, Axiomatic versus constructive procedures in mathematics, *Math. Intelligencer* 7:4 (1985) 10–17, 38.
79. H. Weyl, Mathematics and logic, *Amer. Math. Monthly* 53 (1946) 2–13.
80. O. Zariski, The fundamental ideas of abstract algebraic geometry. In *Proceedings of the Intern. Congr. of Mathematicians*, 1950, Vol. II, Amer. Math. Society, 1952, pp. 77–89.
81. D. Zeilberger, Theorems for a price: tomorrow's semi-rigorous mathematical culture, *Notices of the Amer. Math. Soc.* 40 (1993) 978–981.

Part D
Courses Inspired by History

Chapter 11
Numbers as a Source of Mathematical Ideas

11.1 Introduction

Number systems have been a fruitful source of concepts, results, and theories in the evolution of mathematics. In fact, it has been suggested that much even of modern mathematics has its roots in the study of number and shape [78, 79]. This chapter offers suggestions for introducing various mathematical topics related to, and often originating in, the study of number systems. The material is organized around eight themes, which vary in detail and difficulty, and may serve as source material for courses or topics of varied degrees of sophistication and be addressed to various audiences – for example teachers, mathematics majors, and liberal-arts enthusiasts. The themes deal with algebraic, analytic, geometric, number-theoretic, set-theoretic, cultural, and philosophical issues. Although the themes are interconnected, they can be read independently. In many cases, we sketch the historical origin of the mathematical ideas involved. No attempt is made to be thorough, but references to an extensive bibliography are provided throughout. Readers are invited to come up with their own themes to suit their interests, needs, and objectives. The material in the next chapter may serve as an example of a possible theme.

Our notion of "number" is broad – from the natural through the complex numbers and beyond, to transfinite, p-adic, and hyperreal numbers. The approach within each theme is, when appropriate, historical, or rather genetic (see [111]). Consideration of the origin of mathematical ideas, of the burning questions which the originator of a concept or result tried to answer, lends a useful perspective to the teaching and learning of mathematics. For example, cardinal numbers can of course be studied as a mathematical topic without reference to history, but when viewed in an historical setting as the resolution of centuries of gropings for the meaning of the infinite in mathematics, they acquire special significance. The historical approach also enables us to raise various philosophical issues which lend themselves to classroom discussion. For example:

1. The roles of problems and paradoxes in the genesis of mathematical concepts, results, and theories.

I. Kleiner, *Excursions in the History of Mathematics*,
DOI 10.1007/978-0-8176-8268-2_11, © Springer Science+Business Media, LLC 2012

2. Internal vs. external motivation in the evolution of mathematics.
3. The nature of mathematical existence and proof.
4. The roles of intuition and logic in the creation of mathematics.
5. The relation of mathematics to the physical world.

Now to the themes.

11.2 Beyond the Complex Numbers

This is a good theme with which to begin since it leads rather quickly, and more or less naturally, from the familiar (to students) to the less familiar.

11.2.1 A Brief History of "Standard" Number Systems

A short review of the "standard" number systems, from **N** (the natural numbers) through **Z** (the integers), **Q** (the rationals), and **R** (the reals), to **C** (the complex numbers), sets the scene for what is to come. We focus in this summary on their historical evolution, and especially on *gains and losses* at each transition stage in the evolutionary process. For example, in going from **R** to **C** one gains algebraic closure but loses order, and in going from **Z** to **Q** one gains division but loses divisibility: **Q**, but not **Z**, is closed under division; on the other hand, the notion of divisibility, while fundamental in **Z**, is trivial in **Q**. The transition from **Q** to **R** is accompanied by a loss of "innocence:" while **Z**, **Q**, and **C** can be built up from their predecessors by finitary operations (for example, **Z** can be viewed as consisting of equivalence classes of pairs of elements of **N**), **R** requires infinite processes for its construction (infinite decimals, Cauchy sequences, Dedekind cuts). A point to highlight in this discussion is solvability of polynomial equations: in each number system we can solve more equations than in its predecessor; for example, $2x = 6$ can be solved in **Z** but $2x = 7$ only in **Q**.

In expounding these ideas one can introduce a number of fundamental mathematical notions, such as mathematical induction, closure under an operation, denseness, commensurability, completeness, order, and algebraic closure. See [13,27,29,32,40, 84,87,105,108] for details.

11.2.2 The Quaternions

The introduction of the quaternions **H** follows: $\mathbf{H} = \{a + b_i + c_j + d_k : a, b, c, d \in \mathbf{R}\}$, where i, j, k are arbitrary "units" which satisfy the relations $i^2 = j^2 = k^2 = ijk = -1$. From these relations, other properties of these units,

Fig. 11.1 William Rowan Hamilton (1805–1865)

such as $ij = k = -ji$, can be deducced. All the usual laws of numbers, except for commutativity under multiplication, hold in **H**; technically, **H** is a *skew field* (a *division ring*). Moreover, just as for **C**, a polynomial equation over **H** has a root in **H**. However, in contrast to **C**, an equation of degree n over **H** may have more than n roots in **H**. For example, $x^2 + 1 = 0$ has infinitely many roots: $bi + (1–b^2)^{1/2} j$, where $0 \leq b \leq 1$. See [91,92].

Hamilton's invention of the quaternions in 1843 is well documented. It is a rare instance of the evolving process of mathematical discovery on display, and students can be readily led through it [56, 113]. Although the quaternions did not live up to Hamilton's expectations as a creation rivaling in its applicability the infinitesimal calculus, they proved important in helping "liberate" algebra from arithmetic, that is, from its dependence on the laws of arithmetic, and in helping "liberate" geometry from its restriction to three dimensions. As Poincaré and Hamilton, respectively, put it:

> Hamilton's quaternions give us an example of an operation which presents an almost perfect analogy with multiplication, which may be called multiplication, and yet it is not commutative ... . This presents a revolution in arithmetic which is entirely similar to the one which Lobachevsky effected in geometry [76, p. 29].

> There dawned on me the notion that we must admit, in some sense, a *fourth dimension* of space for the purpose of calculating with triples [40, p. 191].

See [5,31,40,43,56,65,70,105] for details on quaternions.

11.2.3 Other Hypercomplex Systems

Is there a "natural" extension of the quaternions? Cayley and Graves gave an affirmative answer by introducing (in 1844) the *octonions* (*Cayley numbers*) **K**: 8-tuples of reals that form a noncommutative and nonassociative division ring, which is, however, *alternative*: $(aa)b = a(ab)$, and $a(bb) = (ab)b$. See [40, 65].

In introducing the octonions, the problem is (as it was for quaternions) to define multiplication. The difficulty "disappears" if we reconsider the multiplication in **H**. Write $a + bi + cj + dk$ as $(a + bi) + (c + di)j = w + zj$, where $w, z \in \mathbf{C}$, $j^2 = -1$. The quaternions can thus be viewed as pairs of complex numbers. Now define $(w_1 + z_1 j)(w_2 + z_2 j) = (w_1 w_2 - z_2^* z_1) + (z_1 w_2^* + z_2 w_1)j$, where z^* denotes the conjugate of z. (It is important to have the w_i and z_i above in precisely this order.) It is easy to verify that the product in **H** thus defined is the same as the usual product defined in terms of $i,\ j,\ and\ k$.

Let now $\mathbf{K} = \{\alpha + \beta e : \alpha, \beta \in \mathbf{H}\}$, where e is an arbitrary unit with $e^2 = -1$, and define the product in **K** as follows: $(\alpha_1 + \beta_1 e)(\alpha_2 + \beta_2 e) = (\alpha_1\alpha_2 - \beta_2^*\beta_1) + (\beta_1\alpha_2^* + \beta_2\alpha_1)e$ (see the definition above of the product in **H**; the conjugate α^* of the quaternion $\alpha = a + bi + cj + dk$ is $a - bi - cj - dk$). **K** thus becomes an alternative division ring. It is easy to verify that (say) $(ij)e \neq i(je)$, so that **K** is not associative. See [5, 40, 65] for details.

It is tempting to continue in this fashion by defining a "number system" consisting of pairs of octonions. That this is doomed to failure (in the sense below) was shown, independently, by Frobenius and by C. S. Peirce ca 1880:

Theorem. *The only n-tuples of real numbers which form an alternative division ring (which need not be commutative nor associative) are the reals, the complex numbers, the quaternions, and the Cayley numbers.*

See [65] for a not very demanding proof. But to follow the proof one would need some knowledge of the theory of algebras [40, 65]. Such knowledge would, in turn, lead to noncommutative ring theory. See [70].

Incidentally, it is easy to show directly, *once we know what we want to show*, that *triples* of real numbers do not form a division ring extending **C**. For if they did, ij would have to be of the form $a + bi + cj$ $(a, b, c \in \mathbf{R})$. Then $i(ij) = i(a + bi + cj)$.

Multiplying and collecting terms we get $c^2 + 1 = 0$ – a contradiction [83]. There is an *elementary* proof which shows that, for *odd n*, a division ring of n-tuples of reals is possible only for $n = 1$. See [40, p. 190] and [10].

There is, of course, an important product defined on triples, namely the *vector product* $(a_1 i + a_2 j + a_3 k)\mathbf{x}(b_1 i + b_2 j + b_3 k) = (a_2 b_3 - a_3 b_2)i + (a_3 b_1 - a_1 b_3)j + (a_1 b_2 - a_2 b_1)k$. Moreover, the vector product (**x**), the quaternion product $(*)$, and the scalar (inner) product $(\cdot)$ of three-dimensional vectors are related: $\alpha\,\mathbf{x}\,\beta = \alpha * \beta + \alpha \cdot \beta$ [65, p. 29]. The only other Euclidean n-space in which a "cross product" can be defined is when $n = 7$ [82].

11.2.4 What is a Number?

Having defined various number systems with differing properties, one ought to raise the question: "what is a number?," which should quickly lead to the more appropriate question: "how do numbers 'behave'?" The important idea here is that we are often interested in relations among objects rather than in the objects themselves. Another significant point to note is that the notion of number has evolved over time. In the context of this theme, the answer to the question in the title leads to the definitions of ring, field, and division ring. This involves:

1. The consideration of other number-like objects, such as integers modulo n, matrices, (Boolean) rings of sets, polynomials, rational functions, power series, and extended power series. Such examples were instrumental in the emergence of the abstract theories of rings and fields. See [10, 22, 34, 70, 112].
2. Axiomatic characterizations of **Z**, **Q**, **R**, and **C**; that is, doing for these structures what Euclid and Hilbert had done for the Euclidean plane. For example, one can characterize **Z** as the smallest ordered integral domain and **R** as a complete ordered field. These are but instances of the re-emergence of axiomatics at the turn of the twentieth century following a 2,000-year dormancy.

This topic provides a good opportunity to discuss issues of completeness, independence, and consistency of an axiomatic system. As we keep introducing various properties (axioms) in trying to characterize (say) the integers, we ask: Do we have enough such properties (completeness)? Are there now too many (independence)? Perhaps we have picked "incorrect" axioms (consistency)? See [5, 40, 61, 108], and Sect. 14.3.5 for details.

11.3 The Algebraic-Transcendental Dichotomy

11.3.1 Introduction

While the first theme explored number systems beyond **C**, this one investigates such systems, initially fields, between **Z** and **C**. We easily show that there is no field between **Z** and **Q** (**Q** is the smallest field contained in **C**), nor between **R** and **C** (if a field contains **R** and *any* (non-real) complex number, it must be all of **C**). The latter parenthetical observation entails the notion of *adjunction* of an element α to a field F to obtain the field F(α) containing F – a fundamental idea in field theory. It was used by Galois in his development of Galois theory [10, 70]. When applied to $\mathrm{F} = \mathbf{Q}$ and $\alpha = \sqrt{2}$, say, it yields the field $\mathbf{Q}(\sqrt{2})$ of polynomials in $\sqrt{2}$ with coefficients in $\mathbf{Q} : \mathbf{Q}(\sqrt{2}) = \{a + b\sqrt{2} : a, b \in \mathbf{Q}\}$. What if $\alpha = \pi$? In this case polynomials in π do not yield a field – we need rational functions:

$$Q(\pi) = \left\{ \frac{a_0 + a_1\pi + \ldots + a_m\pi^m}{b_0 + b_1\pi + \ldots + b_n\pi^n} : a_i, b_j \in Q \right\}.$$

These two examples highlight the difference between algebraic and transcendental numbers, here $\sqrt{2}$ and π, respectively. We have found this algebraic framework to be a good way to motivate the introduction of these two concepts. That the algebraic/transcendental dichotomy is not unbridgeable is indicated by the remarkable relation $e^{\pi i} + 1 = 0$. See [29,61,67,90].

11.3.2 Algebraic Numbers

An algebraic number is a root of a polynomial equation with rational (equivalently: integer) coefficients. The set of all algebraic numbers forms a field A. For a nice proof based on an elementary result in linear algebra see [90, p. 83]. A is a generalization of $\mathbf{Q}$, which can be viewed as the field of roots of *linear* equations $ax + b = 0$, with $a, b \in \mathbf{Z}$.

Which real algebraic numbers are rational? That is, what are the rational roots among the zeros of $f(x) = a_0 + a_1 + \ldots + a_n x^n$, a_i integers? If s/t is such a root, with s and t relatively prime, it follows readily that s divides a_0 and t divides a_n [89, p. 58]. As a corollary we get that $\sqrt[n]{k}$ is irrational (k and n integers > 1), unless $k = u^n$, $u \in \mathbf{Z}$ (since $x^n - k$ has no rational roots), and that, for example, $\cos 20^o$ is irrational, since it satisfies the equation $8x^3 - 6x - 1 = 0$ which has no *rational* roots. See [89].

For algebraic numbers which are irrational it has been found important to examine how closely they can be approximated by rational numbers. It turns out that such numbers cannot be approximated "too well" by the rationals. Specifically, if α is an irrational algebraic number which is a root of an irreducible polynomial over $\mathbf{Z}$ of degree $n > 1$, then there exists a positive real number c such that for all $p, q \in \mathbf{Z}$ with $q > 0$, $|\alpha - p/q] > c/q^n$. It was this result which enabled Liouville to prove the transcendentality of $\sum_{m=1}^{\infty} 10^{-m!}$. See [29,64,89,104,105] for details.

11.3.3 Transcendental Numbers

Probably the first to define *transcendental numbers*, those complex numbers which are not roots of polynomial equations over $\mathbf{Z}$, was Euler (in the eighteenth century), although he never proved their existence. (The distinction between algebraic and transcendental *functions* was made by Leibniz in the seventeenth century.)

Since the algebraic numbers form a field, it follows that if t is transcendental and a algebraic, $a \neq 0$, then ta is transcendental. Hence there exist infinitely many transcendental numbers if there exists one. The first to prove the existence of transcendental numbers was Liouville, who (as we noted) showed in 1844 that $1/10^{1!} + 1/10^{2!} + 1/10^{3!} + \ldots$ is transcendental. For a nice proof using some basic ideas of calculus see [104, p. 735]; see also [14]. The transcendence of π, hence the impossibility of squaring the circle, was proved by Lindemann in 1882. For a proof see [93]. For the history of π, which involves geometric, analytic, and computational ideas see [11,13,14,23].

An important event occurred at the second International Congress of Mathematicians in Paris in 1900. Hilbert proposed 23 problems which proved to be instrumental in generating much research during the twentieth century (see [18, Chap. 27; 20, 122]). The seventh problem was to determine which complex numbers of the form α^β are transcendental (see [20, p. 242]). In 1934, Gelfond and Schneider proved, independently, that α^β is transcendental if α and β are algebraic, $\alpha \neq 0, 1$, and β is not rational (it may be complex). It follows, of course, that $2^{\sqrt{2}}$ is transcendental, but also that so are, for example, e^π and $\log_{10} 2$. Concerning e^π, note that one of the values of the infinite-valued expression i^{-2i} is e^π (use $e^{\pi i} = -1$). As for $\log_{10} 2$, observe first that it is not rational (let $\log_{10} 2 = a/b$, then $10^a = 2^b$); if $\log_{10} 2$ were algebraic, then $10^{\log_{10} 2} = 2$ would be transcendental. See [89, 90, 122].

The ideas used by Liouville, and by Gelfond and Schneider, have recently been generalized by Roth and Baker et al. Although these new results have not resolved the question of the transcendentality, or even the irrationality, of such numbers as $\pi^2, \pi + e, \pi e, e^e, \pi^\pi, 2^\pi, 2^e, \gamma$ (the Euler constant), they have played a crucial role in the study of a wide variety of Diophantine problems. See [14, 27; 74, Vols. 1 and 2; 80, 90].

11.3.4 Algebraic Numbers and Diophantine Equations

Algebraic numbers arise naturally and importantly in the solution of diophantine equations. In fact, this is where much of their importance lies. For example, to find integer solutions, if any, of $x^2 + y^2 = z^2$ (the Pythagorean triples), $x^2 + 2 = y^3$ (a special case of the Bachet equation), and $x^3 + y^3 = z^3$ (a special case of Fermat's Last Theorem), where $x, y, z \in \mathbf{Z}$, we factor the left side of each equation to obtain, respectively, $x^2 + y^2 = (x + yi)(x - yi)$, $x^2 + 2 = (x + \sqrt{2}i)(x - \sqrt{2}i)$, and $x^3 + y^3 = (x + y)(x + y\omega)(x + y\omega^2)$, where $\omega = (-1 + \sqrt{3}i)/2$. The terms on the right side of each equality are algebraic numbers – in fact, *algebraic integers.* These are roots of *monic* polynomials with integer coefficients; they generalize the ordinary integers, which can be viewed as roots of monic *linear* polynomials $a + x = 0$ over $\mathbf{Z}$.

Take for example the equation $x^2 + y^2 = z^2$. Since $(x + yi)(x - yi) = z^2$, the product of the two "relatively prime algebraic integers" $x + yi$ and $x - yi$ is a square, hence (as for ordinary integers) each is a square (of an algebraic integer). (It has to be demonstrated, of course, that these two algebraic integers are relatively prime.) In particular, $x + yi = (a + bi)^2 = (a^2 - b^2) + 2abi$ $(a, b \in \mathbf{Z})$, hence $x = a^2 - b^2, y = 2ab$, from which we get $z = a^2 + b^2$. This is the formula that yields all Pythagorean triples.

The crucial part of the above argument takes place in the domain D of algebraic integers called "Gaussian integers," $D = \{a + bi : a, b \in \mathbf{Z}\}$. The general idea in dealing with such equations as above is that in order to answer questions having to do with the integers it is often useful to enlarge the domain of integers, in this case to the domain of Gaussian integers. This provides for the "elbow room" not available in **Z**, and enables us to solve the equation. This idea goes back to Euler in the eighteenth century. In the example $x^2 + y^2 = z^2$, the divisibility properties of **Z** carry over to D, making it a *unique factorization domain* and enabling us to justify the above heuristic reasoning (see [97] or [106] for details). A similar approach applies to the equations $x^2 + 2 = y^3$ and $x^3 + y^3 = z^3$. See [1,54] and Chap. 3.

The consideration of unique factorization in various domains of algebraic integers led to the creation of *algebraic number theory* – the study of number-theoretic problems using the tools of abstract algebra. Algebraic number theory was an important source for the introduction into mathematics of the concepts of ring, ideal, field, and module. See [1, 10, 54, 70, 74, Vol. 2, 97, 106] for details.

11.4 Transfinite Numbers

11.4.1 Introduction

> The infinite! No other question has ever moved so profoundly the spirit of man; no other idea has so fruitfully stimulated his intellect; yet, no other concept stands in greater need of clarification than that of the infinite (Hilbert [81, p. vii]).

The clarification sought by Hilbert was initiated by Cantor beginning in the 1870s. This theme deals with some of Cantor's work and some of its consequences. A brief historical survey of manifestations of the infinite in mathematics prior to Cantor's work sets the scene for his ideas. In Greek antiquity, Zeno's paradoxes dealing with the infinite divisibility of space and time confounded contemporary thinkers. Aristotle viewed the natural numbers as a *potential* but not an *actual* infinite. Medieval scholastic speculations on the nature of the infinite included a discussion, without resolution, of the "paradox" that two concentric circles have unequal perimeters but an equal number of points. Galileo pondered the inherent contradiction in comparing (for "size") the positive integers and their squares: on the one hand the former contain the latter, but on the other, one can match up the two collections in a one-one manner. He concluded that the difficulties arise because

> We attempt, with our finite minds, to discuss the infinite, assigning to it those properties which we give to the finite and limited; but this... is wrong, for we cannot speak of infinite quantities as being the one greater or less than or equal to another [101, p. 5].

Even the great Gauss protested "against the use of an infinite quantity as an actual entity." He claimed that "this is never allowed in mathematics," adding that "the infinite is only a manner of speaking" [43, p. 160]. See [13, 27, 81, 101, 105, 115, 109a] for details.

11.4.2 Some Implications of Cantor's Work

The revolution in our understanding of the infinite was brought about by Cantor, almost single-handedly, in the space of a decade. There were philosophical and theological underpinnings to Cantor's creation [33]. The *mathematical* origins of his ideas on set theory had to do with the representation of functions in Fourier series and the consideration of those sets of points for which *unique* representation fails. For a thorough analysis of these matters Cantor realized that he needed a rigorous theory of real numbers. He proceeded to found it using Cauchy sequences. See [30, 33].

Among the standard but thought-provoking topics for discussion associated with Cantor's ideas are: cardinal and ordinal arithmetic, the existence of a nondenumerable infinity of transcendental numbers, the continuum hypothesis, the paradoxes of set theory and the resulting axiomatizations of "naïve" set theory, the axiom of choice and some of its consequences, various philosophies of mathematics, and Gödel's theorems. Among the unusual implications for the student are:

1. The giving up of the fundamental tenet that "the whole is greater than any of its parts." This was one of the "common notions" in Euclid's axiomatization of geometry. The reluctance to part with it stood in the way of progress in the study of the infinite, as shown by the examples of Galileo and of others cited above. See [43, 81, 101, 109, 115, 109a].
2. The existence of "arithmetics" in which the additive and multiplicative cancelation laws, the commutative laws of addition and multiplication, and the right distributive law fail. The first two fail in cardinal arithmetic: for example, $1+\aleph_0 = 2+\aleph_0$ but $1 \neq 2$; $3\cdot\aleph_0 = 4\cdot\aleph_0$ but $3 \neq 4$. The last three fail in ordinal arithmetic: for example, $1+\omega \neq \omega+1$, $1\cdot\omega \neq \omega 1$, and $(1+1)\omega \neq 1\cdot\omega+1\cdot\omega$. See [55, 101, 109a].
3. The existence of an infinity of infinities – cardinals and ordinals – of different "sizes." We have in mind here the infinitely increasing sequence of infinities which is obtained from the result $|A| < |P(A)|$, where $|A|$ denotes the cardinality of an infinite set A, $P(A)$ the set of all subsets of A. If A is the set of all sets then P(A) $\subseteq$ A, hence $|P(A)| \leq |A|$, and in conjunction with $|A| < |P(A)|$ this gives rise to the so-called *Cantor paradox*. If $A = \mathbf{Z}$, then $P(A) = \mathbf{R}$, and since the algebraic numbers are denumerable (as is easy to show), this implies the existence of nondenumerably many transcendental numbers. See [55, 115].
4. The fact that one can have two equally consistent mathematical theories contradicting one another. This is a relatively recent (nineteenth century) realizaiton. Cohen's proof in 1963 of the independence of the continuum hypothesis from the Zermelo–Fraenkel axioms of set theory gave rise to Cantorian and non-Cantorian set theories, in which the continuum hypothesis and its negation, respectively, hold [26]. The other major example of the phenomenon noted above is, of course, Euclidean and non-Euclidean geometries.

5. The idea that "simple" assumptions can have very surprising consequences. The simple assumption we have in mind is the *axiom of choice*. Among its surprising consequences are the Hausdorff and Banach–Tarski paradoxes. The latter says that any three-dimensional object can be cut into a finite number of pieces (ca 10^{50}) and reassembled to produce two objects, each identical to the original object [46, 117]. In a recent result along these lines, it was shown (in 1988) that using the axiom of choice a circle can be "squared" – that is, it can be dissected into a finite number of pieces (not the kind that can be cut out of paper with scissors!) and reassembled to yield a square equal in area to the given circle. See [24, 50, 117].
6. The pluralistic nature of mathematics, namely that mathematicians can differ fundamentally in their views of and approaches to the subject (see Chap. 10). This phenomenon is more prevalent than one would suppose, given the seemingly "deterministic" nature of mathematics. For example, Leibniz (seventeenth century) strongly opposed Descartes' use of algebra in dealing with geometric matters. Hermite (in the nineteenth century) "turn[ed] away with fright and horror from this lamentable evil of functions without derivatives" ([71, p. 973]) introduced earlier by Riemann and Weierstrass. At the beginning of the twentieth century there was a fundamental dispute between the formalists and the intuitionists. Witness the radically different views of Hilbert and Poincaré, respectively, of Cantor's set theory:

> No one shall expel us from the paradise which Cantor created for us [71, p. 1003].
>
> Later generations will regard *Mengenlehre* [Set Theory] as a disease from which one has recovered [71, p. 1003].

See [13, 27, 37, 86], and Chap. 10 for details.

7. The notion that mathematics and theology may have a closer affinity than meets the eye. Gödel's Incompleteness Theorem showed that the consistency of many of the common axiom systems, for example those for number theory or set theory, cannot be formally established. This implies an act of faith on the part of mathematicians in their pursuit of the consequences of axiomatic systems. See Lecture Thirty-Eight in Eves entitled "Mathematics as a branch of theology" [43].

11.5 The Personality of Numbers (The phrase was coined by P.J. Davis [33])

Theorem. *All [natural] numbers are interesting.*

Proof. If not, let n_0 be the least uninteresting number. But that makes n_0 very interesting. □

Fig. 11.2 Geometric representation of some triangular numbers

Further testimony, if any were needed, to the above result may be found in [6, 35, 59, 73, 100, 118]. For example, 5 is the fifth Fibonacci number; it divides at least one of the numbers of every pythagorean triple; it is the smallest integer n for which the general polynomial of degree n is unsolvable by radicals; it is the second Fermat prime; it is the smallest positive integer d for which the ring of integers of the algebraic number field $(Q(\sqrt{-d})$ is not a unique factorization domain; it is the only number of the form $4x^4 + y^4$ which is prime; it is the only prime p for which $p-2$, p and p, $p+2$ are twin primes; and it is the number of regular polyhedra.

Of course, some numbers are more interesting than others. Moreover, it is collections of numbers rather than individual numbers which often command special attention. Among these the *primes* must be singled out as the building blocks of all numbers. Their distribution among the integers is a deep and fascinating story. See [3, 27, 39, 99], and Sect. 1.8.

A much more elementary classification of the integers than into primes and composites is into *even* and *odd* integers. Yet, even that simple idea made possible the profound discovery by the ancient Greeks of the incommensurability of the diagonal and side of a square (what *we* refer to as the irrationality of $\sqrt{2}$). See [68, 71].

In early Greek mathematics, prior to the "crisis" resulting from the proof of the incommensurability of the diagonal and side of a square, geometry and algebra /arithmetic cohabited amicably (cf. Sect. 11.8). An example of this cooperative relationship was the introduction by the Pythagoreans of the *polygonal numbers*. For example, the triangular numbers are 1, 3, 6, 10, ... (The n-th triangular number is $1+2+3+\ldots+n = n(n+1)/2$.) For their geometric representation see Fig. 11.2.

The square numbers 1, 4, 9, ... ($n^2 = 1 + 3 + 5 + \ldots + (2n - 1)$) have geometric representation as squares, and, in general, the k-gonal numbers $n + (n^2 - n)(k - 2)/2, n = 1, 2 3, \ldots$ as (regular) k-sided polygons. (The inclusion of 1 among the polygonal numbers is recent.) There are many interesting relations involving polygonal numbers, some obtained from their geometric configurations (see [12, 27, 41, 64, 68]). The major result, stated in the seventeenth century by Fermat, is that every integer is a sum of ≤ 3 triangular numbers, ≤ 4 square numbers, and, in general, $\leq k$ k-gonal numbers. The case $k = 4$, namely that every integer is a sum of 4 squares, was proved by Lagrange in the eighteenth century, and the general case by Cauchy in the nineteenth. See [88].

Among other "personable" collections of numbers one might investigate the following:

1. *Perfect and amicable numbers.* These, too, date back to the pythagoreans and their numerological notions. See [21, 68, 71, 95] for details.
2. *Fibonacci numbers* and their connection with the golden ratio. See [62, 94, 116].
3. *Large primes*, the factorization of *large numbers* into primes, and their relation to public-key cryptography. This relatively recent topic belies the notion of the "uselessness" of number theory. See [58, 98].

See [27, 28, 38, 72, 84, 94, 96] for more details on this section.

11.6 One, Two, Many

We have mainly in mind here the cultural history of numbers. There is anthropological evidence that counting in prehistoric civilizations was, at one time, of the "one-two-many" variety [120]. Among topics to study are number-words in various societies, number-mysticism, symbolization, the emergence of the abstract notion of number, and notation, both additive and positional, in various bases. (Boyer claims that it is an exaggeration to regard positional notation as a fundamental accomplishment of civilization [19].) See [13, 25, 30, 32, 48, 63, 85, 120] for details.

A related topic for investigation is how various cultures computed, that is, performed the four algebraic operations and root extraction: fingers, pebbles, abacus, algorithms, logarithms, calculating machines, and computers. See [13, 44, 52, 63, 66, 85, 110, 124] for details.

A culminating idea for this theme might be to consider mathematicians' views of the natural numbers in the nineteenth century: from Kronecker's "God made the integers, all the rest is the work of man," through Dedekind's, Frege's, and Peano's "The integers are man-made," to Gödel's "Man-made axioms for the integers are incomplete." See [30, 51, 53, 71, 101].

11.7 Discovery (Invention), Use, Understanding, Justification

The sequence in the title is often the order of evolution of mathematical ideas. It is the reverse of the usual pedagogical order. (Is there a lesson here for pedagogy?) It is especially evident in the evolution of the various number systems: natural, negative, irrational (real), and complex. Each is an absorbing tale. While the previous theme suggests describing the story for the natural numbers, this theme attempts to sketch it for the other number systems. Here we give a very brief outline of the evolution of the *negative numbers.* That of the real and complex numbers is readily available in the mathematical literature, for example in [13, 30, 32, 71, 109], and Chap. 12. These "stories" provide very good vehicles for raising such issues as the role of paradoxes, internal vs. external motivation, and intuition vs. logic as factors in the development of mathematics. See [37, 118], and Chaps. 8 and 12.

Negative numbers entered mathematics as subtrahends (as in $a - b$), as distinct mathematical entities, as coefficients in polynomial equations, and as roots of such equations. At various times, mathematicians accepted or rejected negative numbers in one or another of these settings.

The ancient Babylonian, Egyptian, and Greek civilizations disallowed negative roots of equations and, in general, avoided explicit use of negative numbers, although they were aware, at least *implicitly,* of the rules governing their manipulation; for example, the Greeks had established *geometrically* the law $(a - b)(c - d) = ac - bc - ad + bd$ for $a > b$ and $c > d$ (see [10, 18, 68, 71]). The Chinese (ca. 200 BC) used negative numbers freely in their calculations. The Hindus (ca. 620 AD) gave *explicit* rules for operations with negative numbers, for example, that negative x negative = positive, and allowed for negative roots of equations, for example they stated that every positive number has two square roots. (Diophantus, in his *Arithmetica* (ca 250 AD), also gave rules for manipulating with negative numbers [8].)

European mathematicians of the sixteenth and seventeenth centuries were ambivalent about negative numbers and about their use as coefficients or as roots of polynomials. The Greek tradition of using geometry to solve algebraic problems – their "geometric algebra" (see [10, 68, 71]), no doubt had an impact. Thus Cardano in 1545 considered $x^3 = ax + b$ and $x^3 + ax = b$ $(a, b > 0)$ as distinct equations; replacing the two equations with the single cubic $x^3 + ax + b = 0$ would have meant introducing negative coefficients. Descartes in 1637 determined the number of "true" (positive) and "false" (negative) roots of an equation (see his well known "rule of signs" [68]), but avoided the use of negative coordinates in his development of analytic geometry. Pascal regarded the subtraction of 4 from 0 as yielding zero (!). Wallis "proved" in 1655 that negative numbers are greater than infinity: since $a/0 = \infty$, $a/$a negative no. $> \infty$, for if one decreases the denominator of a fraction one increases its value. Arnauld (seventeenth century) objected to the equality $-1/1 = 1/-1$ on the grounds that the ratio of a smaller to a greater quantity cannot equal the ratio of a greater to a smaller. Leibniz agreed that this was a difficulty but argued that one should tolerate negative numbers because they are useful and lead to consistent results. See [13, 18, 71].

There were exceptions to the above misgivings. Girard in 1623 admitted negative (and complex) roots of equations, and was thus able to state clearly the relation between the coefficients and roots of an equation, as well as the result that every equation of degree n has n roots (no proof was given). Stifel in the sixteenth century used negative numbers as exponents, and Hudde in the seventeenth allowed literal coefficients in an equation to represent any real number, positive or negative. This was a most important idea, for it permitted a unified treatment of polynomial equations of a given degree (recall Cardano having to consider $x^3 + ax = b$ and $x^3 = ax + b$ as distinct equations). See [10, 13, 68, 71, 103].

"Models" of abstract concepts are – and were – an important aid in their understanding and accommodation as bona fide mathematical entities. In the case of negative numbers, the Chinese viewed them as black rods (using red rods for positive numbers), the Hindus thought of them as debts, Fibonacci (early thirteenth century) as losses, and Girard (1620s) anticipated their representation on a number-line by noting that "the negative in geometry indicates a retrogression, where the positive is an advance" [18, p. 343]. These models aided in justifying the *addition* of negative numbers, but not their multiplication,

Textbooks of the eighteenth century, for example Euler's *Algebra*, continued to give detailed rules for manipulation of negative numbers, but some resistance to the very notion of such numbers persisted to the early nineteenth century (see [4, 49, and 71, p. 593]). During this period, however, mathematicians also began to ask *why* such rules should hold. Euler tried to justify the rule $(-a)(-b) = ab$ by noting that since $(-a)b$ equals $-ab$, $(-a)(-b)$ must equal ab (!) (see [4, pp. 31–32]). In the first half of the nineteenth century Peacock (and others) made a bold, though not very successful, attempt to give such justification by creating *symbolical algebra.* This was a precursor of modern axiomatics in algebra – symbols taking on a life of their own, independent of meaning. For example, $a(b - c)$ was "decreed" to equal $ab - ac$ by virtue of its form rather than its content (see [4, 18, Chap. 26, and 70] for details). Finally, in the latter part of the nineteenth century Weierstrass, and independently Peano, gave an abstract definition of negative numbers (integers) as ordered pairs of natural numbers [71, p. 987 ff.]. (Hamilton, about a half century earlier, had defined the complex numbers as ordered pairs of reals [56].) Proofs of $(-a)(-b) = ab$ and of similar results from the axioms of a field or a ring, as done in today's abstract-algebra courses, were given in the early twentieth century during the emergence of the axiomatic movement in algebra [70].

See [4, 13, 49, 66, 68, 71] for further details about the history of the negative numbers.

To summarize the above account: a formal, logical justification of negative numbers came only in the late nineteenth century, although such numbers were used, in one form or another, for over two millennia. Mathematical need played at least as important a role in their evolution as practical utility, and formal manipulations, more often than genuine understanding, guided mathematicians in formulating the rules of operation with negative numbers.

11.8 Numbers and Geometry

Arithmetic and geometry seem, at first sight, to be antithetical. The one deals with the discrete, the other with the continuous. The relations between the two are, however, deep, though often hidden. The tensions between number and geometry, and between the related analytic and synthetic approaches to mathematics, have been very beneficial for the development of the subject. See [27,47,64,84,94a,109] and Chap. 10.

Although the connection between arithmetic and geometry is fundamental, it has not always been amicable. The early Greek harmony between number and shape, given expression in, among other things, the arithmetic development of the Pythagorean theory of similarity, was shattered by the Greek crisis of incommensurability, that is, by the proof of the existence of incommensurable magnitudes [71]. Geometry reigned supreme for roughly the next two millennia, with the notable exceptions of Chinese, Hindu, and Islamic mathematics. The two joined again in the seventeenth century through the emergence of analytic geometry and calculus. With the arithmetization of analysis in the latter part of the nineteenth century (cf. Sect. 11.9), arithmetic gained the upper hand, at least in analysis. But the harmony persisted in other areas, for example, in algebraic geometry. We now give several examples, from different periods, of the collaborative relationship between number and geometry.

1. The notions that geometric relations can be expressed by numbers and that, conversely, relations among numbers have implications in geometry, have their (implicit) roots in ancient Babylonian and Egyptian mathematics (ca. 1500 BC), in the computation of areas and volumes of various geometric figures and in the (apparent) construction of right-angled triangles from relations such as $3^2+4^2=5^2$ (the Egyptian "rope-stretchers" [109, p. 1]). The formal expression of these latter ideas was the (geometric) statement and proof of Pythagoras' theorem and the (arithmetic) determination of all Pythagorean triples, both coming to fruition in ancient Greece (the latter likely even earlier, in Babylon). The geometry suggested number-theoretic questions. Thus, Fermat in the seventeenth century showed that there are no pythagorean triangles whose areas are squares (with integer sides) and hence that $x^4 + y^4 = z^4$ has no nontrivial integer solutions. See [95, p. 199 ff.] and [109, p. 141 ff.].

On the other hand, geometry was used in number-theoretic problems, such as the solution of diophantine equations. For example, finding integer solutions of $x^2 + y^2 = z^2$ is equivalent to finding rational solutions of $u^2 + v^2 = 1$ – that is, finding all points with rational coordinates ("rational points") on the unit circle. If (a, b) is one such point, and if the line with rational slope t passing through (a, b) cuts the unit circle at another point, it too will be a rational point. Conversely, if (c, d) is a second rational point on the unit circle, then the slope of the line joining (a, b) and (c, d) is rational. Thus, one obtains all rational points by letting t take on all rational values. Hence, if we let $(a, b) = (-1, 0)$, a point on the unit circle, the line through $(-1, 0)$ with slope t is $v = t(u + 1)$. Finding its point of intersection with $u^2 + v^2 = 1$ gives all rational points on the unit circle: $u = (1 - t^2)/(1 + t^2)$, $v = 2t/(1+t^2)$ (see [109, p. 4]). From this we obtain all solutions of $x^2 + y^2 = z^2$: $x = p^2–q^2$, $y = 2pq$, $z = p^2 + q^2$, p and q arbitrary integers.

This idea, of "embedding" a diophantine equation in Euclidean space and using the geometry of that space to facilitate the equation's solution, is fundamental in the modern study of diophantine equations, and apparently was already employed in embryonic form by Diophantus in the third century AD [8, 9, 109]. (Cf. Sect. 11.3.4, in which we indicated how the embedding of Diophantine equations in *algebraic* domains facilitates their solution.)

2. Another problem with roots in Greek antiquity is constructions with straightedge and compass. Its resolution, over 2,000 years later, depended on the "arithmetization" of the problem – its transformation to the problem of the "construction" of real numbers. The constructible numbers form a field – a subfield of the algebraic numbers; since π is transcendental, one cannot square a circle. Neither can one double a cube nor trisect an angle. See [29] for a nice, elementary proof using very little field theory, and [18, 71] for historical background.
3. The one–one correspondence between the points on a line and the real numbers "represents a remarkable link between something which is given by our spatial intuition and something that is constructed in a purely logico-conceptual way," observed Weyl [47, p. 159]. This correspondence forms, of course, the basis for analytic geometry, introduced by Descartes and Fermat, independently, in the early seventeenth century. It is a striking example of the fruitful synthetic-analytic tension in mathematics and it suggests the coordinatization of other geometries, for example, projective geometry. See [13].

The idea to pursue here is the assignment of number-like objects to the elements (points, lines) of various geometries. This leads to such algebraic systems as ternary rings, Veblen–Wedderburn systems, division rings, and fields, and to the study of various geometric properties via an analysis of the corresponding algebraic systems. For example, the only known proof of the result that in a finite projective plane Desargues' theorem implies Pappus' theorem follows from the corresponding algebraic result that a finite division ring is a field. See [15, 22] for details.

The geometric–arithmetic correspondence between the points on a line and the real numbers was used by Hilbert in reducing the question of the consistency of Euclidean geometry to that of the consistency of the real-number system, subsequently dealt with by Gödel. See [60; 71, Chap. 42; 109, p. 73; 109a].

4. The important role of the complex numbers in algebra and analysis is well known. Their role in number theory was noted in Sect. 11.3 and will be further explored in Sect. 11.9.3. The prominent part complex numbers play in geometry – in the solution of elementary problems and the formulation of fundamental principles – is perhaps less familiar. This can be explored at various levels via the study of euclidean, hyperbolic, and algebraic geometry. For details see [27, 43, 102, 109, 121].

11.9 Numbers and Analysis

In this theme we will indicate, on the one hand, the roles of the real and hyperreal number systems in the study of the foundations of analysis, and, on the other, the role of analysis – real, complex, and p-adic – in dealing with number-theoretic questions.

11.9.1 The Arithmetization of Analysis

The real numbers are in the foreground or background of much of analysis, yet they were not well understood until the late nineteenth century. Calculus was grounded largely in geometry in the seventeenth century and in algebra in the eighteenth. Even in the first half of the nineteenth century the proofs of several fundamental results in analysis were based on geometric intuition, especially that of the real numbers (see Sect. 4.6). Among these were:

- The existence of the definite integral of a continuous function.
- The convergence of a Cauchy sequence.
- The Intermediate Value Theorem.

The realization that a rigorous foundation of calculus is to be based on an arithmetic rather than a geometric grounding of the real numbers was due mainly to Dedekind and Weierstrass, who, along with Cantor and others, gave "arithmetic" constructions of the reals founded on the rationals. The above three results could now be established rigorously using, in one form or another, the completeness property of the real numbers.

The real numbers were subsequently shown to be constructible from the positive integers, hence analysis was shown to depend logically only on the properties of the natural numbers. The program of building up analysis from the natural numbers was later (1895) called by Felix Klein the "arithmetization of analysis." Since (Euclidean) geometry was also shown to be logically dependent on the properties of the real numbers, hence of the natural numbers, and much of nineteenth-century algebra was dominated by real or complex algebra, it seemed at the end of the nineteenth century that two and a half millennia after the Pythagorean dictum that "all is number," mathematics had come full circle. As Eves put it [43, p. 132]:

> The great edifice of mathematics was shown to be like an enormous inverted pyramid delicately balanced upon the natural number system as a vertex.

The purpose of this subtheme is to impart some insight into these – both technical and conceptual – ideas. See [17, 42, 43, 45, 53, 71, 87] for details.

11.9.2 Nonstandard Analysis

It became evident in the early stages of the development of calculus that infinitely small and infinitely large "numbers" are very useful. The intent of this subtheme is to trace that usefulness, beginning with Archimedes, through Cavalieri, Leibniz, Euler, and Cauchy, and culminating (in the 1960s) in Robinson's nonstandard analysis. The focus is on indicating how the hyperreal numbers – Robinson's formalization of infinitesimals – have given rise to a new foundation of analysis, in the sense of giving alternative answers to the fundamental questions of the subject. For details see [36, 37, 42, 57, 69, 107, 114].

11.9.3 Number Theory

Analysis – be it real, complex, or p-adic – has played a crucial role in number theory in the nineteenth and twentieth centuries, providing yet another example of the interplay between the continuous and the discrete (cf. Sect. 11.8). Here is a glimpse:

1. The Pell equation $x^2–Ny^2 = 1$ can be solved by expanding $\sqrt{N}$ in an (infinite) continued fraction – an analytic process. See [21, 106].
2. Having shown empirically that there is a prime between n and $2n$ for all $n < 6 \times 10^6$, Bertrand conjectured in 1845 that this holds for all positive integers n (the so-called *Bertrand Postulate*). The proof, given by Chebyshev in 1850, uses analytic methods. See [74, Vol. 1, p. 108].
3. Dirichlet's famous result about the infinitude of primes in an arithmetic progression, namely that for all relatively prime positive integers a and b, the arithmetic progression $a,\ a + b,\ a + 2b,\ a + 3b, \ldots$ contains infinitely many primes, is proved using complex analysis, including the Dirichlet L-series. See [3, 16, 54, 74, Vol. 2].
4. Concepts and results used in the study of the distribution of primes among the integers, for example the zeta function, the Riemann hypothesis, the prime-number theorem, are based in a fundamental way on real and complex analysis. See [3, 54; 74, Vol. 2].
5. As the above four examples indicate, to study the arithmetic of Q – that is, number theory – it is useful, at times essential, to "complete" Q to R. Now the field Q of rationals also has, for each prime p, a "p-adic" completion Q_p ("completion" in the sense of analysis). In fact, R and Q_p are *all* the completions of Q (but the Q_p are nonarchimedean fields). Moreover, just as the hyperreal and real numbers can be regarded as being on an equal footing in analysis (this notion will undoubtedly not go unchallenged), so the p-adic and real numbers are on a par in number theory (this is noncontroversial). For example, it is a fundamental result that the diophantine equation $f(x) = 0$ is, for certain f, solvable in Z if and only if it is solvable in R and in Q_p for all primes p [16]. The idea here is similar to those discussed in Sects. 11.3.3 and 11.8 (a): while there one embedded a diophantine equation in, respectively, the ring of integers of an algebraic number field and Euclidean space, in order to bring the *algebra*, respectively, the *geometry* of these structures to bear on the study of the equation, here one embeds the Diophantine equation in the metric spaces R and Q_p in order to bring the *analysis* of these topological fields to bear on the relevant equation. See [2, 7, 16, 75, 77] for details.

References

1. W. W. Adams and L. J. Goldstein, *Introduction to N umber Theory,* Prentice-Hall, 1976.
2. J. Agnew, *Explorations in Number Theory,* Wadsworth, 1972.

3. T. M. Apostol, *Introduction to Analytic Number Theory,* Springer-Verlag, 1976.
4. A. Arcavi, M. Bruckheimer, and R. Ben-Zvi, Maybe a mathematics teacher can profit from the study of the history of mathematics, *For the Learning of Math.* 3:1 (1982) 30–37.
5. B. Artman, *The Concept of Number: From Quaternions to Monads and Topological Fields,* Wiley, 1988.
6. S. Avital, Don't be blue, number two, *Arithm. Teacher* 34 (Sept. 1986) 42–45.
7. G. Bachman, *Introduction to p-adic Numbers and Valuation Theory,* Academic Press, 1964.
8. I. G. Bashmakova, *Diophantus and Diophantine Equations*, Math. Assoc. of Amer., 1997. (Translated from the Russian by A. Shenitzer.)
9. I. G. Bashmakova, Arithmetic of algebraic curves from Diophantus to Poincaré, *Historia Math.* 8:4 (1981) 393–416.
10. I. G. Bashmakova and G. Smirnova, *The Beginnings and Evolution of Algebra,* Math. Assoc. of America, 2000. (Translated from the Russian by A. Shenitzer.)
11. P. Beckmann, *A History of* π, St. Martin's Press, 1971.
12. A. H. Beiler, *Recreations in the Theory of Numbers,* Dover, 1964.
13. W. P. Berlinghoff and F. Q. Gouvea, *Math Through the Ages: A Gentle History for Teachers and Others*, expanded ed., Math. Assoc. of Amer., 2004.
14. L. Berggren, J. Borwein, and P. Borwein, *Pi: A Source Book*, Springer, 1997.
15. L. M. Blumenthal, *A Modern View of Geometry*, W. H. Freeman, 1961.
16. Z. I. Borevich and I. R. Shafarevich, *Number Theory,* Academic Press, 1966.
17. U. Bottazzini, *The Higher Calculus: A History of Real and Complex Analysis from Euler to Weierstrass,* Springer-Verlag, 1986.
18. C. B. Boyer, *A History of Mathematics*, revised by U. C. Merzbach, Wiley & Sons, 1989.
19. C. B. Boyer, Fundamental steps in the development of numeration, *Isis* 35:2 (1944) 153–168.
20. F. E. Browder (ed.), *Mathematical Developments Arising from Hilbert Problems,* 2 Vols., Amer. Math. Soc, 1976.
21. D. M. Burton, *Elementary Number Theory*, 2nd ed., Wm. C. Brown, 1989.
22. D. M. Burton., *A First Course in Rings and Ideals,* Addison-Wesley, 1970.
23. D. Castellanos, The ubiquitous π, *Math. Mag.* 61 (1988) 67–98 and 148–163.
24. B. Cipra, The circle has been squared, *Science* 244: 4904 (May 5 1989) 528.
25. M. P. Closs (ed.), *Native American Mathematics*, Univ. of Texas Press, 1986.
26. P. J. Cohen and R. Hersh, Non-Cantorian set theory, *Scientific Amer.* 217 (Dec. 1967) 104–116.
27. J. H. Conway and R. K. Guy, *The Book of Numbers*, Springer-Verlag, 1996.
28. J. H. Conway and R. K. Guy, Surreal numbers, *Math Horizons* (November 1996) 26–31.
29. R. Courant and H. Robbins, *What is Mathematics*? Oxford Univ. Press, 1941.
30. J. Crossley, *The Emergence of Number,* World Scientific, 1987.
31. M. J. Crowe, *A History of Vector Analysis*, Univ. of Notre Dame Press, 1968.
32. T. Dantzig, *Number: The Language of Science*, 4th ed., Free Press, 1967.
33. J. W. Dauben, *Georg Cantor: His Mathematics and Philosophy of the Infinite*, Harvard Univ. Press, 1979.
34. P. J. Davis, Number, *Sc. Amer.* 211 (Sept. 1964) 51–59.
35. P. J. Davis, *The Lore of Large Numbers,* Random House, 1961.
36. M. Davis and R. Hersh, Nonstandard analysis, *Sc. Amer.* 226 (1972) 78–86.
37. P. J. Davis, R. Hersh, and E. A. Marchisotto, *The Mathematical Experience, Study Edition*, Birkhäuser, 1995 (orig. 1981).
38. U. Dudley, *Numerology, or What Pythagoras Wrought*, Math. Assoc. of Amer., 1997.
39. U. Dudley, Formulas for primes, *Math. Mag.* 56 (1983) 17–22.
40. H. D. Ebbinghaus *et al*, *Numbers,* Springer-Verlag, 1990.
41. A. W. F. Edwards, *Pascal's Arithmetical Triangle*, Oxford Univ. Press, 1987.
42. C. H. Edwards, *The Historical Development of the Calculus*, Springer-Verlag, 1979.
43. H. Eves, *Great Moments in Mathematics: (a) before 1650 and (b) after 1650*, Math. Assoc. of Amer., 1983.
44. G. Flegg, *Numbers: Their History and Meaning,* Andre Deutsch, 1983.

45. C. G. Fraser, Some observations on mathematical analysis in the 18th century, *Arch. Hist. Exact Sci.*, 39:4 (1989) 317–335.
46. R. M. French, The Banach-Tarski theorem, *Math. Intell.* 10:4 (1988) 21–28.
47. A. Gardiner, *Infinite Processes: Background to Analysis,* Springer-Verlag, 1982.
48. M. Gardner, *The Magic Numbers of Dr. Matrix,* Prometheus Books, 1985.
49. M. Gardner, The concept of negative numbers and the difficulty of grasping it, *Scientific Amer.* 236 (1977) 131.
50. J. Gardner and S. Wagon, At long last, the circle has been squared, *Notices of the Amer. Math. Soc.* 36 (1989) 1338–1343.
51. A. Gillies, *Frege, Dedekind, and Peano on the Foundations of Arithmetic,* Van Gorcum, 1982.
52. H. Goldstine, *The Computer from Pascal to Von Neumann, Princeton* Univ. Press, 1972.
53. I. Grattan-Guinness, *From Calculus to Set Theory, 1630–1910: An Introductory History*, Princeton Univ. Press, 2000.
54. E. Grosswald, *Topics from the Theory of Numbers*, 2nd ed., Birkhäuser, 1984.
55. P. R. Halmos, *Naive Set Theory*, Springer-Verlag, 1974 (orig. 1960).
56. T. L. Hankins, *Sir William Rowan Hamilton*, The Johns Hopkins Univ. Press, 1980.
57. V. Harnik, Infinitesimals from Leibniz to Robinson: time to bring them back to school, *Math. Intell.* 8:2 (1986) 41–47, 63.
58. M. E. Hellman, The math of public key cryptography, *Scientific Amer.* 241:2 (Aug. 1979) 146–157.
59. B. Henry, *Every Number is Special,* Dale Seymour, 1985.
60. D. Hilbert, *The Foundations of Geometry,* Open Court, 1959.
61. A. P. Hillman and G. L. Alexanderson, *A First Undergraduate Course in Abstract Algebra*, 4th ed., Wadsworth, 1983.
62. H. E. Huntley, *The Divine Proportion,* Dover, 1970.
63. G. Ifrah, *From One to Zero,* Penguin, 1985.
64. M. C. Irwin, Geometry of continued fractions, *Amer. Math. Monthly* 96 (1989) 696–703.
65. I. Kantor and A. S. Solodovnikov, *Hypercomplex Numbers,* Springer-Verlag, 1989. (Translated from the Russian by A. Shenitzer.)
66. L. C. Karpinski, *The History of Arithmetic,* Russell and Russell, 1965.
67. E. Kasner and J. R. Newman, *Mathematics and the Imagination,* Simon & Schuster, 1967.
68. V. J. Katz, *A History of Mathematics: An Introduction*, 3rd. ed., Addison-Wesley, 2009.
69. J. Keisler, *Elementary Calculus: An Infinitesimal Approach,* 2nd ed., Prindle, Weber & Schmidt, 1986.
70. I. Kleiner, *A History of Abstract Algebra*, *Birkhäuser*, 2007.
71. M. Kline, *Mathematical Thought from Ancient to Modern Times*, Oxford Univ. Press, 1972.
72. D. E. Knuth, *Surreal Numbers*, Addison-Wesley, 1974.
73. F. Le Lionais, *Les Nombres Remarquables,* Hermann, 1983.
74. W. J. LeVeque, *Topics in Number Theory,* 2 Vols., Addison- Wesley, 1965.
75. D. J. Lewis, Diophantine equations and p-adic methods. In *Studies in Number Theory,* ed. by W. J. LeVeque, Math. Assoc. of Amer., 1969, pp. 25–75.
76. C. C. MacDuffee, Algebra's debt to Hamilton, *Scripta Math.* 10 (1944) 25–35.
77. C. C. MacDuffee, The p-adic numbers of Hensel, *Amer. Math. Monthly* 45 (1938) 500–508.
78. S. Mac Lane, *Mathematics: Form and Function,* Springer-Verlag, 1986.
79. S. Mac Lane, Mathematical models: a sketch for the philosophy of mathematics, *Amer. Math. Monthly* 88 (1981) 462–472.
80. E. Maor, *e: The Story of a Number*, *Princeton Univ. Press*, 1994.
81. E. Maor, *To Infinity and Beyond: A Cultural History of the Infinite,* Birkhäuser, 1987.
82. W. Massey, Cross products of vectors in higher dimensional euclidean spaces, *Amer. Math. Monthly* 90 (1983) 697–701.
83. K. O. May, The impossibility of a division algebra of vectors in three dimensional space, *Amer. Math. Monthly* 73 (1966) 289–291.
84. B. Mazur, *Imagining Numbers*, Farrar Straus Giroux, 2003.

85. K. Menninger, *Number Words and Number Symbols: A Cultural History of Numbers*, M.I.T. Press, 1969.
86. G. H. Moore, *Zermelo's Axiom of Choice: Its Origins, Development, and Influence,* Springer-Verlag, 1982.
87. P. J. Nahin, *An Imaginary Tale: The Story of* $\sqrt{-1}$, Princeton Univ. Press, 1998.
88. M. B. Nathanson, A short proof of Cauchy's polygonal theorem, *Proc. Amer. Math. Soc.* 99 (1987) 22–24.
89. I. Niven, *Numbers: Rational and Irrational,* Random House, 1961.
90. I. Niven, *Irrational Numbers*, Math. Assoc. of America, 1956.
91. I. Niven, The roots of a quaternion, *Amer. Math. Monthly* 49 (1942) 386–388.
92. I. Niven, Equations in quaternions, *Amer. Math. Monthly* 48 (1941) 654–661.
93. I. Niven, The transcendence of π, *Amer. Math. Monthly* 46 (1939) 469–471.
94. C. S. Ogilvy and J. T. Anderson, *Excursions in Number Theory*, Oxford Univ. Press, 1966.
94a. C. D. Olds, A. Lax, and G. Davidoff, *The Geometry of Numbers*, Math. Assoc. of Amer., 2000.
95. O. Ore, *Number Theory and its History,* McGraw-Hill, 1948.
96. O. O'Shea and U. Dudley, *The Magic Numbers of the Professor*, Math. Assoc. of Amer., 2007.
97. H. Pollard and H. G. Diamond, *The Theory of Algebraic Numbers*, 2nd ed., Math. Assoc. of Amer., 1975.
98. C. Pomerance, The search for prime numbers, *Scientific Amer.* 247:6 (1982) 136–147.
99. P. Ribenboim, *The Book of Prime Number Records*, 2nd ed., Springer-Verlag, 1989.
100. S. P. Richards, *A Number for Your Thoughts*, S. P. Richards Publ., 1982.
101. R. Rucker, *Infinity and the Mind: The Science and Philosophy of the Infinite,* Birkhäuser, 1982.
102. H. Schwerdtfeger, *Geometry of Complex Numbers,* Dover, 1979.
103. J. Sesiano, The appearance of negative solutions in mediaeval mathematics, *Arch. Hist. Exact Sc.* 32:2 (1985) 105–150.
104. G. F. Simmons, *Calculus with Analytic Geometry,* McGraw-Hill, 1985.
105. E. Sondheimer and A. Rogerson, *Numbers and Infinity: A Historical Account of Mathematical Concepts*, Cambridge Univ. Press, 1981.
106. H. Stark, *An Introduction to Number Theory*, M.I.T. Press, 1978.
107. L. A. Steen, New models of the real-number line, *Scientific Amer*. 225 (1971) 92–99.
108. I. Stewart and D. Tall, *The Foundations of Mathematics*, Oxford Univ. Press, 1977.
109. J. Stillwell, *Mathematics and its History*, 2nd ed., Springer-Verlag, 2002.
109a. J. Stillwell, *Roads to Infinity: The Mathematics of Truth and Proof,* A K Peters, 2010.
110. F. J. Swetz., *Capitalism and Arithmetic,* Open Court, 1987.
111. O. Toeplitz, *The Calculus: A Genetic Approach*, Univ. of Chicago Press, 1963.
112. B. L. Van der Waerden, *A History of Algebra,* Springer-Verlag, 1985.
113. B. L. Van der Waerden, The discovery of quaternions, *Math. Mag.* 49 (1976) 227–234.
114. D. H. Van Osdol, Truth with respect to an ultrafilter or how to make intuition rigorous, *Amer. Math. Monthly* 79 (1972) 355–363.
115. N. Ya. Vilenkin, *In Search of Infinity,* Birkhäuser, 1995. (Translated from the Russian by A. Shenitzer.)
116. N. N. Vorobov, *Fibonacci Numbers,* Blaisdell, 1961.
117. S. Wagon, *The Banach -Tarski Paradox*, Cambridge Univ. Press, 1985.
118. D. Wells, *The Penguin Dictionary of Curious and Interesting Numbers,* Penguin, 1986.
119. R. L. Wilder, *Mathematics as a Cultural System,* Pergamon Press, 1981.
120. R. L. Wilder, *Evolution of Mathematical Concepts: An Elementary Study,* Wiley, 1968.
121. M. Yaglom, *Complex Numbers in Geometry,* Academic Press, 1968.
122. B. H. Yandell, *The Honors Class: Hilbert's Problems and their Solvers*, A K Peters, 2002.
123. S. Yeshurun, Commonly known and less commonly known numbers, *Theta* 3:1 (1989) 28–34.
124. C. Zaslavsky, *Africa Counts,* Prindle, Weber and Schmidt, 1973.

Chapter 12
History of Complex Numbers, with a Moral for Teachers

12.1 Introduction

The usual definition of complex numbers, either as ordered pairs (a, b) of real numbers or as "numbers" of the form $a + bi$, does not give any indication of their long and tortuous evolution, which lasted about 300 years. In this chapter, we will briefly describe that evolution and suggest a number of lessons to be drawn from it. The lessons have to do with the impact of the history of mathematics on our understanding of mathematics and on our effectiveness in teaching it. But more about the moral of this story later.

The chapter may serve as a "unit" – or as a guide – in teaching courses based on the material in Chapter 11 or 14.

12.2 Birth

Our story begins in 1545. What came earlier can be summarized by the following quotation from Bhaskara, a 12th-century Hindu mathematician [6, p. 182]:

> The square of a positive number, also that of a negative number, is positive; and the square root of a positive number is twofold, positive and negative; there is no square root of a negative number, for a negative number is not a square.

In 1545 Cardano, an Italian mathematician, physician, gambler, and philosopher, published a book entitled *Ars Magna (The Great Art)*, in which he described an algebraic method for solving cubic and quartic equations. The publication of this book was a great event in mathematics: The solution of the cubic was the first major achievement in algebra since that time, 3000 years earlier, when the Babylonians showed how to solve *quadratic* equations.

I. Kleiner, *Excursions in the History of Mathematics*,
DOI 10.1007/978-0-8176-8268-2_12,

Cardano too dealt with quadratics in his book. One of the problems he proposed is the following [24, p. 67]:

> If some one says to you, divide 10 into two parts, one of which multiplied into the other shall produce... 40, it is evident that this case or question is impossible. Nevertheless, we shall solve it in this fashion.

He then applied his algorithm (essentially the method of completing the square, but without the use of symbols) to $x+y = 10$ and $xy = 40$ and got the two numbers $5+\sqrt{-15}$ and $5-\sqrt{-15}$ as solutions of the problem. "Putting aside the mental tortures involved," as he expressed it [3, p. 323], he formally multiplied $5+\sqrt{-15}$ by $5-\sqrt{-15}$ and obtained 40. He did not pursue the matter but concluded that the result is "as subtle as it is useless" [18, p. 291]. Judging by past practice in solving qudratic equations (e.g., $x^2+1 = 0$), this view was not unreasonable.

Although Cardano rejected the above solution, the event was nevertheless historic: It was the first time ever that the square root of a negative number was explicitly written down. And as Dantzig observed, "the mere writing down of the impossible gave it a symbolic existence" [6, p. 182].

Cardano's solution of the cubic $x^3 = ax+b$ was given (in modern notation) as

$$x = \sqrt[3]{\frac{b}{2}+\sqrt{\left(\frac{b}{2}\right)^2-\left(\frac{a}{2}\right)^3}}+\sqrt[3]{\frac{b}{2}-\sqrt{\left(\frac{b}{2}\right)^2-\left(\frac{a}{3}\right)^3}},$$

the so-called *Cardano's formula*. When applied (say) to the equation $x^3 = 9x+2$, it yields $x = \sqrt[3]{1+\sqrt{-26}}+\sqrt[3]{1-\sqrt{-26}}$. Cardano claimed – as he had done for the quadratic – that his formula for the solution of the cubic was inapplicable in cases such as this, in which square roots of negative numbers appear. But such square roots could no longer be dismissed, as we now indicate.

The crucial developments are due to Cardano's countryman Bombelli. In his book *L'Algebra* of 1572 he applied Cardano's formula to the now-classic equation $x^3 = 15x+4$, which yielded $x = \sqrt[3]{2+\sqrt{-121}}+\sqrt[3]{2-\sqrt{-121}}$. On the other hand, $x = 4$ is also a solution of the equation, as Bombelli noted by inspection (the two other solutions of $x^3 = 15x+4$, $x = -2\pm\sqrt{3}$, are also real). It now remained to reconcile the formal and "meaningless" solution $x = \sqrt[3]{2+\sqrt{-121}}+\sqrt[3]{2-\sqrt{-121}}$ with the solution $x = 4$.

Bombelli had a "wild thought," namely that since the radicands $2+\sqrt{-121}$ and $2-\sqrt{-121}$ differ only in sign, the same might be true of their cube roots. He thus let $\sqrt[3]{2+\sqrt{-121}} = a+b\sqrt{-1}$, $\sqrt[3]{2-\sqrt{-121}} = a-b\sqrt{-1}$, and proceeded to solve for a and b by manipulating these expressions according to the established rules for real variables. He deduced that $a = 2$ and $b = 1$ and thereby showed that, indeed, $x = \sqrt[3]{2+\sqrt{-121}}+\sqrt[3]{2-\sqrt{-121}} = (2+\sqrt{-1})+(2-\sqrt{-1}) = 4$ [3, p. 327]. Bombelli had given meaning to the "meaningless." As he put it [17, p. 19]:

> It was a wild thought, in the judgment of many; and I too was for a long time of the same opinion. The whole matter seemed to rest on sophistry rather than on truth. Yet I sought so long, until I actually proved this to be the case.

Of course breakthroughs are often achieved in this way – by thinking the unthinkable and daring to present it in public.

The equation $x^3 = 15x + 4$ is an example of the so-called "irreducible case" of the cubic, in which all three solutions are real yet they are expressed by Cardano's formula in terms of complex numbers. In fact, *every* formula for the solution of an irreducible cubic must involve complex numbers. So indeed complex numbers are unavoidable in the solution of cubic equations [2, p. 476]. To resolve the apparent paradox exemplified by this equation, Bombelli developed a calculus of operations with complex numbers. His rules, in our notation, are $(\pm 1)i = \pm i$, $(+i)(+i) = -1$, $(-i)(+i) = +1$, $(\pm 1)(-i) = \mp i$, $(+i)(-i) = +1$, and $(-i)(-i) = -1$. He also considered examples involving addition and multiplication of complex numbers, such as $8i + (-5i) = +3i$ and $(\sqrt[3]{4 + \sqrt{2}i})(\sqrt[3]{3 + \sqrt{8}i}) = \sqrt[3]{8 + 11\sqrt{2}i}$.

Bombelli's work signaled the birth of complex numbers. Birth, however, did not entail legitimacy. It took another 300 years for complex numbers to be accepted as genuine mathematical entities.

Many textbooks, even at the university level, suggest that complex numbers arose in connection with the solution of *quadratic* equations, especially the equation $x^2 + 1 = 0$. But as we have indicated, it was the *cubic* rather than the quadratic that forced their introduction.

12.3 Growth

Bombelli's work was only the beginning of the saga of the complex numbers. Although his book was widely read, complex numbers were shrouded in mystery, little understood, and often entirely ignored. Witness Simon Stevin's remark in 1585 about them [5, p. 96]:

> There is enough legitimate matter, even infinitely much, to exercise oneself without occupying oneself and wasting time on uncertainties.

Similar doubts concerning the meaning and legitimacy of complex numbers persisted for two and a half centuries. Yet during that same period these numbers were used extensively. Here are a number of examples.

As early as 1620 Girard suggested that an equation of degree n may have n roots. Such statements of the fundamental theorem of algebra were however vague and unclear. For example, Descartes, who coined the unfortunate word "imaginary" for the new numbers, stated that although one can imagine that every equation has as many roots as is indicated by its degree, no (real) numbers correspond to some of these imagined roots.

The following quotation, from a letter in 1673 from Huygens to Leibniz, in response to the latter's letter that contained the identity $\sqrt{1+\sqrt{-3}}+\sqrt{1-\sqrt{-3}}=\sqrt{6}$, was typical of the period [5, p. 107]:

> The remark which you make concerning... imaginary quantities which, however, when added together yield a real quantity, is surprising and entirely novel. One would never have believed that $\sqrt{1+\sqrt{-3}}+\sqrt{1-\sqrt{-3}}$ make $\sqrt{6}$ and there is something hidden therein which is incomprehensible to me.

Leibniz, who spent considerable time and effort on the question of the meaning of complex numbers and the possibility of deriving reliable results by applying the ordinary laws of algebra to them, thought of complex roots as "an elegant and wonderful resource of divine intellect, an unnatural birth in the realm of thought, almost an amphibium between being and non-being" [17, p. 159].

Complex numbers were widely used in the 18th century. Leibniz and John Bernoulli employed them as an aid to integration. For example, to evaluate $\int[1/(x^2+a^2)]\mathrm{d}x$, they proceeded as follows:

$$\begin{aligned}\int[1/(x^2+a^2)]\mathrm{d}x &= \int[1/(x+ai)(x-ai)]\mathrm{d}x\\ &= -1/2ai\int[1/(x+ai)(x-ai)]\mathrm{d}x\\ &= -1/2ai\,[\log(x+ai)-\log(x-ai)].\end{aligned}$$

This raised questions about the meaning of the logarithm of complex as well as negative numbers. A heated controversy ensued between Leibniz and Bernoulli, in particular about the meaning of $\log(-1)$. Bernoulli claimed that $\log(-1)=\log i=0$, arguing that $\log(-1)^2=\log 1^2$, hence $2\log(-1)=2\log 1=0$. Thus $\log(-1)=0$, and therefore $0=\log(-1)=\log i^2=2\log i$, from which it follows that $\log i=0$. On the other hand Leibniz argued that $\log(-1)$ is "imaginary" (to him this meant "not real," not necessarily "complex"), putting forward several arguments. Here is one (for others see Chap. 8):

If $\log(-1)$ were real, $\log i$ would be real, since $\log i=\log(-1)^{1/2}=1/2\log(-1)$. But this, according to Leibniz, is absurd.

The controversy was subsequently resolved by Euler, who showed that $\log(-1)=i(\pi+2n\pi)$, n any integer, so that $\log(-1)$ is complex and multivalued. See [13] and Chap. 8.

Complex numbers were used by Lambert for map projection, by d'Alembert in hydrodynamics, and by Euler, d'Alembert, and Lagrange in incorrect proofs of the fundamental theorem of algebra. (Euler, by the way, was the first to designate $\sqrt{-1}$ by i.) See [13, 17, 23].

Euler, who made important use of complex numbers in many fundamental ways, for example, in linking the exponential and trigonometric functions by the formula $e^{ix}=\cos x+i\sin x$, said the following about them [13, p. 594]:

> Because all conceivable numbers are either greater than zero, less than zero or equal to zero, then it is clear that the square root of negative numbers cannot be included among the possible numbers.... And this circumstance leads us to the concept of such numbers, which by their nature are impossible and ordinarily are called imaginary or fancied numbers, because they exist only in the imagination.

Even the great Gauss, who in his doctoral thesis of 1797 gave the first essentially correct proof of the fundamental theorem of algebra, claimed as late as 1825 that "the true metaphysics of $\sqrt{-1}$ is elusive" [13, p. 631].

The desire for a logically satisfactory explanation of complex numbers became manifest in the latter part of the 18th century on philosophical, if not on utilitarian, grounds. With the advent of the Age of Reason in the 18th century, when mathematics was held up as a model to be emulated – not only in the natural sciences but in philosophy as well as in political and social thought – the inadequacy of a rational explanation of complex numbers was disturbing.

The problem of the logical justification of the laws of operation with negative and complex numbers also became a pressing *pedagogical* issue (at, among other places, Cambridge University) at the turn of the 19th century. Since mathematics was viewed by the educational institutions as a paradigm of rational thought, the glaring inadequacies in the logical justification of the operations with negative and complex numbers became untenable. Such questions as "Why is $2 \ x \ i = i \ x \ 2$?" and "Is $\sqrt{ab} = \sqrt{a}\sqrt{b}$ true for negative a and b?" received no satisfactory answers.

Euler, in his text of the 1760s on algebra, claimed $\sqrt{-1}\sqrt{-4} = \sqrt{4} = +2$ as a possible result. Woodhouse opined in 1802 that since imaginary numbers lead to right conclusions, they must have a logic. Around 1830 George Peacock and others at Cambridge set for themselves the task of determining that logic by codifying the laws of operation with numbers. Although their endeavor did not satisfactorily resolve the problem of the logic of the complex numbers, it was perhaps the earliest instance of "axiomatics" in algebra. See [12, Chaps. 1 and 7] and [13, Chap. 32].

By 1831 Gauss had overcome his metaphysical scruples concerning complex numbers and, in connection with a work on number theory, published his results on their geometric representation as points in the plane. Similar representations by Wessel in 1797 and by Argand in 1806 went largely unnoticed. The geometric representation, given Gauss' stamp of approval, dispelled much of the mystery surrounding complex numbers.

In the next two decades further developments took place. In 1833 Hamilton gave an essentially rigorous algebraic definition of complex numbers as pairs of real numbers, and in 1847 Cauchy gave a completely rigorous definition in terms of congruence classes of real polynomials modulo $x^2 + 1$. In this he modeled himself on Gauss' definition of congruences for the integers. See [8, 12, 13, 22].

12.4 Maturity

By the latter part of the 19th century most vestiges of mystery and distrust of complex numbers could be said to have disappeared, although a lack of confidence in them persisted among some textbook writers well into the 20th century.

These authors would often supplement proofs using complex numbers with proofs that did not involve them. Complex numbers could now be viewed in the following ways:

(a) Points or vectors in the plane [14, 17].
(b) Ordered pairs of real numbers [6, 11].
(c) Operators (that is, rotations of vectors in the plane) [8, 22].
(d) Numbers of the form $a + bi$, with a and b real numbers [13, 22].
(e) Polynomials with real coefficients modulo $x^2 + 1$ [8]. (In the language of abstract algebra: the quotient ring of the ring of real polynomials in x modulo the ideal generated by $x^2 + 1$.)
(f) Matrices of the form

$$\begin{bmatrix} a & b \\ -b & a \end{bmatrix},$$

with a and b real numbers [8, 22].

(g) An algebraically closed field containing the reals [2, 8, 13]. (This is an early-20th-century view.)

Although these ways of viewing the complex numbers might seem confusing rather than enlightening, it is important to note that to gain a better understanding of a given concept, result, or theory, it is helpful to consider it in as many contexts and from as many points of view as possible.

The foregoing descriptions of complex numbers are not the end of their story. Various developments in mathematics in the 19th century enable us to gain a deeper insight into the role of complex numbers in mathematics and in other areas. These numbers offer just the right setting for dealing with many problems in mathematics in such diverse areas as algebra, analysis, geometry, and number theory. They have a symmetry and completeness that is often lacking in (say) the integers and the real numbers. Some of the masters who made fundamental contributions to these fields say it best. The following three quotations are by Gauss in 1811, Riemann in 1851, and Hadamard in the 1890s, respectively:

> Analysis... would lose immensely in beauty and balance and would be forced to add very hampering restrictions to truths which would hold generally otherwise, if... imaginary quantities were to be neglected [1, p. 31].

> The original purpose and immediate objective in introducing complex numbers into mathematics is to express laws of dependence between variables by simpler operations on the quantities involved. If one applies these laws of dependence in an extended context, by giving the variables to which they relate complex values, there emerges a regularity and harmony which would otherwise have remained concealed [8, p. 64].

> The shortest path between two truths in the real domain passes through the complex domain [13, p. 626].

We give brief indications of what is involved:

(1) In algebra, the solution of polynomial equations motivated the introduction of complex numbers: Every equation with complex coefficients has a complex

root – this is the celebrated *fundamental theorem of algebra*. Beyond their use in the solution of polynomial equations, the complex numbers offer an example of an algebraically closed field, relative to which many problems in linear algebra and other areas of abstract algebra have their "natural" formulation and solution [8, 13, 17].

(2) In analysis, the 19th century saw the development of a powerful and beautiful branch of mathematics – *complex function theory*. We have already seen how the complex numbers give us deeper insight into the logarithmic, exponential, and trigonometric functions. Moreover, we can evaluate real integrals by means of complex function theory. One indication of the efficacy of the theory is that a function in the complex domain is infinitely differentiable if once differentiable. Such a result is, of course, false in the case of functions of a real variable (cf. $f(x) = x^{4/3}$) [8, 13, 17].

(3) The complex numbers lend symmetry and generality in the formulation and description of various branches of geometry, including Euclidean, inversive, and non-Euclidean. Ample examples can be found in [21] and [26]. For a specific example using complex numbers to solve problems in the real domain we mention Gauss' use of them to show that the regular polygon of 17 sides is constructible with straightedge and compass [13].

(4) In number theory certain diophantine equations can be solved neatly and relatively easily by the use of complex numbers. For example, the equation $x^2 + 2 = y^3$, when expressed as $(x + \sqrt{2}i)(x - \sqrt{2}i) = y^3$, can readily be solved in integers using properties of the complex domain consisting of the set of elements of the form $a + b\sqrt{2}i$, with a and b integers; see [12] and Chaps. 1 and 3.

(5) An elementary illustration of Hadamard's dictum that "the shortest path between two truths in the real domain passes through the complex domain" is supplied by the following proof that the product of sums of two squares of integers is again a sum of two squares of integers; that is, that $(a^2 + b^2)(c^2 + d^2) = u^2 + v^2$. For, $(a^2 + b^2)(c^2 + d^2) = (a + bi)(a - bi)(c + di)(c - di) = [(a + bi)(c + di)][(a - bi)(c - di)] = (u + vi)(u - vi) = u^2 + v^2$, for some integers u, v.

Try to prove this result without the use of complex numbers and without being given the u and v in terms of a, b, c, and d.

In addition to their fundamental uses in mathematics, complex numbers have become a fixture in science and technology. For example, they are used in quantum mechanics and in electric circuitry. The "impossible" has become not only possible but indispensable [13, 17, 23].

12.5 The Moral

In this section we make comments on, and give suggestions for, the use of the history of mathematics in its teaching, in particular with reference to complex

Fig. 12.1 George Polya (1887–1995)

numbers. We ask: What is the history of mathematics good for? Why bother with such "stories" as this one? C. H. Edwards puts it in a nutshell [9, p. vii]:

> Although the study of the history of mathematics has an intrinsic appeal of its own, its chief raison d'être is surely the illumination of mathematics itself.

My colleague Abe Shenitzer says it as follows:

> One can *invent* mathematics without knowing much of its history. One can *use* mathematics without knowing much, if any, of its history. But one cannot *have* a *mature appreciation* of mathematics without a substantial knowledge of its history.

Polya (1962) expresses similar sentiments [19, Introduction]:

> To teach effectively a teacher must develop a feeling for his subject; he cannot make his students sense its vitality if he does not sense it himself. He cannot share his enthusiasm when he has no enthusiasm to share. How he makes his point may be as important as the point he makes; he must personally feel it to be important.

Such "*mature appreciation* of mathematics," and such a "feeling for [the] subject" are essential for teachers to possess. They can provide them with insight, motivation, and perspective – crucial ingredients in the making of a good teacher. As for use in the classroom, it is of course the teacher who can best judge when and how, at what level and in what context (if any) to introduce and relate historical material to the discussion at hand. The introduction of such material can convey to the student the following important lessons, which are usually not imparted in the standard curriculum:

(a) *The meaning of number in mathematics.* Complex numbers do not fit readily into students' notions of what a number is. And of course the meaning of number has changed over the centuries. This story presents some perspective on the issue. It also leads to the question of whether numbers beyond the complex numbers exist; see [8] and Chap. 11.
(b) *The relative roles of physical needs vs intellectual curiosity as motivating factors in the development of mathematics.* The problem of the solution of the cubic, which motivated the introduction of complex numbers, was *not* a practical problem. Mathematicians already knew how to find *approximate* roots of cubic equations. The aim was to find an *exact algebraic* formula – a question without any practical consequences. Yet how useful did the complex numbers turn out to be! This is a recurring theme in the evolution of mathematics [3, 13].
(c) *The relative roles of intuition vs. logic in the evolution of mathematics.* Observation, analogy, induction, and intuition are the initial and often the more natural ways of acquiring mathematical knowledge. Rigor, formalism, and the logical development of a concept or result usually come at the end of a process of mathematical evolution. For complex numbers, too, first came *use* (theoretical rather than practical), then *intuitive understanding*, and finally *formal justification* ([13] and Chaps. 7–10). P. J. Davis has the following take on this issue [7, p. 305]:

> It is paradoxical that while mathematics has the reputation of being the one subject that brooks no contradictions, in reality it has a long history of successful living with contradictions. This is best seen in the extensions of the notion of number that have been made over a period of 2500 years. From limited sets of integers, to infinite sets of integers, to fractions, negative numbers, irrational numbers, complex numbers, transfinite numbers, each extension, in its way, overcame a contradictory set of demands.

(d) *The nature of proof in mathematics.* What was the role of proof in establishing various results about complex numbers (see e.g. the derivation of the value of log i by Bernoulli)? One thing is certain: what was admissible as a proof in the 17th and 18th centuries was no longer acceptable in the 19th and 20th centuries. The practice of proof has evolved over time, as it is still evolving – not necessarily from less to more rigor. See Chaps. 7–10.
(e) *The relative roles of the individual vs. the environment in the creation of mathematics.* We should note first that mathematics is far from a static, lifeless discipline. It is dynamic, constantly evolving, full of failures as well as achievements. That said, what (e.g.) was the role of Bombelli in the creation of complex numbers? Cardano surely had the opportunity to take the great and courageous step of "thinking the unthinkable." Was the time perhaps not ripe for Cardano, but ripe for Bombelli – about 30 years later? Is it the case, as Wolfgang Bolyai, the father of one of the creators of non-Euclidean geometry, stated, that "many things have, as it were, an epoch in which they are discovered in several places simultaneously, just as the violets appear on all sides in the springtime?." [13, p. 861].

This conclusion seems to be borne out by many instances of independent and simultaneous discoveries in mathematics, such as the geometric representation of complex numbers by Wessel, Argand, and Gauss. The complex numbers are an interesting case study of such questions, to which of course we have no definitive answers.

(f) *The genetic principle in mathematics education.* What are the sources of a given concept or theorem? Where did it come from? Why would anyone have bothered with it? These are fascinating questions, and the teacher should at least be aware of the answers to them. When and how he or she uses them in the classroom is another matter. On this Polya says the following [20, p. 132]:

> Having understood how the human race has acquired the knowledge of certain facts or concepts, we are in a better position to judge how the human child should acquire such knowledge.

Can we not have a better appreciation of students' difficulties with complex numbers having witnessed mathematicians of the first rank make mistakes, "prove" erroneous theorems, and often come to the right conclusions for insufficient or invalid reasons?

(g) *Relevance.* Mathematicians usually create their subject without thought of practical applications (see (b) above). The latter, if any, come later, sometimes centuries later. This point relates to "immediate relevance" and to "instant gratification," which students often seek from any given topic presented in class. We must, of course, supply the student with "internal relevance" when introducing a given concept or result.

This brings us to the important and difficult issue of motivation. To some students the applications of a theorem are appealing; to others, the interest is in the logical structure of the theorem. A third factor, useful but often neglected, is the source of the theorem: How did it arise? What motivated mathematicians to introduce it? As for complex numbers, their origin in the solution of the cubic rather than the quadratic should be stressed. Cardano's attempted division of 10 into two parts whose product is 40 reinforces this point. How much further one continues with the historical account is a decision best made by the teacher in the classroom, bearing in mind the lessons that should be conveyed through this or similar historical material.

12.6 Projects

Historical projects arising from the story about the complex numbers can be given to students as subjects for research. Possible topics are the following:

1. The logarithms of negative and complex numbers. Consult [13, 17, 23], and Chap. 8.
2. The evolution of various number systems and the evolution of our conception of number. Consult [5, 6, 8, 10, 16], and Chap. 11.

3. Hypercomplex numbers – the quaternions, the octonions, n-tuples of reals for $n \neq 2, 4, 8$. The discovery (invention?) of the quaternions, in particular, is a fascinating story. Consult [8, 12, 13, 16], and Chap. 11.
4. Gauss' integer congruences and Cauchy's polynomial congruences. The latter led to a new definition (description) of complex numbers. Consult [13, 23].
5. An axiomatic characterization of complex numbers. In this connection one ought to discuss the notion of characterizing a mathematical system, and thus the concept of isomorphism [2, 8]. (Cf. the various equivalent descriptions of complex numbers discussed in Sect. 12.4.)
6. Many elementary and interesting illustrations of Hadamard's comment demonstrate that indeed "the shortest path between two truths in the real domain passes through the complex domain." We are referring to elementary results from various branches of mathematics, results whose statements do not contain complex numbers but whose "best" proofs often use them. For two such examples see Sect. 12.4, items (4) and (5). Others can be found in the References; see for example [4, 11, 18, 21, 26].

References

1. G. Birkhoff, *A Source Book in Classical Analysis*, Harvard Univ. Press, 1973.
2. G. Birkhoff and S. Mac Lane, *A Survey of Modern Algebra*, 4th ed., Macmillan, 1977.
3. D. M. Burton, *The History of Mathematics: An Introduction*, 6th ed., McGraw Hill, 2007.
4. J. W. Cell, Imaginary numbers, *Math. Teacher* 43 (December 1950) 394–396.
5. J. N. Crossley, *The Emergence of Number,* World Scientific, 1987.
6. T. Dantzig, *Number: The Language of Science*, 4th ed., Macmillan, 1954.
7. P. J. Davis, *The Mathematics of Matrices*, Blaisdell, 1965.
8. H.-D. Ebbinghaus et al, *Numbers*, Springer, 1990.
9. C. H. Edwards, *The Historical Development of the Calculus,* Springer, 1974.
10. G. Flegg, *Numbers: Their History and Meaning,* Andre Deutsch, 1983.
11. P. S. Jones, Complex numbers: an example of recurring themes in the development of mathematics, I-III, *Math. Teacher* 47 (February, April, May 1954) 106–114, 257–263, 340–345.
12. I. Kleiner, *A History of Abstract Algebra*, Birkhäuser, 2007.
13. M. Kline, *Mathematical Thought from Ancient to Modern Times*, Oxford Univ. Press, 1972.
14. B. Mazur, *Imagining Numbers*, Farrar Strauss Giroux, 2003.
15. R. B. McClenon, A contribution of Leibniz to the history of complex numbers, *Amer. Math. Monthly* 30 (1923) 369–74.
16. E. Nagel, Impossible numbers: a chapter in the history of modern logic, *Studies in the History of Ideas* 3 (1935) 429–74.
17. P. J. Nahin, *An Imaginary Tale: The Story of* $\sqrt{-1}$, Princeton Univ. Press, 1998.
18. National Council of Teachers of Mathematics, *Historical Topics for the Mathematics Classroom*, Thirty-First Yearbook, The Council, 1969.
19. G. Polya, *Mathematical Methods in Science*, Math. Assoc. of Amer., 1977.
20. G. Polya, *Mathematical Discovery,* Wiley, 1962.
21. H. Schwerdtfeger, *Geometry of Complex Numbers*, Dover, 1979.
22. E. Sondheimer and A. Rogerson, *Numbers and Infinity: A Historical Account of Mathematical Concepts*, Cambridge Univ. Press, 1981.
23. J. Stillwell, *Mathematics and Its History,* Springer, 1989.

24. D. J. Struik, *A Source Book in Mathematics, 1200 – 1800*, Harvard Univ. Press, 1969.
25. G. Windred, History of the theory of imaginary and complex quantities, *Mathematical Gazette* 14 (1930) 533–41.
26. I. M. Yaglom, *Complex Numbers in Geometry,* Academic Press, 1968.

Chapter 13
A History-of-Mathematics Course for Teachers, Based on Great Quotations

13.1 Introduction

Courses in the history of mathematics have been proposed based on great theorems and great problems [15, 29, 33]. Here we outline a course in the history of mathematics with great quotations as points of departure. These three "greats" have in common a number of important pedagogical features: they are interesting, they arouse curiosity, and they display, or lead to, important aspects of the mathematical enterprise. Moreover, the quotations (like the theorems and the problems) cajole, exasperate, stimulate, motivate, seduce, amuse – all welcome didactic traits. Perhaps more importantly, they are guideposts around which one may structure the development of a concept, a result, or a theory.

At my university "History of Mathematics" is a required course in an In-Service Master's Program for secondary-school mathematics teachers. The Program, which has been especially designed for teachers, attempts to give them a broad overview of major mathematical fields and issues, to expand their horizons and deepen their understanding of mathematics, to teach them relatively elementary mathematics from a relatively sophisticated point of view, and to broaden their perspective on the mathematics they teach, so that they can better judge what to emphasize in their teaching and why to emphasize it.

Effective teaching of mathematics requires more than a sound command of the subject matter. In his *Mathematical Methods in Science* Polya explains [39, Introduction]:

> To teach effectively a teacher must develop a feeling for his subject; he cannot make his students sense its vitality if he does not sense it himself. He cannot share his enthusiasm when he has no enthusiasm to share. How he makes his point may be as important as the point he makes; he must personally feel it to be important.

Wise counsel! The history of mathematics can increase teachers' enthusiasm for the subject, promote a sense of its importance, even greatness, and encourage students to ask "why?" in addition to "how?" All these are among the objectives of this course.

I. Kleiner, *Excursions in the History of Mathematics*,
DOI 10.1007/978-0-8176-8268-2_13,

The following two quotations, by C. H. Edwards and O. Toeplitz, respectively, express some of these sentiments:

> Although the study of the history of mathematics has an intrinsic appeal of its own, its chief raison d'être is surely the illumination of mathematics itself. For example, the gradual unfolding of the integral concept – from the volume computations of Archimedes to the intuitive integrals of Newton and Leibniz and finally the definitions of Cauchy, Riemann and Lebesgue – cannot fail to promote a more mature appreciation of modem theories of integration [16, p. vii].

> Regarding all these basic topics in infinitesimal calculus which we teach today as canonical requisites... the question is never raised, 'Why so?' or 'How does one arrive at them?' Yet all these matters must at one time have been goals of an urgent quest, answers to burning questions, at the time, namely, when they were created. If we were to go back to the origins of these ideas, they would lose that dead appearance of cut-and-dried facts and instead take on fresh and vibrant life again [48, p. v].

The focus of the course is on mathematical ideas – their origin and evolution. But the ideas are presented in the context of the mathematics, without which the ideas lack substance. The historical context provides the motivation often lacking in the schools. It also provides an opportunity to do some new mathematics, to fill gaps in the students' mathematical knowledge.

13.2 What Is Mathematics?

To come back to the ideas. The biggest idea of all is undoubtedly the nature of mathematics. I put it to my students as a question: "What is mathematics?" This is the $64,000 question, which of course I do not intend to answer, because I do not know the answer. But raising the question is important. Teachers taking this course have been studying, doing, and teaching mathematics for many years, but have probably reflected little on what the subject is. I do not suggest that this question should be constantly on their minds, but they should have thought about it at least once in their mathematical careers. To paraphrase the conclusion to the preface of Halmos' *Naive Set Theory*, I give my students the following advice in connection with the question "What is mathematics?": Think about it, try to assimilate it, but don't worry too much about it. (Halmos' statement is "Read it, absorb it, and forget it" [24, p. vi].)

The question "What is mathematics?" is a question in the philosophy of mathematics. But it cannot be addressed without an understanding of the subject's history. In general, I heartily endorse Lakatos' remark (paraphrasing Immanuel Kant) that

> The history of mathematics, lacking the guidance of philosophy, [is] blind, while the philosophy of mathematics, turning its back on the most intriguing phenomena in the history of mathematics, [is] empty [32, p. 2].

I begin to address the question "What is mathematics?" by setting down various definitions and descriptions of the subject given over the years. Here are some:

> Mathematics is the study of number and form (Anon).
>
> It is not of the essence of mathematics to be conversant with the ideas of number and quantity (Boole [8, p. 12]).
>
> Mathematics is an art (Anon).
>
> Mathematics is the 'Queen of the Sciences' (Gauss [3, p. 1]).
>
> The profound study of nature is the most fertile source of mathematical discoveries (Fourier [30, p. 1036]).
>
> It is true that Fourier had the opinion that the principal object of mathematics was public use and the explanation of natural phenomena; but a philosopher like him ought to know that the sole object of the science is the honor of the human spirit, and that under this view a problem of [the theory of] numbers is worth as much as a problem on the system of the world (Jacobi [30, p. 813]).
>
> Mathematics is the science which draws necessary conclusions (B. Peirce [37, p. 97]).
>
> The essence of mathematics lies in its freedom (Cantor [30, p. 1031]).
>
> Mathematics, in its widest signification, is the development of all types of formal, necessary, deductive reasoning (Whitehead [53, p. vi]).
>
> Logic merely sanctions the conquests of the intuition (Hadamard [30, p. 1026]).

You will note that I have arranged these quotations in more or less opposing pairs. This may at first seem confusing, nay paradoxical, to the students. But confusion and paradox should be seen not as impediments to learning but rather as opportunities for clarification (see Chap. 8). Although each of the quotations merits considerable discussion, at this point their role is to arouse the students' curiosity and stimulate their interest. The quotations also give them an indication of the subtlety and complexity of the question "What is mathematics?"

But one cannot deal with these quotations in a historical vacuum. So I next give the students a traditional, *very concise*, chronological account of the history of mathematics. The idea is to discuss with them some distinguishing features of various historical periods, and so to give them a brief panoramic view of selected main currents of mathematical thought through the ages – for example, pre-Greek mathematics, the mathematical aspects of the Greek "miracle," the mathematics of the Renaissance, and the advent of "modern" mathematics [6, 47]. We can then return to the quotations and discuss them more meaningfully.

So let us reconsider the initial pair. The first quotation of that pair, "Mathematics is the study of number and form," gives me the opportunity to discuss conjectured origins of mathematics, be they utilitarian or ritualistic [6, 28, 44]; to talk about the relation of number to form, for example, their coexistence in early Greek mathematics, severed by the supposed "crisis of incommensurability"; and to raise the question of whether students are familiar with mathematics *not* dealing with number or form. This question leads to the second quotation, "It is not of the essence of mathematics to be conversant with the ideas of number and quantity." Boole's "heretical" view of mathematics (espoused in 1847) was shared by only some in the nineteenth century. For example:

> Pure mathematics is the theory of forms (Grassmann [11, p. 65]).
>
> Mathematics is concerned only with the enumeration and comparison of relations (Gauss [4, p. 211]).

> [Mathematics is] purely intellectual, a pure theory of forms, which has for its objects not the combination of quantities or their images, the numbers, but things of thought to which there could correspond effective objects or relations, even though such a correspondence is not necessary (Hankel [30, p. 1031]).
>
> Mathematics is the science which draws necessary conclusions (B. Peirce [37, p. 97]).

Grassmann, Hankel, and Peirce – not to speak of Gauss – were leading mathematicians of the nineteenth century, but the perspective expressed in these quotations was then a minority view. To most mathematicians of that time the subject was firmly anchored in "real" entities [28, 30, 47].

Let us now consider the second pair of quotations given in our list of opposed pairs: "Mathematics is an art" and "Mathematics is the 'Queen of the Sciences'." This pair provides the opportunity to discuss aspects of mathematics shared by the sciences and the arts, and to suggest that mathematics possesses characteristics of both [1, 9, 41]. Assuming science is discovered and art created, the question arises: Is mathematics discovered or created (invented)? This brings us face to face with foundational issues: Platonism and formalism [12, 22, 25, 30].

Let me set aside the other three pairs of descriptions of mathematics. The important moral for the students to draw from these apparently contradictory pairs is that they are not mutually exclusive but complementary. Each gives new insights into mathematics; together they illustrate its many facets. But not only are these pairs of quotations not mutually exclusive, they are far from exhaustive. Here are several others, to bring home that point.

> The laws of Nature are written in the language of mathematics... the symbols are triangles, circles and other geometrical figures, without whose help it is impossible to comprehend a single word... (Galileo [30, pp. 328–329]).

Galileo had much to do with the supplanting (in the seventeenth century) of theology by mathematics as the queen of the sciences [28, 30].

> The science of mathematics presents the most brilliant example of how pure reason may successfully enlarge its domain without the aid of experience (Kant [27]).

An eighteenth-century view by one of the foremost philosophers of the Enlightenment [25, 30].

> No mathematician can be a complete mathematician unless he is also something of a poet (Weierstrass [5, p. 432]).

There was undoubtedly "poetry" in the mathematics of Weierstrass [28, 46].

> Mathematics is not the art of computation, but the art of minimal computation (Anon).

Certainly not the average person's view of mathematics!

> In mathematics... we find two tendencies present. On the one hand, the tendency toward *abstraction* seeks to crystallize the *logical* relations inherent in the maze of materials... being studied, and to correlate the material in a systematic and orderly manner. On the other hand, the tendency toward *intuitive understanding* fosters a more immediate grasp of the objects one studies, a live *rapport* with them, so to speak, which stresses the concrete meaning of their relations (Hilbert [26, p. iii]).

So much for Hilbert the formalist! [12, 25].

> The constructs of the mathematical mind are at the same time free and necessary. The individual mathematician feels free to define his notions and set up his axioms as he pleases. But the question is, will he get his fellow mathematicians interested in the constructs of his imagination. We cannot help the feeling that certain mathematical structures which have evolved through the combined efforts of the mathematical community bear the stamp of a necessity not affected by the accidents of their historical birth. Everybody who looks at the spectacle of modem algebra will be struck by this complementarity of freedom and necessity (Weyl [52, pp. 538–539]).

A perceptive statement on the nature of mathematics by one of the twentieth-century's greats [12, 25, 28, 30, 49].

Finally, here is a contemporary "definition" of mathematics:

> Mathematics is what mathematicians do (Anon).

Mathematicians give definitions of mathematics and discuss the nature of mathematics, but they also experiment, visualize, discover, compute, invent, conjecture, formulate, prove, model, apply, and classify. The preceding quotation embodies some of these recent thoughts on the nature of mathematics. The related philosophy of mathematics, given formal expression within the last several decades, is called "quasi-empiricism" [12, 25, 49].

13.3 Non-Euclidean Geometry

Having "settled" the big idea – the nature of mathematics – I can get on with discussing "lesser" matters, such as the evolution of a concept, result, or theory. There are many possibilities here, of course. Two "must" topics for teachers are non-Euclidean geometry and infinity. We begin with the former. It is a fascinating story spanning more than two millennia, and it has fundamental implications in mathematics, philosophy, physics, and pedagogy. Hilbert gives us a possible entrée with his statement [40, p. 240]:

> Every mathematical discipline goes through three periods of development: the *naive*, the *formal*, and the *critical*.

I tell my students that the evolution of a mathematical idea often proceeds in *four* stages: discovery (or invention), use, understanding, and justification. But regardless of whether there are two, three, or more levels, the point to stress is that when it comes to the evolution of mathematical ideas the big bang theory rarely applies.

The following are several quotations around which one can structure some of the major issues in the evolution of non-Euclidean geometry.

> You must not attempt this approach to parallels. I know this way to its very end. I have traversed this bottomless night, which extinguished all light and joy of my life. I entreat you, leave the science of parallels alone.... I thought I would sacrifice myself for the sake of the truth. I was ready to become a martyr who would remove the flaw from geometry and return it purified to mankind. I accomplished monstrous, enormous labors; my creations are

> far better than those of others and yet I have not achieved complete satisfaction. . . . I turned back when I saw that no man can reach the bottom of the night. I turned back unconsoled, pitying myself and all mankind.
> I admit that I expect little from the deviation of your lines. It seems to me that I have been in these regions; that I have travelled past all reefs of this infernal Dead Sea and have always come back with broken mast and torn sail.
> The ruin of my disposition and my fall date back to this time. I thoughtlessly risked my life and happiness [36, pp. 31–32].

A wonderfully evocative quotation, from Wolfgang Bolyai, a friend of Gauss, to his son Janos, one of the inventors of non-Euclidean geometry. Mathematical passions! (cf. Chap. 10.)

> The assumption that the sum of the three angles is less than 180° leads to a curious geometry, quite different from ours [the Euclidean], but thoroughly consistent, which I have developed to my entire satisfaction, so that I can solve every problem in it with the exception of the determination of a constant. . .
> The theorems of this geometry appear to be paradoxical and, to the uninitiated, absurd; but calm, steady reflection reveals that they contain nothing at all impossible. . . . All my efforts to discover a contradiction, an inconsistency, in this noneuclidean geometry have been without success. . . .
> I do not fear that any man who has shown that he possesses a thoughtful mathematical mind will misunderstand what has been said above, but in any case consider it a private communication of which no public use or use leading in any way to publicity is to be made. Perhaps I shall myself, if I have at some future time more leisure than in my present circumstances, make public my investigations [55, pp. 46–47].

Unmistakably Gauss, in an 1824 letter to Taurinus, who had also been working on the theory of parallels.

> Mathematical discoveries, like springtime violets in the woods, have their season which no human can hasten or retard [4, p. 263].

The season was upon them, and Wolfgang Bolyai admonished his son to publish his discoveries in non-Euclidean geometry lest others claim priority. This raises an interesting question: If mathematical discoveries have their season, what role does the individual play in the development of mathematics? For example, are the near misses of Galileo on the infinite and of Saccheri on non-Euclidean geometry due to the fact that the "season" for these two areas of mathematics had not yet arrived? [23, 28, 30, 54].

> If geometry were an experimental science, it would not be an exact science. It would be subjected to continual revision. . . . *The geometrical axioms are therefore neither synthetic a priori intuitions nor experimental facts.* They are conventions. Our choice among all possible conventions is guided by experimental facts; but it remains *free*, and is only limited by the necessity of avoiding every contradiction, and thus it is that postulates may remain rigorously true even when the experimental laws which have determined their adoption are only approximate. In other words, *the axioms of geometry* (I do not speak of those of arithmetic) *are only definitions in disguise.* What then are we to think of the question: Is Euclidean geometry true? It has no meaning. We might as well ask if the metric system is true and if the old weights and measures are false; if Cartesian coordinates are true and polar coordinates false. One geometry cannot be more true than another; it can only be more convenient [38, pp. 49–50].

This is Poincaré's pronouncement about the newly emergent view of the nature of geometry and its relation to the physical world.

To finish the geometry segment, here is a celebrated quotation which makes it starkly clear how axiomatics of geometry à la Euclid differs from axiomatics of geometry à la Hilbert:

> It must be possible to replace in all geometric statements the words point, line, plane by table, chair, mug (Hilbert [51, p. 14]).

Surely Euclid and his contemporaries would have found this view shocking!

See [23, 28, 30, 36, 46, 55] for various aspects of the material in this section.

13.4 The Infinite

> No other question has ever moved so profoundly the spirit of man; no other idea has so fruitfully stimulated his intellect; yet no other concept stands in greater need of clarification than that of the infinite (Hilbert [35, p. vii]).

We are still far from having clarified the concept of the infinite, a century after Hilbert's challenge.

Here are several more quotations which help focus the discussion.

> [The difficulties in the study of the infinite arise because] we attempt, with our finite minds, to discuss the infinite, assigning to it those properties which we give to the finite and limited; but this... is wrong, for we cannot speak of infinite quantities as being the one greater or less than or equal to another [20, p. 31].

Galileo, resigned, following an unsuccessful attempt to compare for size the positive integers and their squares.

> I see it, but I don't believe it [30, p. 997].

This is Cantor's expression of bewilderment, conveyed in a letter to Dedekind, following his proof that the real numbers (a one-dimensional domain) and the complex numbers (two-dimensional) have the same cardinality.

> Later generations will regard set theory as a disease from which one has recovered (Poincaré [30, p. 1003]).
>
> No one shall expel us from the paradise which Cantor has created for us (Hilbert [30, p. 1003]).

Who said there is no democracy in mathematics! Of course the idea of "democracy" in this subject is hard for students to accept, but it is a much more common phenomenon than might appear (see Chap. 10).

Although Cantor's set theory is standard fare, its implications for the students are far from standard. Here are some:

(a) The whole need not be greater than its parts [35].
(b) Infinity comes in different sizes [42].

Fig. 13.1 Georg Cantor (1845–1918)

(c) There are "arithmetics" in which the additive and multiplicative cancelation laws, the commutative laws of addition and multiplication, and one of the two distributive laws fail (cardinal and ordinal arithmetic) [24].
(d) One can have two equally consistent mathematical theories contradicting one another (Cantorian and non-Cantorian set theories) [10].
(e) "Simple" assumptions can have formidable consequences (the axiom of choice as the assumption and the Banach–Tarski paradox as a consequence) [50].

We conclude the discussion of the infinite with the following quotation by Weyl [51, p. 12]:

> Mathematics has been called the science of the infinite. Indeed, the mathematician invents finite constructions by which questions are decided that by their very nature refer to the infinite. This is his glory.

That is one of the paradoxes about mathematics which make the subject so alluring (see Chap. 8).

For details about the infinite see [24, 30, 35, 42, 46a] and Sect. 11.4.

13.5 The Twentieth Century: Foundational Issues

As a final topic for consideration, it is important to give high school teachers a sense of at least *some* twentieth-century developments in mathematics. Among other things, this will demonstrate that mathematics has not stopped growing and prospering. The quotations below provide entry points into a number of central ideas of the mathematics of the twentieth century, including foundational issues, Gödel's work, and the role of the computer. The first is Russell's provocative, perhaps facetious, description of mathematics.

> Mathematics is the subject in which we do not know what we are talking about nor whether what we are saying is true [30, p. 1196].

Russell's portrayal of the subject raises teachers' eyebrows.

> The great edifice of mathematics was shown to be like an enormous inverted pyramid delicately balanced upon the natural number system as a vertex [18, p. 132].

This quotation, from Eves' *Great Moments in Mathematics*, recalls the arithmetization of analysis in the late nineteenth century and points to a useful insight which students should be aware of – a latter-day pythagoreanism. (Recall the Pythagorean decree that "all is number" [30].) See also Sects. 4.6.3 and 11.9.1.

Before I state the next quotation, I need a definition of religion (attributable to the contemporary mathematician De Sua) [13, p. 305]:

> Religion is any discipline whose foundations rest on an element of faith, irrespective of any element of reason which may be present.

Now the quotation, also from De Sua [13, p. 305]:

> Mathematics is the only branch of theology possessing a rigorous demonstration of the fact that it should be so classified.

De Sua is referring here to Gödel's revolutionary work. An awareness of Gödel's ideas should be part of every student's mathematical culture (see [25, 42, 46]). Here is another way of saying what De Sua asserts:

> Gödel gave a formal demonstration of the inadequacy of formal demonstrations (Anon).

The next quotation is from Dieudonné [14, p. 19]:

> Now... the basic principle of modern mathematics is to achieve a complete *fusion* [of] 'geometric' and 'analytic' ideas.

The terms "geometric" and "analytic" are to be construed broadly, referring to both method and subject. For examples of that fusion note such areas of mathematics as algebraic geometry, algebraic topology, topological algebra, and diophantine geometry, as well as the use of metric notions in number theory (p-adic numbers), of topology in algebra (the Zariski topology), and of algebra in geometry (Klein's Erlangen Program). Some sense of this unity-in-diversity of mathematics can and should be conveyed to students (see [28, 30, 46] and Chap. 11). For example:

(a) To solve $x^2+y^2=z^2$ (nontrivially) in integers is to find the points with rational coordinates on the unit circle $u^2+v^2=1$ [46].
(b) To prove the nonconstructibility with straightedge and compass of the three Greek classical construction problems one must resort to abstract algebra [21].

(c) The only known proof of the fact that in a finite projective plane Desargues' theorem implies Pappus' theorem (both are theorems in geometry) involves showing that a finite division ring is a field [7].

No account of twentieth-century mathematics is adequate if it does not mention the computer. Here, then, is a quotation from Lynn Steen which fills the bill admirably [45, p. 34]:

> The intruder has changed the ecosystem of mathematics, profoundly and permanently.

Steen has in mind not so much the routine use of computers in mathematics – now commonplace – as the ways in which computers have affected the direction of mathematics: its problems, its methods, and its practitioners' conception of their subject. Two examples will suffice: computers have been used – indispensably – in the proofs of several long-outstanding conjectures, for instance the Four-Color conjecture and the Kepler conjecture (see Chaps. 7 and 10); and computers are central in "experimental mathematics" – a recently founded field [2].

13.6 Conclusion

The history of mathematics can be studied chronologically, thematically, topically, and biographically. I have used in this course elements of each approach. The quotations have played a pivotal function in all of them. It is perhaps not inappropriate to conclude this chapter with several more quotations – historical and pedagogical.

> The Divine intellect indeed knows infinitely more propositions [than we can ever know]. But with regard to those few which the human intellect does understand, I believe that its knowledge equals the Divine in objective certainty... (Galileo [19, p. 103]).
>
> I have had my results for a long time, but I do not yet know how to arrive at them (Gauss [32, p. 9]). See Sect. 10.10.
>
> If I only had the theorems! Then I should find the proofs easily enough (Riemann [32, p. 9]).
>
> The utmost abstractions are the true weapons with which to combat our thought of concrete fact (Whitehead [31, p. 466]).
>
> God exists since mathematics is consistent and the devil exists since we cannot prove the consistency (Weyl [30, p. 1206]). See [18].
>
> Education is that which remains when one has forgotten everything learned in school (Einstein [17, p. 63]).
>
> Being a language, mathematics may be used not only to inform but also, among other things, to seduce (Mandelbrot [34, p. 20]).
>
> To teach creatively is not to cover, but to uncover the syllabus (Bowden [43, back jacket]).

References

1. M. Atiyah, Mathematics: art and science, *Bull. Amer. Math. Soc.* 43 (2005) 87-88.
2. D. Bailey, J. Borwein et al, *Experimental Mathematics in Action*, A K Peters, 2007.

3. E. T. Bell, *Mathematics: Queen and Servant of Science*, G. Bell & Sons, 1952.
4. E. T. Bell, *The Development of Mathematics*, 2nd ed., McGraw-Hill, 1945.
5. E. T. Bell, *Men of Mathematics*, Simon and Schuster, 1937.
6. W. P. Berlinghoff and F. Q. Gouvea, *Math Through the Ages: A Gentle History for Teachers and Others*, expanded ed., Math. Assoc. of Amer., 2004.
7. L. Blumenthal, *A Modem View of Geometry*, W.H. Freeman & Co., 1961.
8. G. Boole, *An Investigation of the Laws of Thought*, Dover, 1951 (orig. 1854).
9. A. Borel, Mathematics: art and science, *Math. Intelligencer* 5:4 (1983) 9-17.
10. P. Cohen and R. Hersh, Non-Cantorian set theory, *Scientific Amer.* 217 (Dec. 1967) 104-116.
11. M. Crowe, *A History of Vector Analysis*, Univ. of Notre Dame Press, 1967.
12. P. J. Davis and R. Hersh, *The Mathematical Experience*, Birkhäuser, 1981.
13. F. De Sua, Consistency and completeness: a resumé, *Amer. Math. Monthly* 63 (1956) 295-305.
14. J. Dieudonné, Should we teach modem mathematics? *Amer. Scientist* 61 (Jan./Feb. 1973) 16-19.
15. W. Dunham, *Journey Through Genius: The Great Theorems of Mathematics*, Wiley, 1990.
16. C. H. Edwards, *The Historical Development of the Calculus*, Springer-Verlag, 1979.
17. A. Einstein, *Ideas and Opinions*, Crown Publ., 1954.
18. H. Eves, *Great Moments in Mathematics (After 1650)*, Math. Assoc. of Amer., 1981.
19. Galileo Galilei, *Dialogues Concerning the Two Chief World Systems,* tr. by S. Drake, 2nd ed., Univ. of California Press, 1970.
20. Galileo Galilei, *Two New Sciences*, tr. by H. Crew and A. de Salvio, Dover, 1954.
21. J. Gallian, *Contemporary Abstract Algebra*, 2nd ed., D.C. Heath & Co., 1990.
22. N. Goodman, Mathematics as an objective science, *Amer. Math. Monthly* 86 (1979) 540-551.
23. M. J. Greenberg, *Euclidean and Non-Euclidean Geometries: Development and History*, 2nd ed., W.H. Freeman & Co., 1980.
24. P. Halmos, *Naive Set Theory*, Springer-Verlag, 1960.
25. R. Hersh, *What is Mathematics, Really?*, Oxford Univ. Press, 1997.
26. D. Hilbert and S. Cohn-Vossen, *Geometry and the Imagination*, Chelsea, 1952.
27. I. Kant, *Critique of Pure Reason,* transl. by M. Müller, 2nd ed., Macmillan,1927.
28. V. J. Katz, *A History of Mathematics: An Introduction*, 3rd ed., Addison-Wesley, 2009.
29. I. Kleiner, Famous problems in mathematics: an outline of a course, *For the Learning of Mathematics* 6:1 (1986) 31-38.
30. M. Kline, *Mathematical Thought from Ancient to Modern Times*, Oxford Univ. Press, 1972.
31. M. Kline, *Mathematics in Western Culture*, Oxford Univ. Press, 1953.
32. I. Lakatos, *Proofs and Refutations*, Cambridge Univ. Press, 1976.
33. R. Laubenbacher and D. Pengelley, Great problems of mathematics: a course based on original sources, *Amer. Math. Monthly* 99 (1992) 313-317.
34. B. Mandelbrot, *Fractals: Form, Chance, and Dimension*, W.H. Freeman & Co., 1977.
35. E. Maor, *To Infinity and Beyond*, Birkhäuser, 1987.
36. H. Meschkowski, *Noneuclidean Geometry*, Academic Press, 1964.
37. B. Peirce, Linear associative algebras, *Amer. Jour. Math.* 4 (1881) 97-215.
38. H. Poincaré, *Science and Hypothesis*, Dover, 1952.
39. G. Polya, *Mathematical Methods in Science*, Math. Assoc. of Amer., 1977.
40. R. Remmert, *Theory of Complex Functions*, Springer-Verlag, 1989.
41. G. – C. Rota, The phenomenology of mathematical beauty, *Synthese* 111 (1997) 171-182.
42. R. Rucker, *Infinity and the Mind*, Birkhäuser, 1982.
43. M. M. Schiffer and L. Bowden, *The Role of Mathematics in Science*, Math. Assoc. of Amer., 1984.
44. A. Seidenberg, The origin of mathematics, *Arch. Hist. Exact Sci.* 18 (1978) 301-342.
45. L. Steen, Living with a new mathematical species, *Math. Intelligencer* 8:2 (1986) 33-40.
46. J. Stillwell, *Mathematics and Its History*, Springer-Verlag, 1989.
46a. J. Stillwell, *Roads to Infinity: The Mathematics of Truth and Proof*, A K Peters, 2010.
47. D. Struik, *A Concise History of Mathematics*, 4th rev. ed., Dover, 1987.
48. O. Toeplitz, *The Calculus: A Genetic Approach*, Univ. of Chicago Press, 1963.

49. T. Tymoczko, *New Directions in the Philosophy of Mathematics*, Birkhäuser, 1986.
50. S. Wagon, *The Banach-Tarski Paradox*, Cambridge Univ. Press, 1985.
51. H. Weyl, Axiomatic versus constructive procedures in mathematics, *Math. Intelligencer* 7:4 (1985) 10-17.
52. H. Weyl, A half-century of mathematics, *Amer. Math. Monthly* 58 (1951) 523-553.
53. A. N. Whitehead, *A Treatise on Universal Algebra*, Hafiier, 1960.
54. R. Wilder, *Evolution of Mathematical Concepts: An Elementary Study*, John Wiley & Sons, 1968.
55. H. Wolfe, *Introduction to Non-Euclidean Geometry*, Holt, Rinehart & Winston, 1945.

Chapter 14
Famous Problems in Mathematics

14.1 Introduction

"Famous Problems in Mathematics" is the title of a one-semester course at the third-year level offered in the department of mathematics at my University. The course has a significant historical component, but it is not a course in the history of mathematics. The historical perspective is, however, essential. One of the objectives of the course is to make students aware that mathematics *has* a history, and that it may be interesting, useful, and important to bring history to bear on the study of mathematics.

The course tries to legitimize in the eyes of students that it makes sense to *talk* about mathematics in addition to *doing* mathematics, and that it makes sense to deal with ideas in mathematics in addition to dealing with "mathematical technology." In brief, the course attempts to make students "mathematically civilized" [109, p. 603]. (Some technical details about the course are given at the end of the chapter.)

14.2 The Themes

Before dealing with the "famous problems," let me list some themes which I try to pursue in the course, with brief indications of intent.

14.2.1 The Origin of Concepts, Results, and Theories

A major theme of the course is that "concrete" problems often give rise to "abstract" concepts and theories. In fact, Dieudonné has argued that "the history

I. Kleiner, *Excursions in the History of Mathematics*,
DOI 10.1007/978-0-8176-8268-2_14,

of mathematics shows that a theory almost always originates in efforts to solve a specific problem" [29, Introduction]. Our problems 1, 2, 3, 6, and 7 illustrate this point. For a discussion see [7, pp. 229–230;39,40,69,79,119,134].

14.2.2 The Roles of Intuition vs. Logic

Students often see only the logical side of the mathematical enterprise. But in the view of Hadamard, "logic merely sanctions the conquests of the intuition" [73, p. 1026]. History often bears him out (see [69] and Chaps. 4, 8–10). On the other hand, there were times in the evolution of mathematics when logical rather than intuitive thinking was the creative force. The discovery/invention of non-Euclidean geometry and of set theory are prime examples. For the working mathematician, there is an ongoing interplay between intuition and logic. See [5, 8, 19, Vol. 2; 25,68,131,132].

14.2.3 Changing Standards of Rigor

The concepts of "proof" and "rigor" change with time. Moreover, the change is not necessarily from the less to the more rigorous – there are fluctuations in standards of rigor. An entire issue of the *Two Year College Mathematics Journal* (v. 12, no. 2, 1981) is devoted to the question of what a proof is. See also [25,27,49,50,77,85, 99,130], and Chaps. 7–10.

What we have likely witnessed during the last decades of the twentieth century – both pedagogically and professionally – is a reaction against the strict rigor and abstraction which have dominated mathematics for much of that century. Rigor is of course essential in mathematics, but "it ought to suit the occasion" [110, p. ix].

14.2.4 The Roles of the Individual vs. the Environment

A sociological theory concerning the development of mathematics can be summarized succinctly and poetically by the following statement of W. Bolyai (the father of one of the creators of non-Euclidean geometry): "Mathematical discoveries, like springtime violets in the woods, have their season which no human can hasten or retard" [7, p. 263]. Against this note Cantor's decree that "mathematics is entirely free in its development... The essence of mathematics lies in its freedom" [73, p. 1031]. We explore this "complementarity of freedom and necessity" [126, p. 539]. See [19, v. 1; 89,108,128,129].

The human drama inherent in the creation of mathematics is often ignored when we teach. Even if there is a certain inevitability in mathematical creations, they are made by people – people with personalities, passions, and prejudices, which sometimes have a bearing on the mathematics they create. Cantor is a case in point. (For an analysis of the significance of Cantor's personality on the creation of his transfinite set theory see [21].) The intent, then, is to pay attention to the creators as well as the creations of mathematics. See [6,47,92,95], and Chap. 10.

14.2.5 Mathematics and the Physical World

The relationship between mathematics and the physical world is longstanding. It has enriched our understanding of both. Moreover, our view of this relationship has changed over time (especially in the nineteenth century). Witness the following words of Whitehead: "The paradox is now fully established that the utmost abstractions are the true weapons with which to control our thought of concrete fact" [75, p. 466]. For elaboration see [16, 19, Vol. 3, 57, 75, 104, 117, 127] and Chap. 5.

14.2.6 The Relativity of Mathematics

Mathematical truths are not absolute – they are context-dependent. For example, the statement "If $a + b = a + c$ then $b = c$" is true in the domain of, say, real or complex numbers but false in the domain of transfinite numbers. Again, the equation $x^2 + 1 = 0$ has no solutions in the domain of real numbers, two solutions in the domain of complex numbers, and infinitely many solutions in the domain of quaternions. See [70] and Chap. 11.

14.2.7 Mathematics: Discovery or Invention?

This question arises more or less naturally in connection with various mathematical developments in the nineteenth century which are dealt with in the course. Moreover, one need not opt for one characterization or the other. Davis and Hersh suggest that the typical working mathematician is a Platonist on weekdays and a formalist on weekends – thus viewing mathematics as *both* a discovery *and* an invention [25, p. 321]. See also [5,71,108,116].

The above themes are of major importance in the history and philosophy of mathematics, and one cannot treat them exhaustively in a one-semester course. They are however central to the course. Moreover, they are not dealt with as separate topics, but are discussed in the course of dealing with the various problems. So much for the underlying themes. Now to the "famous problems."

14.3 The Problems

The content of the course is flexible and one can choose the problems more or less as one pleases, keeping in mind the level and objectives of the course. Here are some of my choices. They are dictated by personal taste, by the level of the course, by the fact that the subject matter of the problems is usually not dealt with in the standard courses, and by the relevance of the problems to the themes which I am trying to expound.

Herein I have described nine problems – some in detail, others very sketchily. The problems are independent of each other (although some reinforce one another) and can be dealt with in any order.

14.3.1 Problem 1: Diophantine Equations

These are equations in two or more variables with integer coefficients in which the solutions sought are integers. Diophantine equations are fundamental in number theory.

I begin this topic with the equations $x^2 + y^2 = z^2$ and $x^2 + 2 = y^3$. The first goes back to Euclid, c. 300 BC (and in one form or another to the Babylonians, c. 1600 BC), whose work apparently inspired Fermat's conjecture about $x^n + y^n = z^n$ (see Chap. 2). The second equation is a special case of another famous Diophantine equation, $x^2+k = y^3$ – the so-called Bachet equation, studied by Fermat and others (Chap. 2). I proceed to solve these two equations "formally" – and analogously – as follows:

(a) Factor the left-hand side of the equation $x^2 + y^2 = z^2$, which gives $(x + yi)(x - yi) = z^2$. This is now an equation in so-called "Gaussian integers" G, "numbers" of the form $a + bi$, where a and b are ordinary integers. Now the set Z of (ordinary) integers has the property that if $ab = c^2$ $(a, b, c \in Z)$, and a and b are relatively prime, then $a = u^2$ and $b = v^2$ for some u, $v \in Z$ (the result also holds with squares replaced by cubes and higher powers). Applying this result in the set G, it follows from $(x + yi)(x - yi) = z^2$ that each of $x + yi$ and $x - yi$ is a square in G. In particular, $x + yi = (a + bi)^2$ for some a, $b \in Z$. This gives $x+yi = (a^2-b^2)+2abi$, so that $x = a^2-b^2$, $y = 2ab$. Since $z^2 = x^2+y^2$, $z = a^2+b^2$. Conversely, we easily verify that $x = a^2-b^2$, $y = 2ab$, $z = a^2+b^2$ is a solution of $x^2 + y^2 = z^2$ for every a, $b \in Z$. Thus, we have found all solutions of $x^2 + y^2 = z^2 : x = a^2 - b^2, y = 2ab, z = a^2 + b^2$, where a and b are arbitrary integers.

(b) Now to the equation $x^2 + 2 = y^3$. Employing the same idea as in (a), we factor its left-hand side, which yields $(x + \sqrt{2}i)(x - \sqrt{2}i) = y^3$. This is an equation in the domain $D = \{a + b\sqrt{2}i : a, b \in Z\}$ of "quadratic integers." Since the product of the two elements $x + \sqrt{2}i$ and $x - \sqrt{2}i$ of D is a cube, it follows that each is a cube. In particular, $x + \sqrt{2}i = (a + b\sqrt{2}i)^3$ for some a, $b \in Z$.

Thus $x + \sqrt{2}i = (a^3 - 6ab^2) + (3a^2b - 2b^3)\sqrt{2}i$, and equating real and imaginary parts, $x = a^3 - 6ab^2$ and $1 = (3a^2b - 2b^3) = b(3a^2 - 2b^2)$. Since $a, b \in Z$, we must have $b = \pm 1$ and $3a^2 - 2b^2 = \pm 1$. Substituting $b = \pm 1$ into the last equation, and performing some elementary algebraic manipulation, we get $x = a^3 - 6ab^2 = \pm 1 \mp 6 = \pm 5$. Since $x^2 + 2 = y^3$, $y^3 = 27$ and $y = 3$. We note that $x = 5$, $y = 3$ and $x = -5$, $y = 3$ are indeed solutions of $x^2 + 2 = y^3$, hence they are its *only* solutions.

14.3.1.1 Examination of the Above "Solutions"

Once the two equations in (a) and (b) have been "solved," we examine the solutions carefully in order to justify the various steps. The basic questions that must be answered are as follows: What are the properties of the (ordinary) integers that carry over to the other two types of "integers" (G and D), and how can this be done? To answer these questions we introduce the concept of a *unique factorization domain* and develop enough machinery relevant to such domains to close the logical gaps in the formal solutions. For details see [3, 11, 44, 69, 97, 115].

The above procedure is the reverse of what is done in standard courses, in which we would first define a unique factorization domain and then (perhaps) give an application to the solution of Diophantine equations. However, I have found the above approach to be a good way of motivating the introduction of the concept of unique factorization domain. It is my experience that students are much more receptive to digesting abstract concepts when their introduction is motivated by concrete problems.

If an instructor wants to spend more time on this topic, the following interesting Diophantine equations pursue similar themes: (c) $n = x^2 + y^2$, (d) $n = x^2 + y^2 + z^2 + w^2$, (e) $x^3 + y^3 = z^3$, and more generally, (f) $x^n + y^n = z^n$. To elaborate:

(c) This is Fermat's problem of determining which integers can be represented as sums of two squares (see Chap. 2). Since $n = x^2 + y^2 = (x + yi)(x - yi)$, one applies some of the results developed above for Gaussian integers to resolve this problem rather quickly. For details see [3, 11, 44, 62, pp.112–113].

(d) Here we want to show, following Lagrange, that every positive integer is a sum of four squares. We proceed as follows: $n = x^2 + y^2 + z^2 + w^2 = (x + yi + zj + wk)$, where $x + yi + zj + wk$ and $x - yi - zj - wk$ are conjugate quaternions. The problem can be solved in a manner analogous to that in (c) above. We require some knowledge of quaternions (see Sect. 14.3.4). For details see [11, pp. 127–133; 62, pp. 329–335]. An alternative, short, and interesting algebraic proof of Lagrange's theorem conceptually related to (c) is given in [117].

(e) To show that $x^3 + y^3 = z^3$ has no nontrivial integer solutions, we factor $x^3 + y^3$ into $(x + y)(x + wy)(x + w^2y)$, where $w = 1/2(-1 + \sqrt{3}i)$ (a primitive cube root of 1), and use the fact that $B = \{a + bw : a, b \in Z\}$ is a unique factorization domain. For details see [3, 30, 55, 59].

(f) If we consider Fermat's equation $x^p + y^p = z^p$ for an arbitrary prime p (it suffices to consider $x^n + y^n = z^n$ for n prime), we can "prove" in a similar manner to the case $p = 3$ that the equation has no nontrivial integral solutions (!). We have to assume that the domain of cyclotomic integers $D_p = \{a_0 + a_1 w + a_2 w^2 + \ldots + a_{p-2} w^{p-2} : a_i \in Z,\, w$ a primitive p-th root of $1\}$ is a unique factorization domain. (See [3, p. 103] or [13, pp. 160–163] for details of such a "proof.") It is precisely this assumption which Lamé made in 1843 when he announced that he had proved Fermat's Last Theorem. He was of course unaware that D_p is not a unique factorization domain for every p. See [3, 34, 35], and Chap. 3.

It is such equations as the above, especially (f), which have been instrumental in the rise of a new branch of mathematics – *algebraic number theory* – and in particular in the introduction of such concepts as *unique factorization domain, ring, field, and ideal.* They provide a very good illustration of our theme that concrete problems often give rise to abstract concepts and theories (Sect. 14.2.1). See [3, 11, 34, 35, 55, 69, 100, 106].

14.3.2 *Problem 2: Distribution of Primes Among the Integers*

The study of prime numbers has fascinated and challenged some of the greatest mathematicians of all time, from Greek antiquity to the present. The purpose of this problem is to give students a sense of that fascination and that challenge. It is also to show that important questions about natural numbers cannot be settled by restricting one's attention to the natural numbers – an idea already encountered in Problem 1. The basic difficulty is that the integers have too little structure. Thus in Problem 1 we enlarged the domain of (ordinary) integers to that of "algebraic integers" so as to be able to employ algebraic methods, and in this problem we extend the domain of integers to that of real or complex numbers in order to be able to use analytic tools.

In the mid-eighteenth century Euler stated that

> Mathematicians have tried in vain to this day to discover some order in the sequence of prime numbers and we have reason to believe that it is a mystery into which the human mind will never penetrate [37, p. 241].

Some of the facts which attest to the lack of "order," the "mystery," are as follows:

(a) Numerical evidence suggests that there are infinitely many primes that are as close together as possible – the so-called "twin primes:" 11, 13; 107, 109; and 10006427, 10006429 are three such pairs.
(b) At the same time, there are arbitrarily large sequences of consecutive composite integers: for example, $10^6!+2, 10^6!+3, \ldots, 10^6!+10^6$ is a sequence of 999,999 consecutive composite integers ($n!$ denotes n factorial).
(c) Yet, the next prime after any given prime cannot be "too far removed" from it: there exists a prime between n and $2n$ for any integer n. This is the so-called Bertrand Postulate.

Euler's apparent pessimism did not prove to be entirely justified. For although we find no regularity in the distribution of the primes when considered *individually*, Gauss found regularity in their distribution when considered *collectively*. Thus Gauss tried to describe not "how" but "how often" the primes occur in the integers. We have in mind the Prime Number Theorem which Gauss (and independently Legendre) conjectured but was unable to prove, namely that $\pi(x)$ is asymptotic to $x/\log x$, where $\pi(x)$ is the number of primes $\leq x$; that is, $\lim_{x\to\infty} \pi(x)/(x/\log x) = 1$. Davis and Hersh said of this Theorem that "it is one of the finest examples of the extraction of order from chaos in the whole of mathematics" [25, p. 210].

Attempts to prove the Prime Number Theorem stimulated the development of the branch of analysis called complex function theory, and this in turn led Hadamard and de la Vallée Poussin (independently) to a proof of the theorem (in 1896). The starting point for the use of analysis in number theory, which eventually led to a new branch of the latter subject – *analytic number theory* – was Euler's work.

In 1737 Euler showed that $\sum_1^\infty 1/n^s = \Pi(1-p^{-s})^{-1}$, where s is any real number >1, and p ranges over the primes. Euler was a master of formal manipulation of series. Inspired by Leibniz's result that $1-1/3+1/5-1/7+\ldots = \pi/4$, he proved in 1736 that $1+1/2^2+1/3^2+\ldots = \pi^2/6$, and soon thereafter that $1+1/2^{2n}+1/3^{2n}+\ldots = \pi^{2n}q$, q a rational number, n any positive integer. This apparently led him to the series $\sum_1^\infty 1/n^s$ and to the discovery of the identity $\sum_1^\infty 1/n^s = \Pi(1-p^{-s})^{-1}$.

Euler noted that this identity gives a new proof that there are infinitely many primes (assume the contrary and take limits of both sides as $s \to 1$), and that it can be used to show that the series $\sum 1/p$ diverges (take the log of both sides). Moreover, an elementary argument, based on a similar idea, proves (which Euler did) that there are infinitely many primes in the two arithmetic progressions $\{4n+1\}$ and $\{4n+3\}$ (cf. (b) below). See [4, 48, 59].

In 1859 Riemann attempted to prove the Prime Number Theorem by introducing the *zeta function* $\zeta(s) = \sum_1^\infty 1/n^s$, where now s was a *complex* variable with real part > 1, and noted that Euler's identity extends to this complex domain. This led him to the celebrated – and still undecided – *Riemann Hypothesis* concerning the roots of $\zeta(s)$. In this course we discuss some of the known results about the Riemann Hypothesis, and the relationship of the Hypothesis to the Prime Number Theorem. See [4, 59].

Another aspect of the problem of the distribution of primes has to do with prime-producing formulas. Since, as the above evidence suggests, it is very unlikely that a formula can be found which will produce *all the primes and only the primes*, what about formulas which produce a subset of the primes (and possibly also composites)? For example:

(a) $f(n) = 2n^2+29$ is prime for $n \leq 28$; $f(n) = n^2+n+41$ is prime for $n \leq 39$; and $f(n) = n^2 - 79n + 1{,}601$ is prime for $n \leq 79$. These formulas illustrate the failure of "scientific induction" in mathematics.

The second formula is an instance of the formula $f_q(n) = n^2 + n + q$, which takes on primes for all $n \leq q - 2$ if and only if $q = 3, 5, 11, 17$, and 41 (a result due to Euler). These are precisely the values of q for which the domain A_d of "integers" is a unique factorization domain, where A_d equals $\{a + b\sqrt{d} : a, b \in Z\}$ if $d \equiv 1 \pmod 4$, and $\{(a + b\sqrt{d})/2 : a, b \in Z, a$ and b both even or both odd$\}$ if $d \equiv 3 \pmod 4$, where $d = 1-4q$ is the "discriminant" of $f_q(n)$. See [31,43] for details.

(b) As we noted, Euler showed that $f(n) = 4n + 1$ and $f(n) = 4n + 3$ yield infinitely many primes as n ranges over the positive integers. In 1837 Dirichlet effected a grand generalization of this result by showing, using fairly deep analytic tools, that $f(n) = an + b$ yields an infinite number of primes for *any* relatively prime positive integers a and b, $n = 1, 2, 3, \ldots$. See [4].

(c) It is not known whether $f(n) = 2^{2^n} + 1$ and $f(n) = 2^n - 1$ produce infinitely many primes as n ranges over the positive integers. The former are the famous Fermat numbers, the latter the equally famous Mersenne numbers. The Fermat numbers are related to the question of the construction of regular polygons with straightedge and compass, the Mersenne numbers to the determination of even perfect numbers. See [98].

(d) W.H. Mills showed in 1947 that there exists a real number a such that $[a^{3^n}]$ is a prime for every integer n, where for any real number x, $[x]$ denotes the greatest integer $\leq x$. It was subsequently shown that there are infinitely many such a's (in fact, their cardinality is that of the continuum), although not a single value for a is known. See [98].

(e) It is known that no polynomial with integer coefficients will produce only primes. It was a remarkable achievement when Matijasevich produced (in 1970) a polynomial which assumes all the primes and only primes for its *positive* values. The polynomial, of degree 37 in 24 variables, was found as the result of deep insights into Hilbert's Tenth Problem. See [64] for details.

Despite the many interesting and powerful results on the distribution of primes obtained since Euler's statement on the topic (see the top of this section), it is quite fitting to conclude with the following quote from Weyl, who more than 200 years after Euler echoed the latter's sentiments [126, p. 532]:

> The notion of prime number is of course as old and as primitive as that of the multiplication of natural numbers. Hence it is most surprising to find that the distribution of primes among all natural numbers is of such a highly irregular and almost mysterious character.

For references on various aspects of Problem 2 see [4,31,46,48,56,93,101,102,137].

14.3.2.1 Remarks on Problems 1 and 2

In addition to providing illustrations of some of the themes mentioned at the beginning of this chapter (in particular, themes (a), (b), and (e)), the study of number theory as exemplified in the first two problems sheds light on the following:

(a) "Simplicity" in mathematics is complex: there is an abundance of "simple" questions to which there are as yet no answers.
(b) To study problems formulated in a given system (in this case, the integers), it is often helpful to enlarge the system – a recurrent theme in mathematics.
(c) The computer can be a useful device in the study of various branches of mathematics.

Moreover, number theory (I find) is a good topic with which to begin a course such as outlined here. It is intrinsically interesting to students, and it lends itself, perhaps more than many other topics, to student participation. This sets a good tone (one hopes) for the remainder of the course.

14.3.3 Problem 3: Polynomial Equations

The Babylonians knew how to solve quadratic equations, essentially by the method of completing the square, about 4000 years ago. Little progress was achieved in the theory of algebraic solution of equations for the next 3,500 years, until the sixteenth-century Italian school of algebra made a fundamental breakthrough by giving an algebraic solution of the cubic equation, and soon thereafter the quartic equation. We focus on this breakthrough, which is intimately related to the discovery of complex numbers.

Students think that it was the *quadratic* equation (in particular $x^2+1=0$) which led to the introduction of complex numbers. This is not the case. It was the *cubic* which gave rise to them. The "why" and "how" of this interesting story are explored. The subsequent evolution of the complex numbers is briefly dealt with. The complex numbers are an interesting case study of the genesis, evolution, and acceptance of a mathematical system. Their story is related in some detail in Chap. 12. See also [17, 21, 23, 73, 83, 114].

Some indication is given of the theory of polynomial equations beyond the quartic; in particular, how the permutations of the roots of a polynomial equation aid in its solution – an important source of the rise of *group theory*. See [2, v. 1; 3, 69, 73, 119].

This problem illustrates themes (a), (d), (e), and (g).

14.3.4 Problem 4: Are There Numbers Beyond the Complex Numbers?

The answer depends on what we mean by "numbers." We explore the historical evolution of the various number systems – from the natural numbers through the complex numbers – indicating gains and losses at each stage of the evolutionary process. Following this we introduce the *quaternions* and the *octonions* (Cayley numbers), indicating how these led to the study of *noncommutative algebra*. For details see [33, 66, 69, 73, 78, 87, 114, 122], and Chap. 11.

This problem illustrates themes (a), (b), (d), (e), (f), and (g).

14.3.5 Problem 5: Why Is (−1)(−1)=1?

This is an instance of the problem of rigorous justification of the laws of operation with negative numbers. It became a pressing problem – for both pedagogical and professional reasons – at Cambridge University around 1830. In fact, the very existence of negative numbers came into question. Peacock and others set themselves the task of resolving this problem by codifying the laws of operation with numbers. This was perhaps the earliest instance of *axiomatics in algebra*. The seeds of "abstract algebra" that emerge here are as follows:

(i) The manipulation of symbols for their own sake – the so-called "symbolical algebra;" interpretation comes later.
(ii) Some freedom to choose the laws obeyed by the symbols.

We discuss some of these issues, focussing on the following:

(a) Reasons why the problem of the negative numbers became a burning issue at the time.
(b) Some proposed solutions, especially Peacock's, embodied in his "principle of permanence of equivalent forms."
(c) Reactions to the symbolical approach to algebra.
(d) Implications of the symbolical approach for subsequent developments in algebra (cf. the works of De Morgan, Hamilton, Boole, Cayley).

Pointing out some of the limitations in Peacock's development, we next take a more modern, Hilbertian approach to the problem of negative numbers. Just as Hilbert "defined" (characterized) the real numbers axiomatically as a complete ordered field, so we characterize the integers as an ordered integral domain in which the positive elements are well ordered. Once this is done we can prove such laws as $(-1)(-1) = 1$, and more generally, $(-a)(-b) = ab$), $a \times 0 = 0$, and others.

The following are some issues which we discuss in this context:

1. How can we prove a law such as $(-1)(-1) = 1$? This question leads to the concept of *axioms*. We cannot prove everything.
2. What axioms do we set down in order to *characterize* the integers? This question enables us to introduce the concepts of ring, integral domain, ordered structure.
3. How do we know when we have enough axioms? This question permits us to introduce the concept of *completeness* of a set of axioms (to be elaborated in Problem 9).

Fig. 14.1 George Peacock (1791–1858)

4. What does it mean to characterize the integers? This question sets the stage for the introduction of the concept of *isomorphism*. We have characterized the integers by means of a set of axioms when any two systems satisfying these axioms are isomorphic. Thus, for example, the axioms for an ordered integral domain do not characterize the integers since the rationals are also an ordered integral domain, and the integers and rationals are not isomorphic, as can readily be shown.
5. Having characterized the integers, do we now have perhaps too many axioms? In fact the commutativity of addition can be derived from the other axioms for an integral domain. Here we come face to face with the concept of *independence* of a set of axioms (see Problem 9).
6. Are we at liberty to pick and choose axioms as we please? This leads us to the concept of *consistency* of a set of axioms (again, to be elaborated in Problem 9) and, more broadly, to the question of "freedom of choice" in mathematics.

For details on symbolical algebra see [15, 69, 103–105, 112]; for the Hilbertian approach see [10, 33, 88].

This problem illustrates themes (a), (b), (c), and (g).

14.3.5.1 Remark on Problems 3, 4, and 5

These three problems come from algebra and give an indication of the transition from "classical" algebra – the study of polynomial equations and laws of operation with numbers, to "modern" algebra – the study of axiomatic systems. In fact, I often begin teaching a course in abstract algebra with Problem 5. Moreover, a "nonstandard" course in abstract algebra, in which "concrete" problems motivate the introduction of abstract concepts, can be structured around Problems 5, 1, 3, and 4. See [69].

14.3.6 Problem 6: Euclid's Parallel Postulate

This problem gave rise to the creation of *non-Euclidean geometry*, the reevaluation of the *foundations of Euclidean geometry*, and the study of *axiomatics*. It is an excellent problem for raising many interesting issues (e.g., "what is mathematics?") and, in particular, addressing *all* the themes (a) to (g) in Sect. 14.2. For details see [5, 12, 39, 41, 54, 67, 90, 91, 133].

14.3.7 Problem 7: Uniqueness of Representation of a Function in a Fourier Series

The study of *Fourier series* had a great impact on subsequent developments in mathematics (see Sect. 14.4 (e) below). The problem of "unique representation" was addressed by Cantor. This led him to the creation of *set theory* and the clarification of the concept of the (actual) infinite. For the origin of Cantor's set theory in the study of Fourier series see [24, 51].

In this problem we are not concerned so much with Cantor's technical achievements in set theory as with his conceptual breakthrough in coming to grips with the actual infinite, and the consequences of this for mathematics and beyond. On the technical side, we study some *cardinal arithmetic*, and touch on *algebraic and transcendental numbers*. (Recall Cantor's proof that there is a continuum of transcendental numbers.) For details see [20, 39, 96, 107, 114, 124, 138].

This is an excellent topic for illustrating themes (a), (b), (d), (f), and (g).

14.3.8 Problem 8: Paradoxes in Set Theory

Various approaches to resolving Russell's paradox concerning the set $S = \{x : x \notin x\}$ led in the early twentieth century to different *axiomatizations of set theory*. For example, Russell's theory of types forbids asking if $S \epsilon S$; the Zermelo–Fraenkel theory forbids the formation of S; the Von Neumann-Gödel-Bernays theory classifies S as a class but not as a set. Among other reasons, these axiomatizations led to various philosophies of mathematics: logicism, formalism, and intuitionism. For details see [5, 8, 19, Vols. 1 & 2; 25, 39, 61, 90, 107, 133].

The problem helps illustrate themes (a), (b), (c), (d), and (f).

14.3.9 Problem 9: Consistency, Completeness, Independence

Here we study the *continuum hypothesis* and especially *Gödel's theorems* – one of the greatest mathematical and intellectual achievements of the twentieth century – and their impact on mathematics and beyond. For details see [39, 61, 63, 90, 94, 107, 113, 133].

These matters illustrate themes (b), (c), (f), and (g).

14.3.9.1 Remark on Problems 6, 7, 8, and 9

In addition to illustrating the various themes as indicated, these problems relate to questions in the philosophy of mathematics, and especially to the fundamental question about the nature of mathematics. See Chap. 13.

14.4 Other Problems

Here are nine more problems, technically somewhat more demanding, which may be considered in such a course.

(a) The Königsberg Bridge Problem; classification of regular polyhedra; the Four-Colour Theorem. These problems helped motivate the development of graph theory and topology. See [9, 20, 30, 84, 121, 136].
(b) Measurement: length, area, and volume. These motivated the development of the integral. See [45, 60, 81].
(c) "Exotic" functions; space-filling curves. Such examples motivated the rigorization and arithmetization of analysis. See [73, 84, 124], and Chaps. 4, 5, and 8.
(d) Isoperimetric problems; other maxima and minima problems. These motivated the creation of the calculus of variations. See [30, 110, 120].
(e) Aspects of Fourier series – led to a reevaluation of a number of fundamental concepts of analysis such as function, integral, and convergence. See [14, 52, 80, 84, 123].
(f) The logarithms of negative and complex numbers. This problem demonstrates the early use of complex numbers and of analysis by some of the seventeenth- and eighteenth-century masters of the subject. See [14, 18, 82, 83, 86].
(g) The Vibrating-String problem, the Heat-Conduction problem, and their relation to the evolution of the function concept. See [14, 45, 52, 84], and Chap. 5.
(h) The arithmetization of analysis. See [14, 39, 51].
(i) The Erlangen Program – influential in the clarification of the nature of geometry and in the rise of the group concept. See [53, 54, 73, 118].

14.5 General Remarks on the Course

(1) The technical elements of the course are not very demanding. Many students, however, find the intellectual aspects challenging. To deal with ideas in mathematics, to be asked to read independently in the mathematical literature, and to write mini-essays are tasks which mathematics students are not – but should become – accustomed to.
(2) No textbook is used. However, *many* references are given and students are expected to *read* some of them! (see the extensive list of *References* below).
(3) The prerequisites for the course are any two mathematics courses. Students with only this minimum prerequisite are asked to take concurrently at least one or two more mathematics courses. One is looking for the elusive quality of "mathematical maturity" rather than for specific technical proficiency.
(4) In a one-semester course one can deal adequately with only some of the above nine problems.

References

1. A. A. Albert (ed.), *Studies in Modern Algebra*, Math. Assoc. of Amer., 1963.
2. A. D. Aleksandrov et al., *Mathematics: Its Content, Methods, and Meaning*, 3 vols., M.I.T. Press, 1963.
3. R. B. J. T. Allenby, *Rings, Fields and Groups*, Edward Arnold Publ., 1983.
4. T. M. Apostol, *Introduction to Analytic Number Theory*, Springer, 1976.
5. S. F. Barker, *Philosophy of Mathematics*, Prentice-Hall, 1964.
6. E. T. Bell, *Men of Mathematics*, Simon & Schuster, 1965.
7. E. T. Bell, *The Development of Mathematics*, 2nd ed., McGraw-Hill, 1945.
8. P. Benacerraf & H. Putnam, *Philosophy of Mathematics: Selected Readings*, Prentice-Hall, 1964 (reprinted by Cambridge Univ. Press, 1983).
9. N. L Biggs, E. K. Lloyd, and R. J. Wilson, *Graph Theory: 1736–1936*, Oxford Univ. Press, 1986.
10. G. Birkhoff and S. Mac Lane, *A Survey of Modern Algebra*, 4th ed., Macmillan, 1977 (orig. 1941).
11. E. D. Bolker, *Elementary Number Theory: An Algebraic Approach*, W.A. Benjamin, 1970.
12. R. Bonola, *Non-Euclidean Geometry*, Dover, 1955 (orig. 1911).
13. Z. I. Borevich and I. R. Shafarevich, *Number Theory*, Academic Press, 1966.
14. U. Bottazzini, *The Higher Calculus: A History of Real and Complex Analysis from Euler to Weierstrass*, Springer, 1986.
15. C. B. Boyer, *A History of Mathematics*, 2nd ed., revised by U. Merzbach, Wiley, 1989.
16. F. E. Browder, Does pure mathematics have a relation to the sciences? *Amer. Scientist* 64 (Sept./Oct.1976) 542–549.
17. D. M. Burton, *The History of Mathematics: An Introduction*, 6th ed., McGraw-Hill, 2007.
18. F. Cajori, History of exponential and logarithmic concepts, *Amer. Math. Monthly* 20 (1913) 5–14, 35–47, 75–84, 107–117, 148–151, 173–182, and 205–210.
19. D. M. Campbell and J. C. Higgins (eds.), *Mathematics: People, Problems, Results,* 3 Vols., Wadsworth, 1984.
20. R. Courant and H. Robbins, *What is Mathematics?* Oxford Univ. Press, 1941.
21. J. N. Crossley, *The Emergence of Number*, World Scientific,1987.

22. J. N. Crossley et al., *What is Mathematical Logic?* Oxford Univ. Press, 1972.
23. T. Dantzig, *Number: The Language of Science,* 4th ed., Free Press, 1967 (orig. 1930).
24. J. W. Dauben, *Georg Cantor: His Mathematics and Philosophy of the Infinite*, Harvard Univ. Press, 1979
25. P. J. Davis and R. Hersh, *The Mathematical Experience,* Birkhäuser, 1981.
26. H. De Long, *A Profile of Mathematical Logic,* Addison Wesley, 1970.
27. R. A. De Millo et al, Social processes and proofs of theorems and programs, *Math. Intelligencer* 3 (1980) 31–40.
28. L. E. Dickson, Fermat's Last Theorem and the origin and nature of the theory of algebraic numbers, *Ann. Math.* 18 (1916–17) 161–187.
29. J. Dieudonné, *A Panorama of Pure Mathematics,* Academic Press, 1982.
30. H. Dörrie, *100 Great Problems of Elementary Mathematics: Their History and Solution*, Dover, 1965.
31. U. Dudley, Formulas for primes, *Math. Mag.* 56 (1983) 17–22.
32. W. Dunham, *Journey Through Genius: The Great Theorems of Mathematics*, Wiley, 1990.
33. H.-D. Ebbinghaus et al, *Numbers*, Springer, 1990.
34. H. M. Edwards, Fermat's Last Theorem, *Scientific Amer.* 239 (Oct. 1978) 104–122.
35. H. M. Edwards, *Fermat's Last Theorem: A Genetic Introduction to Algebraic Number Theory,* Springer, 1977.
36. H. M. Edwards, *Riemann's Zeta Function,* Academic Press, 1974
37. L. Euler, Discovery of a most extraordinary law of numbers concerning the sum of their divisors, *Opera Omnia*, Ser. 1, Vol. 2, pp. 241–253. (E175 in the Eneström index.)
38. H. Eves. *An Introduction to the History of Mathematics*, 6th ed., Saunders College Publ., 1990 (orig. 1953).
39. H. Eves, *Great Moments in Mathematics (after 1650)*, Math. Assoc. of Amer., 1981.
40. H. Eves, *Great Moments in Mathematics (before 1650)*, Math. Assoc. of Amer., 1980.
41. H. Eves and C. Newsom, *An Introduction to the Foundations and Fundamental Concepts of Mathematics,* Holt, Rinehart & Winston, 1958.
42. S. Feferman, What does logic have to tell us about mathematical proofs? *Math. Intelligencer* 2 (1979) 20–24.
43. D. Fendel, Prime producing polynomials and principal ideal domains, *Math. Mag.* 58 (1985) 204–210.
44. H. Flanders, A tale of two squares – and two rings, *Math. Mag.* 58 (1985) 3–11.
45. A. Gardiner, *Infinite Processes: Background to Analysis*, Springer, 1982.
46. M. Gardner, Patterns in primes are a clue to the strong law of small numbers, *Scientific Amer.* 243 (Dec. 1980) 18–28.
47. C. C. Gillispie (ed.), *Dictionary of Scientific Biography*, 16 vols., Scribner's, 1970–1980.
48. J. R. Goldman, *The Queen of Mathematics: A Historically Motivated Guide to Number Theory*, A K Peters, 1998.
49. J. Grabiner, Changing attitudes toward mathematical rigor: Lagrange and analysis in the eighteenth and nineteenth centuries. In *Epistemological and Social Problems of the Sciences in the Early 19th Century*, H. Jahnke & M. Otte (eds.), D. Reidel, 1981, pp. 311–330.
50. J. Grabiner, Is mathematical truth time-dependent? *Amer. Math. Monthly* 81 (1974) 354–365.
51. I. Grattan-Guinness (ed.), *From the Calculus to Set Theory, 1630- 1910: An Introductory History,* Princeton Univ. Press, 2000.
52. I. Grattan-Guinness, *The Development of the Foundations of Mathematical Analysis from Euler to Riemann,* MIT Press, 1970.
53. J. Gray, *Worlds out of Nothing: A Course in the History of Geometry in the 19th Century*, Springer, 2007.
54. M. J. Greenberg, *Euclidean and Non-Euclidean Geometries: Development and History*, 2nd ed., W. H. Freeman & Co., 1980.
55. E. Grosswald, *Topics from the Theory of Numbers*, 2nd ed., Birkhäuser, 1984.
56. R. K. Guy, Conway's prime producing machine, *Math. Mag.* 56 (1983) 2–33.

57. R. W. Hamming, The unreasonable effectiveness of mathematics, *Amer. Math. Monthly* 87 (1980) 81–90.
58. G. Hanna, *Rigorous Proof in Mathematics Education*, Ontario Institute for Studies in Educ. Press, 1983.
59. G. H. Hardy & E. M. Wright, *An Introduction to the Theory of Numbers*, 4th ed., Oxford Univ. Press, 1960 (orig. 1938).
60. T. Hawkins, *Lebesgue's Theory of Integration: Its Origins and Development*, Chelsea, 1970.
61. R. Hersh, *What is Mathematics, Really*?, Oxford Univ. Press, 1997.
62. I. N. Herstein, *Topics in Algebra*, Blaisdell Publ. Co., 1964.
63. D. R. Hofstadter, Analogies and metaphors to explain Gödel's Theorem, *Two Yr. Coll. Math. Jour.* 13 (1982) 98–114.
64. J. P. Jones et al., Diophantine representation of the set of primes, *Amer. Math. Monthly* 83 (1976) 449–464.
65. M. Kac & S. M. Ulam, *Mathematics and Logic: Retrospect and Prospects*, The New Amer. Library, 1969.
66. I. L. Kantor and A. S. Solodovnikov, *Hypercomplex Numbers: An Elementary Introduction to Algebras*, Springer-Verlag, 1989.
67. E. Kasner & J. R. Newman, *Mathematics and the Imagination*, Simon & Schuster, 1967 (orig. 1940).
68. J. G. Kemeny, Rigor vs. intuition in mathematics, *Math. Teacher* 54 (1961) 66–74.
69. I. Kleiner, *A History of Abstract Algebra,* Birkhäuser, 2007.
70. I. Kleiner & S. Avital, The relativity of mathematics, *Math. Teacher* 77 (1984) 554–558, 562.
71. M. Kline, *Mathematics: The Loss of Certainty*, Oxford Univ. Press, 1980.
72. M. Kline (ed.), *Mathematics: An introduction to its Spirit and Use*, W.H. Freeman & Co., 1979.
73. M. Kline, *Mathematical Thought from Ancient to Modern Times*, Oxford Univ. Press, 1972.
74. M. Kline (ed.), *Mathematics in the Modern World*, W. H. Freeman & Co., 1968.
75. M. Kline, *Mathematics in Western Culture*, Oxford Univ. Press, 1953.
76. G. Kolata, Does Gödel's Theorem matter for mathematics? *Science* 218 (Nov. 1982) 779–780.
77. G. Kolata, Mathematical proofs: the genesis of reasonable doubt, *Science* 192 (June 1976) 989–990.
78. A.G. Kurosh, *Lectures on General Algebra*, Chelsea Publ. Co., 1963.
79. I. Lakatos, *Proofs and Refutations: The Logic of Mathematical Discovery*, Cambridge Univ. Press, 1976.
80. R. E. Langer, Fourier series: the genesis and evolution of a theory, *Amer. Math. Monthly* 54 Supplement (1947) 1–86.
81. R. Laugenbacher and D. Pengelley, *Mathematical Expeditions: Chronicles by the Explorers*, Springer, 1999.
82. D. Laugwitz, Controversies about numbers and functions. In *The Growth of Mathematical Knowledge*, E. Grosholz and H. Berger (eds), Kluwer, 2000, pp. 177–198.
83. Leapfrogs, *Imaginary Logarithms*, Leapfrogs, 1978.
84. J. R. Manheim, *The Genesis of Point Set Topology*, Pergamon Press, 1964.
85. Yu. I. Manin, How convincing is a proof? *Math. Intelligencer* 2 (1979) 17–18.
86. P. Marchi, The controversy between Leibniz and Bernoulli on the nature of the logarithms of negative numbers. In *Akten das II Inter. Leibniz-Kongress,* Bnd II, 1974, pp. 67–75.
87. K. O. May, The Impossibility of a division algebra of vectors in three dimensional space, *Amer. Math. Monthly* 73 (1966) 289–291.
88. N. H. McCoy, *Introduction to Modern Algebra,* Allyn & Bacon, 1968.
89. R. K. Merton, Singletons and multiples in scientific discovery: a chapter in the sociology of science, *Proc. Amer. Philos. Soc.* 105 (1961) 470–486.
90. H. Meschkowski, *Evolution of Mathematical Thought,* Holden-Day, 1965.
91. H. Meschkowski, *Noneuclidean Geometry,* Academic Press, 1964.
92. H. Meschkowski, *Ways of Thought of Great Mathematicians,* Holden-Day, 1964.

93. H. L. Montgomery, Zeta zeros on the critical line, *Amer. Math. Monthly* 86 (1979) 43–45.
94. E. Nagel and J. R. Newman, *Gödel's Proof*, New York Univ. Press, 1958.
95. J. R. Newman (ed.), *The World of Mathematics*, 4 vols., Simon &. Schuster, 1956.
96. I. Niven, *Numbers: Rational and Irrational,* Random House, 1961.
97. C. S. Ogilvy and. J. T. Anderson, *Excursions in Number Theory*, Oxford Univ. Press, 1966.
98. O. Ore, *Number Theory and Its History*, McGraw-Hill, 1948.
99. J. Pierpont, Mathematical rigor, past and present, *Bull. Amer. Math. Soc.* 34 (1928) 23–53.
100. H. Pollard and. H. G. Diamond, *The Theory of Algebraic Numbers*, 2nd ed., Math. Assoc. of Amer., 1975.
101. C. Pomerance, The search for primes, *Scientific Amer.* 247 (Dec. 1982) 136–147.
102. C. Pomerance, Recent developments in primality testing, *Math. Intelligencer* 3 (1981) 97–105.
103. H. Pycior, Augustus De Morgan's algebraic work: the three stages, *Isis* 74 (1983) 211–226.
104. H. Pycior, Historical roots of confusion among beginning algebra students: a newly discovered manuscript, *Math. Mag.* 55 (1982) 150–156.
105. H. Pycior, George Peacock and the British origins of symbolical algebra, *Hist. Math.* 8 (1981) 23–45.
106. F. Richman, *Number Theory: An Introduction to Algebra*, Brooks/Cole Publ. Co., 1971.
107. R. Rucker, *Infinity and the Mind: The Science and Philosophy of the Infinite,* Birkhäuser, 1982.
108. W. L. Schaaf (ed.), *Our Mathematical Heritage,* Macmillan, 1963
109. O. Shisha, Mathematically civilized, *Notices Amer. Math. Soc.* 30 (1983) 603.
110. G. F. Simmons, *Differential Equations, with Applications and Historical Notes,* McGraw-Hill, 1972.
111. C. Small, A simple proof of the Four-Square theorem, *Amer. Math. Monthly* 89 (1982) 59–61.
112. G. C. Smith, De Morgan and the laws of Algebra, *Centaurus* 25 (1981) 50–70.
113. R. M. Smullyan, *What is the Name of this Book? The Riddle of Dracula and Other Logical Puzzles,* Prentice-Hall, 1978.
114. E. Sondheimer and A. Rogerson, *Numbers and Infinity: A Historical Account of Mathematical Concepts*, Cambridge Univ. Press, 1981.
115. H. Stark, *An Introduction to Number Theory*, Markham Publ. Co., 1970.
116. L. A. Steen (ed.), *Mathematics Today: Twelve Informal Essays,* Springer, 1978.
117. I. Stewart, The science of significant form, *Math. Intelligencer* 3 (1981) 50–58.
118. J. Stillwell, *The Four Pillars of Geometry,* Springer, 2005.
119. H. Tietze, *Famous Problems of Mathematics,* Graylock Press, 1965.
120. V. M. Tikhomirov, *Stories About Maxima and Minima*, Amer. Math. Society, 1990.
121. R. J. Trudeau, *Introduction to Graph Theory*, Dover, 1993.
122. B. L. Van der Waerden, Hamilton's discovery of quaternions, *Math. Mag.* 49 (1976) 227–234.
123. E. B. Van Vleck, The influence of Fourier Series upon the development of mathematics, *Science* 39 (1914) 113–124.
124. N. Ya. Vilenkin, *Stories About Sets,* Academic Press, 1968. (Translated by A. Shenitzer.)
125. A. Weil, *Number Theory: An approach Through History,* Birkhäuser, 1984.
126. H. Weyl, A half-century of mathematics, *Amer. Math. Monthly* 58 (1951) 523–553.
127. E. P. Wigner, The unreasonable effectiveness of mathematics in the natural sciences, *Comm. Pure &. Appl. Math.* 13 (1960) 1–14.
128. R. L. Wilder, *Mathematics As a Cultural System,* Pergamon Press, 1981.
129. R. L. Wilder, *Evolution of Mathematical Concepts: An Elementary Study,* Wiley, 1968.
130. R. L. Wilder, Relativity of standards of mathematical rigor, *Dict. of the Hist. of Ideas* 3 (1968) 170–177.
131. R. L. Wilder, The role of intuition, *Science* 156 (1967) 605–610.
132. R. L. Wilder, The role of the axiomatic method, *Amer. Math. Monthly* 74 (1967) 115–127.
133. R. L. Wilder, *Introduction to the Foundations of Mathematics,* Wiley, 1965.

134. R. L. Wilder, The origin and growth of mathematical concepts, *Bull. Amer. Math. Soc.* 59 (1953) 423–448.
135. R. L. Wilder, The nature of mathematical proof, *Amer. Math. Monthly* 51 (1944) 309–323.
136. R. Wilson, *Four Colours Suffice: How the Map Problem was Solved*, Penguin, 2002.
137. D. Zagier, The first 50 million prime numbers, *Math. Intelligencer* 0 (1977) 7–19.
138. L. Zippin, *Uses of Infinity*, Random House, 1962.

Part E
Brief Biographies of Selected Mathematicians

Chapter 15
The Biographies

15.1 Richard Dedekind (1831–1916)

15.1.1 Introduction

The nineteenth century was a golden age in mathematics. Entirely new subjects emerged – for example, abstract algebra, non-Euclidean geometry, set theory, complex analysis; and old ones were radically transformed – for example, real analysis, number theory. Just as important, the spirit of mathematics, the way of thinking about it and doing it, changed fundamentally, even if gradually.

Mathematicians turned more and more for the genesis of their ideas from the sensory and empirical to the intellectual and abstract. Witness the introduction of noncommutative algebras, non-Euclidean geometries, continuous nowhere differentiable functions, space-filling curves, n-dimensional spaces, and completed infinities of different sizes. Cantor's dictum that "the essence of mathematics lies in its freedom" became a reality – though one to which many mathematicians took strong exception.

Other pivotal changes were the emphasis on *rigorous* proof and the acceptance of nonconstructive existence proofs, the focus on concepts rather than on formulas and algorithms, the stress on generality and abstraction, the resurrection of the axiomatic method, and the use of set-theoretic modes of thinking. Dedekind was an exemplary practitioner of many of these new approaches; in fact, he initiated several of them – as we shall see. See [14].

15.1.2 Life

He was born in Brunswick, Germany (also the birth place of Gauss). His father was a lawyer and a professor at the Collegium Carolinum (an educational institution

I. Kleiner, *Excursions in the History of Mathematics*,
DOI 10.1007/978-0-8176-8268-2_15,

Fig. 15.1 Richard Dedekind (1831–1916)

between a high school and a university), and his mother the daughter of a professor at the same college. The youngest of four children, he never married, living for many years with his sister until her death in 1914.

Between the ages of seven and sixteen Dedekind attended the local gymnasium, studying physics and chemistry. However, he found these subjects unsatisfactory since they lacked logical structure! In 1848, at sixteen, he entered the Collegium Carolinum (which Gauss had earlier attended). There he mastered the elements of analytic geometry, calculus, algebra, and mechanics. He was thus well prepared when he entered the University of Göttingen two years later.

He got his doctorate under Gauss in 1852 (at the age of twenty-one) on the topic of Eulerian integrals. Gauss noted about the dissertation that "the author evinces not only a very good knowledge of the relevant field, but also such an independence as augurs favorably for his future achievement" [2, p. 517].

Riemann came to Göttingen in 1851 to pursue doctoral studies with Gauss, and Dirichlet came in 1855 to succeed Gauss upon his death. Dedekind formed lasting friendships with both Riemann and Dirichlet, and was influenced both by their mathematics and by their approach to the subject, which focused on getting at the underlying concepts of a theory rather than the computations. Dirichlet in particular made a "new man" out of him, he said [3, p. 2].

In 1858 Dedekind was appointed professor at the prestigious Zürich Polytechnic (now the ETH). He was recommended for this position by Dirichlet, who, in addition to praising his mathematical abilities, called him "an exceptional pedagogue." He stayed at the Zürich Polytechnic four years, and in 1862 became professor at the Brunswick Polytechnic in his home town, where he spent the last fifty years of his life.

We focus on three of Dedekind's important contributions: the founding of algebraic number theory (1871), the definition of the real numbers in terms of what are now known as Dedekind cuts (1872), and the definition of the natural numbers in terms of sets (1888).

15.1.3 Algebraic Numbers

Algebraic number theory is the study of number theory using the tools of abstract algebra. Pioneering work in the subject was done (in the 1840s) by Kummer, who showed that in the domain of cyclotomic integers every "ideal number" is a unique product of (ideal) primes. Inspired by Kummer's work, Dedekind extended it very significantly by showing that in the ring of integers of an algebraic number field every ideal is a unique product of prime ideals. The concepts "field," "ring," and "ideal" (among others) were introduced by him. Here are his definitions of field and ideal, which were important technical and *methodological* achievements.

A system [set] F of real or complex numbers is called a *field* if the sum, difference, product, and quotient of any two numbers of F belong to F.

A subset I of the integers R of an algebraic number field K is an *ideal* of R if it has the following two properties: (a) If $a, b \in$ I, then $a \pm b \in$ I. (b) If $a \in$ I, c $\in$ R, then $ac \in$ I.

Dedekind introduced here two fundamental innovations:

1. Use of the axiomatic method in algebra, although in the concrete setting of the complex numbers (all that was needed for his algebraic number theory). This approach influenced Hilbert and especially Noether, and became a staple of twentieth-century mathematics.
2. Use of set-theoretic language and of the completed infinite. (Observe that Dedekind's fields and ideals are infinite sets.) This predated Cantor's work on sets later in the 1870s.

As we mentioned, the conceptual focus adopted by Dedekind was promoted earlier by his two colleagues and friends, Dirichlet and Riemann. Speaking of Dirichlet's work, and noting the famous "Dirichlet Principle" in analysis, Minkowski referred to "the other principle of Dirichlet" as the view that mathematical problems should be solved through a minimum of blind calculation and a maximum of forethought [5, p. 68].

Dedekind's work founded a new subject – algebraic number theory. It also embodied a breakthrough in the evolution of abstract algebra. His approach and methods were revolutionary. Bourbaki called the work "magisterial," Landau said it "brought order to chaos, and light to the deepest darkness," and Noether noted that "its style of thought now [the 1920s] permeates the entirety of modern algebra" [12, p. 763]. Despite the high praise from such distinguished quarters, Dedekind's ideal theory/algebraic number theory did not get a positive reception until the 1890s. Most nineteenth-century mathematicians were not prepared for its modern spirit. See [15, Chap. 3] for details.

15.1.4 Real Numbers

The real numbers were viewed throughout history variously as magnitudes, ratios of magnitudes, quantities, infinite decimals, or points on a line. None of these definitions was rigorous, nor were the properties of the real numbers explicitly formulated. This became a pressing issue for Dedekind already in 1858, when he began to teach calculus upon his appointment to the Zürich Polytechnic. The following quotation reveals the prevailing state of affairs and Dedekind's thinking on the matter [8, pp. 1–2]:

> As professor in the Polytechnic school in Zürich I found myself for the first time obliged to lecture upon the elements of the differential calculus and felt more keenly than ever before the lack of a really scientific formulation for arithmetic. In discussing the notion of the approach of a variable magnitude to a fixed limiting value, and especially in proving the theorem that every magnitude which grows continually, but not beyond all limits, must certainly approach a limiting value, I had recourse to geometric evidence. Even now such resort to geometric intuition in a first presentation of the differential calculus, I regard as exceedingly useful from the didactic standpoint .... But that this form of introduction into the differential calculus can make no claim to being scientific, no one will deny. For myself this feeling of dissatisfaction was so overpowering that I made the fixed resolve to keep meditating on the question till I should find a purely arithmetic and perfectly rigorous foundation for the principles of infinitesimal analysis.

Find it he did. To provide some context: Cauchy gave a rigorous presentation of the calculus based on the concept of limit in a seminal work begun in 1821. But he left unresolved a number of foundational issues. Since the real numbers are in the foreground or background of much of analysis, and were viewed *geometrically* by Cauchy and his contemporaries, these mathematicians resorted to intuitive geometric arguments in order to establish a number of the fundamental results of analysis, for example the Intermediate Value Theorem. This Dedekind found unacceptable. (So did several other mathematicians around 1872, in particular Cantor, Weierstrass, and Heine. Each gave a rigorous but different presentation of the real numbers.)

Dedekind's definition of the reals was in terms of "cuts," as is well known. Briefly, a cut (now called "Dedekind cut") is a partition (A, B) of the rationals into two sets A and B such that every element of A is less than each element of B. The real numbers are defined to be the totality of all such cuts. Note that a cut is a pair of infinite sets. In fact, the entire development of *Continuity and Irrational Numbers* (Dedekind's monograph on this topic [8]) is in the language of sets. He used the final section of the work "to explain the connection between the preceding investigations and certain fundamental theorems of infinitesimal analysis" [8, p. 24].

15.1.5 Natural Numbers

Dedekind published his definition of the natural numbers in his pamphlet of 1882, *Was sind und was sollen die Zahlen* (*What are Numbers and What Should They Be*, mistranslated in [8] as *The Nature and Meaning of Numbers*). The essence of his approach was to reduce the properties of the integers to those of sets and mappings (a presentiment of the logicist school). Thus the first twenty or so pages of the tract deal exclusively with the latter topics. The first sentence reads: "In what follows I understand by a *thing* [an element] every object of our thought" [8, p. 44]. Dedekind went on to define sets: "It very frequently happens that different things, $a, b, c, \ldots$ for some reason can be considered from a common point of view, can be associated in the mind, and we say that they form a *system* [set] S; we call the things $a, b, c, \ldots$ *elements* of the system S" [8, p. 45]. He also defined equality of sets, subsets, unions, intersections, mappings of sets, and composition of maps – all in the modern spirit in which they are presented today. Among the theorems he proved are the following:

1. There exist infinite sets (!). (Dedekind defined an infinite set as one which has a proper subset of the same cardinality (our language) as the set itself.)
2. Every infinite set contains a copy of the natural numbers.
3. Up to isomorphism, the set of natural numbers is unique.

See [8] for details.

The monograph did not win universal praise from contemporary mathematicians. Even Dedekind anticipated potential misgivings about the unorthodox, abstract nature of the presentation (see [8, p. 33]). It was the most explicit of his works in its use of set-theoretic notions – rare in nineteenth-century mathematics but central in the twentieth century. It inspired Peano in his axiomatic definition (in 1889) of the natural numbers and Zermelo in his search (in the 1900s) for an axiom system for sets [12, pp. 787–790; 13].

15.1.6 Other Work

Here we mention briefly another three of Dedekind's contributions – to algebraic geometry, lattices, and the zeta function.

1. *Algebraic geometry* Dedekind collaborated with Weber on editing Riemann's collected works. This was likely the inspiration for their groundbreaking joint paper of 1882, "Theory of algebraic functions of a single variable," in which they put part of Riemann's work on abelian functions, which depended on the unproved Dirichlet Principle, into rigorous algebraic language. The fundamental idea of their approach was to carry over to algebraic function fields the ideas which Dedekind had earlier introduced for algebraic number fields, thus pointing to the strong analogy between algebraic number theory and algebraic geometry. This analogy would prove extremely fruitful for both theories [12, 15, Chaps. 3 and 4; 16, pp. 157–162].
2. *Lattices* In two papers, in 1897 and 1900, Dedekind introduced the notion of a lattice. The motivation came from number theory, in particular properties possessed by various operations on ideals and modules (sums, products). The definition of a lattice was axiomatic [1, p. 130]:

> If two operations on two arbitrary elements A, B of a (finite or infinite) system [set] G generate two elements $A \pm B$ of the same system G that satisfy the conditions (1–3) [below], then, regardless of the nature of these elements, G is called a dual group [lattice] with respect to the operations $\pm$: (1) $A+B = B+A$, $A-B = B-A$; (2) $(A+B)+C = A+(B+C)$, $(A-B)-C = A-(B-C)$; (3) $A+(A-B) = A$, $A-(A+B) = A$.

He derived various results from these identities, including the idempotent laws $A + A = A$ and $A - A = A$. His work on lattices inspired Ore and Birkhoff when they founded lattice theory as an independent subject in the 1930s [5].

3. *The zeta function*: The zeta function $\zeta(s)$ and its extensions and generalizations have been most important tools in analytic number theory since Riemann introduced $\zeta(s)$ in 1859. Among Dedekind's significant contributions to mathematics was his generalization (in 1879) of Riemann's zeta function to algebraic number fields. He defined the zeta function of such a field K to be $\zeta_K(s) = \sum 1/N(I)^s$, where s is a real number greater than 1, the summation is over all the ideals I of the ring of integers of K, and N(I) denotes the norm of I. Dedekind found a formula giving the number of classes h(K) of K, the so-called "class number" of K (it is finite for all K) [9, p. 208]. Just as Riemann's zeta function turned out to be important in the study of integer primes, so Dedekind's was instrumental in the study of the primes in the ring of integers of an algebraic number field.

15.1.7 Conclusion

It is time to conclude our account of Dedekind. In his Supplements to Dirichlet's *Zahlentheorie* he founded algebraic number theory and brought about a turning point in the evolution of abstract algebra, and in his works on the real and natural numbers he tamed the continuous by reducing it to the discrete (the arithmetization of analysis). But beyond the fundamental concepts that he introduced and the important results that he proved, were the methods that he inaugurated. He was guided by philosophical principles in introducing many of his important innovations. "He does seem to be a great and true philosopher of the subject – a genuine philosopher, of and in mathematics," notes Stein [17, p. 249]. One of his philosophical principles was a focus on intrinsic, conceptual properties over formulas, calculations, or concrete representations. Another was the acceptance of nonconstructive definitions and proofs as legitimate mathematical methods – an attitude rare at the time.

His two very significant methodological innovations were the use of the axiomatic method outside of geometry and the institution of set-theoretic modes of thinking. The axiomatic method was just beginning to surface after 2,000 years of dormancy. Dedekind was instrumental in pointing to its mathematical power and pedagogical value. His use of set-theoretic formulations, including that of the completed infinite – taboo at the time – preceded by about ten years Cantor's seminal work on the subject. Edwards refers to his Supplement X (1871) as "the 'birthplace' of the modern set-theoretic approach to the foundations of mathematics" [11, p. 9].

Not everyone was pleased with Dedekind's way of doing mathematics. Even among his mathematical soul-mates there was discomfort. When Weber wrote to Frobenius in 1893 about the forthcoming publication of his *Lehrbuch der Algebra*, the latter responded as follows [5, p. 128]:

> Your announcement of a work on algebra makes me very happy. ... Hopefully you will follow Dedekind's way, yet avoid the highly abstract approach that he so eagerly pursues now. ... It is indeed unnecessary to push abstraction so far. I am therefore satisfied that you write the *Algebra* and not our venerable friend and master, who had also once considered that plan.

But of course Dedekind's "highly abstract approach" became commonplace in the twentieth century. Among the early converts were Hilbert, Steinitz, and Emmy Noether. The latter, who edited Dedekind's works, used to say modestly that all she had done could already be found in his researches. Dedekind himself was modest and retiring and did not seek honors. But they came his way. He was elected to the Göttingen, Berlin, Rome, and Paris Academies, and received numerous other scientific honors on the occasion of the 50th anniversary of his doctorate.

The mathematician and historian of mathematics Harold Edwards, who was a great admirer of Kronecker's approach to mathematics, which was antithetical to Dedekind's, paid him a singular honor [11, p. 20]:

> Dedekind's legacy ... consisted not only of important theorems, examples, and concepts, but of a whole *style* of doing mathematics that has been an inspiration to each succeeding generation.

References

1. I. G. Bashmakova and A. N. Rudakov, Algebra and algebraic number theory. In *Mathematics of the* 19th *Century*, ed. by A. N. Kolmogorov and A. P. Yushkevich, Birkhäuser, 2001, pp. 35–135. (Translated from the Russian by A. Shenitzer, H. Grant, and O. B. Sheinin.)
2. E. T. Bell, *Men of Mathematics*, Simon and Schuster, 1937.
3. K. S. Biermann, Dedekind, Richard. In *Dictionary of Scientific Biography*, ed. by C. C. Gillispie, Charles Scribner's Sons, Vol. 4, 1981, pp. 1–5.
4. U. Bottazzini, *The Higher Calculus: A History of Real and Complex Analysis from Euler to Weierstrass*, Springer-Verlag, 1986.
5. L. Corry, *Modern Algebra and the Rise of Mathematical Structures*, Birkhäuser, 1996.
6. R. Dedekind, *Theory of Algebraic Integers*, translated and introduced by John Stillwell, Cambridge University Press, 1996.
7. R. Dedekind, Letter to Keferstein. In *From Frege to Gödel: A Source Book in Mathematical Logic*, 1879–1931, Harvard Univ. Press, 1977, pp. 98–103.
8. R. Dedekkind, Essays on the Theory of Numbers, Dover, 1963. (The book consists of the two essays: *Continuity and Irrational Numbers* and *The Nature and Meaning of Numbers*.)
9. J. Dieudonné (ed.), Abrégé d'histoire des mathtématiques, 1700–1900, Vol. 1, Hermann, 1978.
10. H. M. Edwards, Mathematical ideas, ideals, and ideology, *Mathematical Intelligencer* 14:2 (1992) 6–19.
11. H. M. Edwards, Dedekind's invention of ideals, *Bull. London Math. Soc.* 15 (1983) 8–17.
12. W. Ewald (ed.), *From Kant to Hilbert: A Source Book in the Foundations of Mathematics*, Vol II, Oxford University Press, 1996, pp. 753–837.

13. J. Ferreiros, Traditional logic and early history of sets, *Arch. for Hist. of the Exact Sci.* 50 (1996/97) 5–71.
14. J. Gray, *Plato's Ghost: The Modernist Transformation of Mathematics*, Princeton Univ. Press, 2008.
15. I. Kleiner, *A History of Abstract Algebra*, Birkhäuser, 2007.
16. D. Laugwitz, *Bernhard Riemann, 1826–1866*, Birkhäuser, 1999. (Translated from the German by A. Shenitzer.)
17. H. Stein, Logos, logic and *Logistiké*. In *History and Philosophy of Modern Mahematics*, ed. by W. Aspray and P. Kitcher, Univ. Minnesota Press, 1988.

15.2 Leonhard Euler (1707–1783)

15.2.1 Introduction

Euler was the most productive mathematician ever, and one of the greatest. Seventy-six volumes of his collected works have been published to date as well as three volumes of his correspondence and several of his books. He made seminal contributions to all of the then-existing areas of mathematics as well as to mechanics, dynamics, optics, and astronomy. There was, in fact, little distinction in the eighteenth century between mathematics, especially analysis, and its fields of applications. And the very men who fashioned the infinitesimal concepts and methods (the Bernoullis, Euler, Lagrange, d'Alembert, and others) also formulated and derived the laws governing the motions of fluids and of rigid bodies, the bending of beams, and the vibrations of elastic bodies.

Not only did Euler contribute to all extant fields of mathematics, he pointed to the creation of new ones: complex functions, elliptic functions, algebraic number theory, analytic number theory, partition theory, graph theory, calculus of variations, differential equations (ordinary and partial), differential geometry, and topology.

Calculation/computation is an important experimental tool of the mathematician. Euler was a superb calculator – and we have in mind at least as much *symbolic* as numerical computation – who often arrived at beautiful results inductively and heuristically. He "calculated without apparent effort, as men breathe or as eagles sustain themselves in the wind," noted the scientist François Arago [2, p. 139]. To him – as to most mathematicians of his time – algorithms were at least as important as abstract proofs, and the solution of special problems at least as weighty as the formulation of general theories.

But Euler must not be characterized as "merely" a problem-solver. He sought the general in the particular, and introduced general methods to solve specific problems.

Since he was employed by the academies rather than the universities (see Sect. 15.2.2 below), he did no formal teaching. However, he wrote books on various topics which were models of insight and clarity, and which inspired students and teachers. Among such books are the two-volume *Introduction to Analysis of the Infinite* (known more commonly as the *Introductio*, which he considered as a precalculus book), *Basic Principles of the Differential Calculus*, *Basic Principles of the Integral Calculus* (3 vols.), *Elements of Algebra* (which contains much on number theory), *Mechanics* (which made the subject analytical), *A Method for Finding Curved Lines Enjoying Properties of Maximum or Minimum* ... (a book on the calculus of variations), and *Letters to a German Princess* (composed to give lessons on various topics in science and philosophy to Frederick the Great's fifteen-year-old niece! The letters became very popular and were published in book form in seven languages). See [17].

15.2.2 Life

Euler was born in Basel, Switzerland. His father graduated in theology from the University of Basel and became a pastor. He was proficient in mathematics and as a student attended lectures by Jakob Bernoulli. He taught his precocious son mathematics, among other subjects. At thirteen Euler entered the University of Basel to get a general education in the humanities before specializing. One of his instructors was Johann Bernoulli, who taught him mathematics and physics. More importantly, Euler says, "he gave me ... valuable advice to start reading more difficult mathematical books on my own. ... This, undoubtedly, is the best method to succeed in mathematical subjects" [17, p. 468].

Euler had broad interests – in mathematics and beyond – even as a student. At fifteen he got a Bachelor of Arts degree, giving a speech in praise of temperance. That same year he was a respondent at the defense of two theses – on logic and on the history of law. The following year he received a Master's degree in philosophy, presenting a talk comparing the philosophical ideas of Descartes and Newton. To please his father, he studied theology with the goal of becoming a minister, but he soon gave up that idea in favor of mathematics. He remained, however, a believer throughout his life.

At eighteen he published his first paper (in *Acta Eroditorum*) on isochronous curves in a resistant medium, soon to be followed by a second paper in the same journal on reciprocal algebraic trajectories.

Scientific activity in the eighteenth century centered around academies rather than universities, and Euler spent all his professional life at the academies of St. Petersburg and Berlin. Nineteenth-century mathematician Michel Chasles explains why invitations by enlightened rulers for mathematicians to be employed in their kingdoms was of benefit to them [2, p. 139]:

> History shows that those heads of empires who have encouraged the cultivation of mathematics, the common source of all the exact sciences, are also those who have been the most brilliant and whose glory is the most durable.

At nineteen Euler was recommended by Johann Bernoulli's sons, Nikolaus II and Daniel, for a position at the St. Petersburg Academy, recently established by Catherine the Great. He was part of a group of eminent scientists, including Daniel

Bernoulli, Jakob Hermann, and Christian Goldbach, and thrived in these surroundings. He became professor of physics at age twenty-four and of mathematics two years later (replacing Daniel Bernoulli, who returned to Basel).

Some of his duties in the Academy were to carry out a study of Russian territory and to find solutions to various technological problems. He worked on map-making, shipbuilding, and navigation, the last being a most important problem for contemporary empires. But of course his main interests and efforts were in mathematical research. He lost the sight of his right eye in 1738, at age thirty.

In 1741, after fourteen years in St. Petersburg, and because of a negative climate in the Academy and political turmoil in Russia, Euler moved to Berlin, at the invitation of Frederick the Great. He was appointed director of the mathematical sciences at the Berlin Academy and substituted for its president, Maupertuis, when the latter was away. Among many duties, Euler wrote elementary textbooks for the schools, supervised the observatory and the botanical gardens, managed the publication of various calendars and geographical maps, advised the government on the organization of state lotteries, problems of insurance, annuities and widows' pensions, and oversaw the works on pumps and pipes at the hydraulic system of the royal summer palace.

He stayed at the Berlin Academy for twenty-five years, but following disputes with King Frederick and disagreements with Voltaire, the other luminary of the Academy (who was favored by Frederick), he moved back to St. Petersburg, at the invitation of Catherine the Great. He became completely blind at age sixty-four, but continued productive mathematical work till the day he died, twelve years later.

Euler was kind and generous. He had a happy and prosperous family life, married to his first wife for over forty years. Three years after her death, he wed her half-sister, with whom he stayed for the last seven years of his life. He had thirteen children but only five survived beyond infancy. And he loved to play with his grandchildren.

See [2, 4, 5, 7, 17] for further details about Euler's life.

We now give a glimpse of some of Euler's contributions to mathematics. Given his voluminous output, we can barely scratch the surface. We are fortunate, however, to have, among other useful accounts, several collections of books and articles which appeared c. 2007, the 300th anniversary of his birth (see [3–8, 11–13, 17]).

15.2.3 *Analysis*

Euler's publications in analysis exceeded by far those in any other field. Little wonder: The seventeenth century bequeathed to the eighteenth a marvelous and powerful subject – calculus – which mathematicians eagerly explored and applied to the solution of scientific problems. Known to his colleagues as "analysis incarnate," Euler advanced and systematized the differential and integral calculus through his influential textbooks. And he contributed to what later came to be independent fields of analysis (see Sect. 15.2.1 above). We focus on two topics.

Fig. 15.2 Leonhard Euler (1707–1783)

(a) The function concept

The evolution of the concept of function was intimately tied to the development of analysis in the eighteenth and nineteenth centuries. The calculus of Newton and Leibniz is *not* a calculus of functions, it is a calculus of *curves*. The eighteenth century witnessed a gradual "algebraization" of calculus – the replacement of the concept of variable, applied to geometric objects, with the concept of function as an algebraic formula. In this spirit, the first definition of function was given in 1718 by Johann Bernoulli: "One calls here Function of a variable a quantity composed in any manner whatever of this variable and of constants" (see Chap. 5). Bernoulli did not explain what "composed in any manner whatever" meant, but he had in mind an algebraic formula.

It was Euler who took the lead in this process of algebraization, and who had a crucial role in shaping the function concept in the eighteenth century. In his influential text *Introductio in Analysin Infinitorum* of 1748 functions play a central role: the calculus, he asserted, is about functions, not curves. He defined a function as an "analytic expression" (a formula): "A function of a variable quantity is an analytical expression composed in any manner from that variable quantity and numbers or constant quantities." He did not define the term "analytic expression," but gave it meaning by explaining that admissible "analytic expressions" involve the four algebraic operations, roots, exponentials, logarithms, trigonometric functions, differentials, and integrals. It is important to keep in mind that an analytic expression was taken to be a *single* formula valid over the *entire* real line. Thus, for example, neither $f(x) = \begin{cases} 1, & \text{if } x > 0 \\ 2, & \text{if } x \leq 0 \end{cases}$ nor $g(x) = x^2$ if $-1 \leq x \leq 1$ were considered to be functions.

Euler's view of functions was to evolve soon thereafter, following his solution at mid-century of the *vibrating-string problem*: to describe the motion of a taut elastic string fixed at both ends (0 and l say) and released to vibrate. The motion, d'Alembert and Euler showed, is governed by the *wave equation* $\partial^2 y/\partial t^2 = a^2(\partial^2 y/\partial x^2))$ (a constant). Soon a fierce controversy arose between d'Alembert and Euler about the nature of the solution. The goal was to find the *most general* solution. D'Alembert claimed that the solution, hence the initial shape of the string, given by $y = f(x)$, must be an analytic expression – a formula – since these were the only permissible functions. In fact $f(x)$ must be twice differentiable, claimed d'Alembert, since it satisfies the wave equation.

Euler disagreed that this solution is the most general. From physical considerations, he argued that the initial shape of the string can be given by several analytic expressions in different subintervals of $(0, l)$, or, more generally, by a curve drawn free hand. But neither of these was an analytic expression – a *single* formula. According to Grattan-Guinness, the debate between Euler and d'Alembert brought "the whole of eighteenth-century analysis ... under inspection: the theory of functions, the role of algebra, the real line continuum and the convergence of series..." [11, p. 2].

Euler's view of functions evolved over a period of several years. Compare the definition he gave in his 1748 *Introductio* (see above) with the following definition given in 1755, in which the term "analytic expression" does not appear:

> If ... some quantities depend on others in such a way that if the latter are changed the former undergo changes themselves then the former quantities are called functions of the latter quantities. ... If, therefore, x denotes a variable quantity then all the quantities which depend on x in any manner whatever or are determined by it are called its functions. ...

The saga of the nature of function continued for another two centuries. See Chap. 5.

(b) Infinitely small and infinitely large quantities

Power series were an important tool in the algebraization of analysis in the eighteenth century. They were manipulated as polynomials, with little if any attention paid to convergence. Euler claimed that every function could be represented by a power series, with possibly negative or fractional exponents. To substantiate it, he gave a hands-on reason: "If anyone doubts this, this doubt will be removed by the expansion of every function" [3, p. 10].

We present Euler's expansion of $\cos x$ in a power series. Essential tools in this derivation are infinitely small and infinitely large numbers, as well as complex numbers, all of which he used – here and elsewhere – unhesitatingly, and with great success, although none had rigorous backing:

Use the binomial theorem to expand the left-hand side of the identity $(\cos z + i \sin z)^n = \cos nz + i \sin nz$ and equate the real part to $\cos nz$. This yields $\cos nz = (\cos z)^n - [n(n-1)/2!](\cos z)^{n-2}(\sin z)^2 + [n(n-1)(n-2)(n-3)/4!](\cos z)^{n-4}(\sin z)^4 - \cdots$ Let now n be an infinitely large integer and z an infinitely small number. Then $\cos z = 1, \sin z = z, n(n-1) = n^2, n(n-1)(n-2)(n-3) = n^4, \ldots$.

The above equation becomes $\cos nz = 1 - n^2z^2/2! + n^4z^4/4! - \cdots$.

Letting $nz = x$ (Euler claims that nz is finite since n is infinitely large and z infinitely small) we get $\cos x = 1 - x^2/2! + x^4/4! - \cdots$.

What artistry! What brilliant use of symbolic computation! It took another century for this "artistry" to be rigorously explained. See [3, 8, 11, 15], and Chaps. 4 and 7.

15.2.4 Number Theory

Fermat is regarded as the founder of modern number theory, but he could not get his mathematical colleagues in the seventeenth century interested in the subject. Euler was the first to take up the study of number theory in close to 100 years. He took to the subject with "passionate addiction," according to Legendre [16, p. 325].

Euler's interest in number theory was stimulated by his friend Goldbach, with whom he carried on a correspondence over several decades. This began with a letter in 1729, when Euler was twenty-two, in which Goldbach asked Euler's opinion of Fermat's claim that the Fermat numbers $F_k = 2^{2^k} + 1$ are all prime. Euler was skeptical, but it was only two years later that he came up with the counterexample F_5, which he showed to be divisible by 641. This set him on a lifelong study of Fermat's number-theoretic work.

We describe several examples which in due course gave rise (along with other developments) to major branches of number theory: analytic and algebraic number theory.

(a) Analytic number theory

The overriding reason why there was little interest in number theory in the seventeenth and eighteenth centuries was that the period saw the ascendance of calculus as the predominant mathematical field, so mathematicians turned their attention to the exploration of that subject. A major topic was the summation of series. Leibniz' summation $1 - 1/3 + 1/5 - 1/7 + \cdots = \pi/4$ fascinated mathematicians. The summation $1 + 1/4 + 1/9 + 1/16 + \ldots$ baffled them. Leibniz and the Bernoulli brothers (Jakob and Johann) failed to find the sum of this series.

In 1735 Euler succeeded. He showed that $1+1/4+1/9+1/16+\cdots = \pi^2/6$ [7]. This was a spectacular achievement for the young mathematician. It helped establish his growing reputation. He next studied the series $\sum_{n=1}^{\infty} 1/n^{2k}$ for arbitrary k and proved the following beautiful result: $\sum 1/n^{2k} = (2^{2k-1}\pi^{2k}|B_{2k}|)/(2k)!$, where B_i are the "Bernoulli numbers," the coefficients in the power series expansion $x/(e^x - 1) = \sum_1^{\infty} B_n x^n/n!$ [5, 7].

The next natural problem for Euler was to sum the series $\sum 1/n^{2k+1}$. This, however, proved to be a mystery – to him and to his successors. (Only in 1978 was it shown that $\sum 1/n^3$ is irrational; in 2000 it was announced that $\sum 1/n^{2k+1}$ is irrational for infinitely many k, but the status of, for example, $\sum 1/n^5$ is still unknown.)

It was probably the lack of knowledge about the series $\sum 1/n^{2k+1}$ that persuaded Euler to study the function $\zeta(s) = \sum_{n=1}^{\infty} 1/n^s$ for real $s > 1$ (for which the series converges). The complex analogue of this function came to be known as the (Riemann) *zeta function*. It turned out to be a pivotal function in number theory. See [1,5,7,10,16].

In 1737 Euler proved the following about the above function: $\sum 1/n^s = \Pi(1 - p^{-s})^{-1}$, where the product is taken over all the primes p. This very important identity, known as the *Euler product formula*, may be viewed as an analytic counterpart of the fundamental theorem of arithmetic (FTA), the unique representation of the integers as products of primes. (The identity is, in fact, equivalent to the FTA [14, p. 41].) Using his product formula he proved the following two corollaries bearing on the distribution of primes among the integers: (a) There are infinitely many primes, and (b) $\sum 1/p$ diverges, where the sum ranges over all the primes. (Since $\sum 1/n^2$ converges, this shows that there are "more" primes than squares among the integers.) See [1].

The introduction of analysis – the study of the continuous – into number theory – the study of the discrete – may have appeared paradoxical at the time, but it was a crucial development, extensively exploited in subsequent centuries. Here is how Euler viewed these matters [16, p. 176]:

> One may see how closely and wonderfully infinitesimal analysis is related to the theory of numbers, however repugnant the latter may seem to that higher kind of calculus.

Building bridges between different mathematical fields is an important and powerful idea. Euler's work led in the first decades of the nineteenth century to the rise of a new subject – *analytic number theory*. See [1,7,10,16], and Chap. 1.

(b) Algebraic number theory

Here is another example of bridge-building, this time between number theory and abstract algebra. A major motivation for this development was the study of Diophantine equations. We consider the classic equation $x^2+2 = y^3$, a special case of the important *Bachet equation* $x^2 + k = y^3$, k an integer (see Chap. 2). Fermat claimed to have solved it, but gave no indication how. Euler gave the solution in his *Elements of Algebra* of 1770 by introducing a new – and what turned out to be a very important – technique: He factored the left side of $x^2 + 2 = y^3$, which transformed the equation to$(x + \sqrt{-2})(x - \sqrt{-2}) = y^3$. This was now an equation in the domain of "complex integers" of the form $\mathrm{Z}(\sqrt{-2}) = \{\mathrm{a} + b\sqrt{-2} : a, b \in \mathrm{Z}\}$. We call them "integers" because they possess many of the number-theoretic properties of the ordinary integers. Euler exploited this analogy to solve $x^2 + 2 = y^3$. The solution is $x = \pm 5$, $y = 3$. See Sect. 9.5.1.

He had taken the audacious step of introducing "foreign objects" – complex integers – into number theory. Weil claimed that this was a "momentous event" [16, p. 242].

Euler solved the equation $x^3 + y^3 = z^3$ in a similar way, but now embedding it in the domain $Z(\sqrt{-3}) = \{a + b\sqrt{-3} :\ a, b \in \mathrm{Z}\}$ [9, pp. 41–44].

In these examples he introduced a most important idea into number theory: embedding problems about integers in algebraic domains, such as $Z(\sqrt{-2})$ and $Z(\sqrt{-3})$. A major issue that arose was to investigate unique factorization (in some sense) in such domains, which was needed for the solution of the corresponding equations. This was the start of a fruitful interaction between number theory and abstract algebra. It found its expression in the remarkable achievements of Dedekind and Kronecker, who in the 1870s introduced such fundamental algebraic concepts as unique factorization domain, ideal, ring, field, and Dedekind domain, giving rise to a new and important subfield of number theory – *algebraic number theory*. See [9,10,16], and Chaps. 1 and 3. For the significance of Euler's work in number theory on modern developments in the field see [16].

15.2.5 Conclusion

Euler was an outstanding universalist, one of the very few. There is hardly an area of mathematics or its applications to which he did not make significant contributions. André Weil put it well [7, p. 171]:

> No mathematician ever attained such a position of undisputed leadership in all branches of mathematics, pure and applied, as Euler did for the best part of the eighteenth century.

Condorcet, in his eulogy of Euler, noted that "all mathematicians are his disciples" [7, p. xxviii], and Johann Bernoulli referred to him as the "incomparable Leonhard Euler" [17, p. 47]. Laplace admonished us to "read Euler, read Euler, he is the master of us all" [15, p. 124], and Gauss, who was not given to excessive praise, asserted that

> The study of Euler's works will remain the best school for the different fields of mathematics and nothing else can replace it [15, p. 124].

References

1. T. M. Apostol, *Introduction to Analytic Number Theory*, Springer-Verlag, 1976.
2. E. T. Bell, *Men of Mathematics*, Simon and Schuster, 1937.
3. U. Bottazzini, *The Higher Calculus: A History of Real and Complex Analysis from Euler to Weierstrass*, Springer-Verlag, 1986.
4. R. E. Bradley and C. E. Sandifer (eds.), *Leonhard Euler: Life, Work, and Legacy*, Elsevier, 2007.
5. W. Dunham (ed.), *The Genius of Euler: Reflections on his Life and Work*, Math. Assoc. of Amer., 2007.
6. W. Dunham, *The Calculus Gallery: Masterpieces from Newton to Lebesgue*, Princeton Univ. Press, 2005.
7. W. Dunham, *Euler, the Master of Us All*, Math. Assoc. of Amer., 1999.
8. C. H. Edwards, *The Historical Development of the Calculus*, Springer-Verlag, 1979.
9. H. M. Edwards, *Fermat's Last Theorem: A Genetic Introduction to Algebraic Number Theory*, Springer-Verlag, 1977.
10. J. R. Goldman, *The Queen of Mathematics: A Historically Motivated Guide to Number Theory*, A K Peters, 1998.

11. I. Grattan-Guinness, *The Development of the Foundations of Mathematical Analysis from Euler to Riemann*, M.I.T. Press, 1970.
12. C. E. Sandifer, *The Early Mathematics of Leonhard Euler*, Math. Assoc. of Amer., 2007.
13. C. E. Sandifer, *How Euler Did It*, Math. Assoc. of Amer., 2007.
14. J. Stillwell, *Elements of Number Theory*, Springer-Verlag, 2003.
15. D. J. Struik, *A Concise History of Mathematics*, 4th rev. ed., Dover, 1987.
16. A. Weil, *Number Theory: An Approach through History, from Hammurapi to Legendre*, Birkhäuser, 1984.
17. A. P. Youschkevitch, Euler, Leonhard. In *Dictionary of Scientific Biography*, ed. by C. C. Gillispie, Charles Scribner's Sons, 1981, Vol. 4, pp. 467–484.

15.3 Carl Friedrich Gauss (1777–1855)

15.3.1 Life

Gauss was born in Brunswick, Germany, the only son of working-class parents. His father was "worthy of esteem [but] domineering, uncouth and unrefined," according to Gauss [11, p. 298]. His mother was intelligent and of strong character, but only semiliterate. Gauss was a most precocious child and joked later in life that he could count before he could talk. At age eight he astonished his teacher by finding, almost instantly, the sum of the first hundred integers. When he was fourteen, the Duke of Brunswick, who had heard of his reputation, became his patron, and went on to support his education for about ten years.

With his mother's – but not his father's – encouragement, he entered in 1792 the Collegium Carolinum, studying classical languages and, on his own, the works of Newton, Euler, and Lagrange. In 1795, when he enrolled at the University of Göttingen, he was still undecided about which of his two intellectual loves – philology or mathematics – he would pursue as a career.

He opted for mathematics the following year, when he managed to prove that the regular polygon of 17 sides is constructible with straightedge and compass. This was not just a personal triumph; it was the first discovery of a constructible regular polygon in over 2,000 years (the ancient Greeks knew how to construct regular polygons of 3, 4, 5, and 15 sides). Another early landmark was Gauss' proof in 1799 of the Fundamental Theorem of Algebra, which eluded d'Alembert, Euler, and Lagrange. He considered that theorem so important that he gave four proofs of it during his lifetime. The one in 1799 earned him a Ph.D. degree from the University of Helmstedt.

Gauss married happily in 1805. He remarried, unhappily, a year after the death of his first wife in 1809, from which he never fully recovered. He had three children with each of his two wives. He achieved a peaceful home life only in 1831, following the death of his second wife, at which time his younger daughter took over the household duties and "became the intimate companion of his last twenty-four years" [11, p. 302]. See also [2].

Fig. 15.3 Carl Friedrich Gauss (1777–1855)

15.3.2 Disquisitiones Arithmeticae

Gauss made groundbreaking contributions in all areas of mathematics to which he turned: algebra, analysis (both real and complex), geometry (differential and non-Euclidean), number theory, probability, and statistics. He was the Prince of Mathematicians to his contemporaries and is, by universal acknowledgment, one of the three foremost mathematicians of all time (the other two are Archimedes and Newton).

Number theory, the Queen of Mathematics according to Gauss, was his first and greatest mathematical love. The *Disquisitiones Arithmeticae*, arguably his best work, was completed in 1798, when he was twenty-one (!), but was not published till 1801 [5]. In the seventeenth and eighteenth centuries, number theory consisted of a collection of isolated, though brilliant, results, pioneered principally by Fermat, Euler, and Lagrange. In the *Disquisitiones* Gauss systematized the subject, solved a number of its difficult and central problems, and pointed directions for future researchers. But the work was austere and demanding, and it had few readers until Dirichlet, in his *Vorlesungen über Zahlentheorie* of 1863, made it accessible to the mathematical public.

The *Disquisitiones* begins with the definition of congruence – another first. This offers an excellent example of the power of a felicitous notation: the idea of divisibility is expressed in algebraic form, thereby lending the suggestive power of algebraic expressions to arithmetical investigations. A major achievement is the proof of one of the central theorems of number theory, the *quadratic reciprocity law*, already conjectured by Euler and Legendre and rediscovered by Gauss at age seventeen. It describes the relationship between the solvability of $x^2 \equiv p \pmod q$ and $x^2 \equiv q \pmod p$ for odd primes p and q [8]. "This theorem has inspired some deep ideas of modern algebra and is of great importance throughout number theory and in other branches of mathematics" (Stewart [13, p. 125]). Gauss called it "the golden theorem" (theorema aureum) and during his lifetime proved it in eight different ways, hoping to extend it to higher reciprocity laws [8].

Another fundamental accomplishment in the *Disquisitiones* is the comprehensive but subtle theory of *binary quadratic forms*, $f(x, y) = ax^2 + bxy + cy^2$, which studies the representation of integers by such expressions. Fermat in the seventeenth century began to make inroads into this subject by showing that every positive integer is a sum of two squares, $n = x^2 + y^2$. He also found those integers which are sums of $x^2 + 2y^2$ or $x^2 + 3y^2$. In the eighteenth century the problem was intensively investigated by Euler and Lagrange. But the crucial breakthroughs were made by Gauss. Most important was his definition of the composition of binary quadratic forms and his proof that the equivalence classes of such forms with a given discriminant are (in our language) an abelian group under this composition. This result inspired, among others, Dirichlet, Kummer, and Dedekind to try to gain conceptual insight into Gauss' composition of forms. Dedekind succeeded by means of his theory of ideals. See [10, Chap. 3].

The final section of the *Disquisitiones* – an outstanding blend of algebra, geometry, and number theory – deals with *cyclotomy*: the division of a circle into n equal parts. Algebraically, it asks for the solution of $x^n - 1 = 0$. Gauss showed that this so-called cyclotomic equation is solvable by radicals for every positive integer n. This was an important result in the program, initiated by Lagrange in 1770 and brought to fruition by Galois about 1830, of determining which polynomial equations are solvable by radicals. See [10, Chap. 2].

An important by-product of Gauss' results on cyclotomy was the characterization of regular polygons constructible with straightedge and compass: a regular n-gon is so constructible if and only if $n = 2^k p_1 p_2 \dots p_s$, where the p_i are distinct primes of the form $2^{2^t} + 1$, so-called Fermat primes. Gauss proved the sufficiency of his condition for constructibility (the harder part) but only asserted its necessity. This was shown in 1837 by Wantzel.

Little wonder that the *Disquisitiones* made Gauss an instant celebrity. Its wealth and profundity of ideas are still being mined. See [1–3, 7, 13] for further details.

15.3.3 Biquadratic Reciprocity

Gauss returned to number theory in 1831, introducing another groundbreaking idea. This appeared in a paper on *biquadratic reciprocity*, which investigates the relation between the solvability of $x^4 \equiv p \pmod q$ and $x^4 \equiv q \pmod p$, p and q primes. He found that just to *state* a law of biquadratic reciprocity he needed to enlarge the domain of arithmetic. He put it thus [8, p. 108]:

> The previously accepted laws of arithmetic are in no way sufficient for the foundations of a general theory [of reciprocity] .... Such a theory necessarily demands that ... the domain of higher arithmetic needs to be endlessly enlarged.

A prophetic statement indeed. Gauss was calling (in modern terms) for the founding of an arithmetic theory of algebraic numbers. He took the first step by enlarging the domain of arithmetic with the introduction of what came to be known as the *Gaussian integers*, defined as $\mathrm{G} = \{a + bi : a, b \in \mathrm{Z}\}$. He carefully analyzed the arithmetical structure of G, showing that its nonzero, noninvertible elements can be written uniquely as products of "primes" (in G), that is, that G is a unique factorization domain. This extension enabled him to state and prove the biquadratic reciprocity law – an important first in the deep study of reciprocity laws in the nineteenth and twentieth centuries [8]. The article was also a noteworthy contribution to the founding of a new subject – *algebraic number theory*, which flourished in the nineteenth century following work by Kummer, Dedekind, Kronecker, and others. See [10] and Chap. 3.

In the 1831 paper cited above Gauss also defined the complex numbers as points in the plane, an idea which he had formed 30 years earlier. Although these numbers had been used for a century, it was Gauss' sanction that at long last made them respectable as bona fide mathematical entities. Gauss used them extensively and importantly, for example in elliptic function theory.

15.3.4 Differential Geometry

Students of astronomy and physics salute Gauss as one of their own. In 1801, as the *Disquisitiones* was coming off the presses, he correctly predicted, with very little observational data, the location of a new "planet" – Ceres. This was a brilliant achievement, denied to his contemporaries, and it established Gauss as a first-rate scientist.

For economic reasons he accepted in 1807 the directorship of the Göttingen Observatory and a professorship in astronomy, positions he held for the next forty-seven years, until his death. He thenceforth made important contributions to both the theoretical and observational aspects of astronomy and to various branches of physics, including mechanics, optics, acoustics, and geomagnetism. But he always sought the mathematical connection, and in one instance in particular his efforts bore exceptional fruit.

In 1820 Gauss was asked by the Kingdom of Hanover (to which Göttingen belonged) to supervise a geodesic survey, which lasted several years. A major task was the precise measurement of large triangles on the earth's surface. The stimulus (presumably) thus provided to Gauss' fertile mind gave birth in 1827 to his famous paper on curved surfaces, "Disquisitiones generales circa superficies curvas," in which he formulated the fundamental notion of (Gaussian) *curvature* and founded the study of the *intrinsic (differential) geometry* of curved surfaces (cf. the intrinsic geometry on the surface of a sphere) [11, p. 304]. Riemann built on these ideas in the 1850s to found the theory of n-dimensional manifolds, which later proved indispensable in Einstein's general theory of relativity [2, 3].

15.3.5 Probability and Statistics

Related to Gauss' astronomical work, particularly his calculation of orbits of asteroids and comets, were achievements in probability and statistics. In 1809, in a paper on the "Theory of motion of heavenly bodies," he introduced the *method of least squares* (independently found by Legendre) for obtaining the "best fit" to a series of experimental observations. In this connection he devised what came to be known as *Gaussian elimination* for the solution of a system of linear equations. In the same work he also showed that the distribution of errors when using the least squares method is "normal." This is the source of the *Gaussian (normal) distribution*, represented graphically by a bell-shaped curve [2, 7]. Motivated by surveying problems, he made further significant contributions to statistics in an 1823 paper entitled "Theoria combinationis observationum erroribus minimis obnoxiae" [11].

15.3.6 The Diary

Ideas no less profound and far-reaching than those present in Gauss' *published* works (of which we have discussed only some) were found in his *mathematical diary* [6]. This is a remarkable 19-page document of 146 very brief, often cryptic, entries dealing with discoveries, mainly in number theory, algebra, and analysis, covering the years 1796–1814. The diary became public only in 1898. The first entry, dated March 30, 1796, notes Gauss' discovery of the constructibility of the regular 17-gon: "The principles upon which the division of the circle depend, and geometrical divisibility of the same into 17 parts, etc." [6, p. 106]. Entry 72 reads: "I have demonstrated the possibility of a plane." Gray, who translated and commented on the diary, states [6, p. 117]: "This refers to Gauss' interest in the foundations of Euclidean geometry. In a letter to W. Bolyai ... Gauss indicated that the usual definition of a plane presumed too much."

The publication of many of the 146 entries would have made mere mortals famous. Some of the entries anticipated major creations of nineteenth-century mathematics: complex analysis, elliptic function theory, and non-Euclidean geometry. Had Gauss published these in his lifetime, it has been suggested, the development of mathematics would have been advanced by decades (see [1, 11, 13]). Speaking of Gauss' influence on his successors, E.T. Bell said in 1937 that "he lives everywhere in mathematics" [1, p. 269], a tribute even more applicable today.

One might speculate why Gauss did not publish the discoveries listed in his diary: He had so many original ideas at any given time that, being a perfectionist, he had little time to put them in a form sufficiently satisfactory (to him) for publication. His motto "pauca sed matura" (few but ripe) likely guided his attitude to publication, as did his fear of controversy. For more on this issue see [2, 13].

15.3.7 Personality

A comment about aspects of Gauss' personality. He was aloof, worked in isolation, and had no mathematical collaborators, perhaps because he believed he had no mathematical equals. "Jacobi complained in a letter to his brother ... that in twenty years Gauss had not cited any publication by him or by Dirichlet" [11, p. 308].

Some of the events surrounding non-Euclidean geometry are perhaps instructive in this respect. There is no doubt that Gauss was in possession of the elements of non-Euclidean geometry about two decades prior to its publication in the 1830s by J. Bolyai and by Lobachevsky. (For reasons why he did not publish his work on non-Euclidean geometry see [1, 2, 11, 13].) When J. Bolyai's father, W. Bolyai, wrote to his friend Gauss about his son's discovery, Gauss responded that he could not praise the work since he had done all that years earlier. The younger Bolyai was greatly disappointed and died without proper recognition of his great achievement. Gauss did, however, praise the number-theoretic work of the young Eisenstein, "who had been one of the few to tell Gauss anything he did not already know" [11. p. 307]. And he "often gave practical assistance to his friends and to deserving young scientists" [11, p. 301]. I will leave to readers to decide to what extent Gauss was prepared to recognize mathematical talent in others. To pursue this, and other matters having to do with his personality, see [1, 2, 11, 13].

15.3.8 Conclusion

Gauss was a transitional figure in the evolution of mathematics. Stewart put it well [13, p. 130]:

> In many ways Gauss stood at the crossroads. He can be viewed equally well as either the first of the modern mathematicians or the last of the great classical ones. The paradox can easily be resolved: his methods were modern in spirit but his choice of problems was classical.

The nineteenth century witnessed fundamental transformations in mathematics, among them a growing insistence on rigor. Gauss was a leading exponent of this emerging spirit, which began to permeate all areas of mathematics. For example, in his important 1812 work on the *hypergeometric series* he was the first to insist on a rigorous treatment of the convergence of series [2, 3]. "It is demanded of a proof that all doubt become impossible," he wrote to a friend. And he practiced what he preached. His proofs were elegant and polished, often to the point where all traces of his method of discovery were removed. "He is like the fox, who erases his tracks in the sand with his tail," deplored Abel [13, p. 125].

The finished product of Gauss' researches gives no indication of his great skill in, and love of, computation. Some of his deepest theorems in number theory, for example the quadratic reciprocity law, were inspired by calculation. He conjectured the *Prime Number Theorem*, namely that $\pi(x) \sim x/(\log x)$, where $\pi(x)$ is the number of primes $\leq x$ ("$\sim$" denotes "asymptotic," that is, $\lim_{x\to\infty} \pi(x)/[x/(\log x)] = 1$) by first putting together a table of all primes up to 3,000,000. Calculation also enabled him to (re)discover at age fifteen the binomial theorem for rational exponents and the arithmetic–geometric mean [11, p. 298]. Already at this young age

> his lifelong heuristic pattern had been set: extensive empirical investigation leading to conjectures and new insights that guided further experiment and observation [11, p. 298].

We conclude with the following quotation (source unknown), which captures important features of Gauss' mathematical genius:

> It was likely a striking, and possibly unique, combination of remarkable insight, formidable computing ability, and great logical power that produced a mathematician whose ideas are still bearing rich fruit today, two centuries after he burst on the mathematical scene.

References

1. E. T. Bell, *Men of Mathematics*, Simon and Schuster, 1937.
2. W. K. Bühler, *Gauss: A Biographical Study*, Springer-Verlag, 1981.
3. G. W. Dunnington, *Carl Friedrich Gauss: Titan of Science*, Hafner Publ., 1955.
4. W. Ewald (compiler), *From Kant to Hilbert: A Source Book in the Foundations of Mathematics*, Oxford Univ. Press, 1996.
5. C. F. Gauss, *Disquisitiones Arithmeticae*, translated by A.A. Clark, Springer-Verlag, 1986.
6. J. J. Gray, A commentary on Gauss' mathematical diary, 1796–1814, with an English translation, *Expositiones Mathematicae* 2 (1984) 97–130.
7. T. Hall, *Carl Friedrich Gauss: A Biography*, M.I.T. Press, 1970.
8. K. Ireland and M. Rosen, *A Classical Introduction to Modern Number Theory*, Springer, 1982.
9. I. James, *Remarkable Mathematicians: From Euler to Von Neumann*, Math. Assoc. of America, 2002.
10. I. Kleiner, *A History of Abstract Algebra*, Birkhäuser, 2007.
11. K. O. May, Gauss, Carl Friedrich. In *Dictionary of Scientific Biography*, edited by C. C. Gillispie, Charles Scribner's Sons, 1981, Vol. 5, pp. 298–315.
12. G. M. Rassias (ed.), *The Mathematical Heritage of C. F. Gauss*, World Scientific, 1991.
13. I. Stewart, Gauss, *Scientific Amer.* 237 (July 1977) 122–131.

15.4 David Hilbert (1862–1943)

15.4.1 Introduction

Hilbert was arguably the foremost mathematician of the first half of the twentieth century. He made important contributions to invariant theory, algebraic number theory, foundations of geometry, analysis, theoretical physics, and metamathematics. Moreover, he stimulated the development of mathematics in the twentieth century by presenting 23 open problems at the International Congress of Mathematicians held in Paris in 1900. And he was one of the moving spirits in the abstract, axiomatic approach so characteristic of the mathematics of the first half of the twentieth century.

15.4.2 Life

Hilbert was born in Königsberg, Germany. His father was a judge, his mother "an unusual woman" for the times – interested in astronomy, mathematics, and philosophy [9, p. 2]. Mathematics appealed to Hilbert at an early age because, he said later, "it was easy, effortless. It required no memorization. He could always figure it out again for himself" [9, p. 6]. But to study mathematics he first had to obtain a diploma from a gymnasium focusing on Greek and Latin.

In 1880 he took the examination for admission to Königsberg University, and he studied mathematics there for the next four years. He got his PhD in 1885, writing his thesis on invariants, and the following year became "Privatdozent" (an *unpaid* position at German universities entitling appointees to teach; they were given nominal pay by students attending their classes). Hilbert rose at Königsberg to the ranks of associate professor in 1892 and full professor a year later. The university had a number of stars, past and present: Kant, Jacobi, Weber, Lindemann, Minkowski, and Hurwitz. The latter two had great influence on Hilbert's mathematical development and interests.

On the recommendation of Felix Klein, Hilbert was appointed in 1895 to a full professorship at Göttingen University, where such luminaries as Gauss, Dirichlet, and Riemann had held sway. He married in 1892 and had one child. He retired from the University in 1930.

He was in Germany during the two World Wars. Weyl gives an evaluation of aspects of Hilbert's personality relevant to these events [11, p. 612]:

> Hilbert was singularly free from national and racial prejudices; in all public questions, be they political, social or spiritual he stood forever on the side of freedom, frequently in isolated opposition against the compact majority of his environment. He kept his head clear and was not afraid to swim against the current, even amidst the violent passions aroused by the first world war that swept so many other scientists off their feet. It was not mere chance that when the Nazis "purged" the German universities in 1933 their hand fell most heavily on the Hilbert school and that Hilbert's most intimate collaborators left Germany either voluntarily or under pressure of Nazi persecution. He himself was too old, and stayed behind; but the years after 1933 became for him years of ever deepening tragic loneliness.

Weyl himself left Germany for the US at this time.

See [3, 7, 9, 11] for details on this section.

We now describe briefly Hilbert's major contributions in six areas: invariants, algebraic numbers, geometry, analysis, physics, and foundations of mathematics. These are ordered according to the periods in which the corresponding work was done (which we indicate in the headings). His modus operandi was to focus on a

Fig. 15.4 David Hilbert (1862–1943)

given area at a given time and not to come back to it. At the same time, he claimed that "the science of mathematics as I see it is an indivisible whole, an organism whose ability to survive rests on the connection between its parts" [11, p. 617].

15.4.3 Invariants (1885–1893)

The notion of invariance is fundamental in mathematics (as it is in science). Gauss was among the first to explicitly recognize invariance, in his number-theoretic investigations of binary quadratic forms. Invariants also proved important in geometry, especially projective and algebraic geometry, which sought properties of figures invariant under projective and birational transformations, respectively. In the mid-nineteenth century invariant theory became an independent field of study, divorced from its number-theoretic and geometric origins. In fact, between the 1840s and the 1880s it became an important branch of *algebra*. See [8, 11], and Chaps. 1 and 9.

An important problem of the abstract theory of invariants was to discover invariants of various forms (e.g., binary quadratic forms, ternary cubic forms; a "form" is a homogeneous polynomial of any degree in any number of variables). Many of the major mathematicians of the second half of the nineteenth century, among them Cayley, Sylvester, Jordan, Hermite, Clebsch, Gordan, and Hesse worked on the computation of invariants of specific forms. This led to the major problem of invariant theory, namely to determine a complete system of invariants – a "basis" – for an arbitrary form; that is, to find invariants of the form – it was conjectured that finitely many would do – such that every other invariant could be expressed as a combination of these.

Cayley showed in 1856 that the finitely many invariants he had found earlier for binary quartic forms (forms of degree four in two variables) are a complete system. About ten years later Gordan – the so-called King of Invariants – proved that every *binary* form, of any degree, has a finite basis. Gordan's proof of this important result was difficult and computational; he *exhibited* a complete system of invariants.

In 1888 Hilbert astonished the mathematical world by announcing a new, conceptual approach to the problem of invariants. The idea was to rephrase the problem in terms of the newly emerging concepts of rings and ideals: to consider, instead of invariants, expressions in a finite number of variables, in short, the polynomial ring in those variables. Hilbert then proved what came to be known as the *Hilbert Basis Theorem*, namely that every ideal in the ring of polynomials in finitely many variables with coefficients in a field has a finite basis. A corollary was that every form, of any degree, in any number of variables, has a finite complete system of invariants. This result seemed to have "killed" invariant theory. But it reemerged, with vigor, in the second half of the twentieth century.

Gordan's reaction to Hilbert's proof, which did not explicitly exhibit the complete system of invariants, was that "this is not mathematics; it is theology" [8, p. 930]. When Hilbert later gave a constructive proof of this result (which he, however, did not consider significant), it elicited the following response from Gordan: "I have convinced myself that theology also has its advantages" [8, p. 930].

The Hilbert Basis Theorem was a most important result in the newly emerging abstract algebra. It was also fundamental in the modern approach to algebraic geometry. So was Hilbert's *Nullstellensatz*, which deals with the one–one correspondence between ideals and varieties, and which he discovered in connection with his work on invariants. The "theology" of the 1880s became the mathematical gospel of the early twentieth century. See [3, 8, 9, 11].

15.4.4 Algebraic Numbers (1893–1898)

Algebraic number theory is the study of number-theoretic problems using the concepts and results of abstract algebra, mainly those of groups, rings, fields, modules, and ideals. In fact, some of these abstract concepts were *invented* in order to deal with number-theoretic problems. The initial inroads in the subject were made in the eighteenth century by Euler and Lagrange, but the fundamental breakthroughs were achieved in the nineteenth century. Two basic problems provided the early stimulus for these developments: reciprocity laws and Fermat's Last Theorem. See Chaps. 1 and 3.

The strategy that began to emerge was to embed the domain of integers, within which these problems were formulated, in domains of what came to be known as algebraic integers. Early important examples of such domains of "integers" were the Gaussian integers $G = \{a + bi : a, b \in Z\}$ and the cyclotomic integers $C_p = \{a_0 + a_1 w + a_2 w^2 + \ldots + a_{p-1} w^{p-1} : a_i \in Z\}$, p prime, w a primitive p-th root of 1. A crucial issue became to establish unique factorization – in some sense – in such domains. This was done partially by Kummer using ideal numbers, but the grand theory of factorization was established by Dedekind using ideals (and in a less accessible way by Kronecker using divisors). He showed that every ideal in the ring of integers of an algebraic number field (of which the above domains are examples) is a unique product of prime ideals. See Chap. 1.

At their 1893 meeting, the German Mathematical Society asked Hilbert and Minkowski to prepare a report on the current state of number theory, bringing to order the different approaches and methods of Kummer, Kronecker, Dirichlet, Dedekind, and others. Minkowski soon withdrew from the project and in 1897 Hilbert produced the masterful *Zahlbericht* (*Report on Number Theory*). It was "infinitely more" than a "report." It was "a jewel of mathematical literature" [11, p. 626]. Hilbert rephrased and extended what had been done in the subject, introduced new fundamental concepts and results, and "handed over to his pupils a complex of problems of such fascination as that of the relation between number theory and modular functions" [11, p. 634]. Here is a quote from the preface of the *Report* [11, p. 626]:

> The theory of number fields [algebraic number theory] is an edifice of rare beauty and harmony. The most richly executed part of this building ... is the theory of Abelian fields which Kummer by his work on higher laws of reciprocity, and Kronecker by his investigations on the complex multiplication of elliptic functions, have opened up to us.

Two fundamental, related, topics in the *Report* to which Hilbert made important contributions are reciprocity laws and class fields. (In fact, it turned out that reciprocity laws could be phrased within class field theory.) Gauss and Eisenstein in the early nineteenth century began the study of reciprocity (see Chap. 1). Kummer in the 1840s investigated higher reciprocity laws. Hilbert formulated a general reciprocity law, introducing the very important *norm residue symbol*, which generalized the Legendre symbol so useful in quadratic reciprocity [4, p. 143]. Hilbert's reciprocity law was extended by Artin, Hasse, Tagaki, and others in the early decades of the twentieth century. The *Artin Reciprocity Law* is said to be the most general such law.

Kronecker, Weber, and Hilbert were the first to realize that to study factorization in algebraic number fields K (finite extensions of the rationals Q) it is important to consider their Galois groups over Q (cf. classical Galois theory). In fact, it is important, more generally, to study extensions L of K whose Galois groups are abelian. Such extensions are called *abelian extensions. Class field theory* attempts to describe all abelian extensions of K. For example, Kronecker and Weber showed that when $K = Q$, every abelian extension of Q is a subfield of some cyclotomic field. If K is an imaginary quadratic field, "the theory of *complex multiplication* uses certain elliptic curves to give an explicit description of the Abelian extensions of K and their Galois groups" [2, p. 504]. In his study of class fields, Hilbert made an important conjecture – proved in 1907 by Furtwängler – that every algebraic number field K has a unique abelian extension L whose structure as reflected in the Galois group of L over K is identical to the structure of the class group of K. (The "class group" of a field is a measure of its departure from being a unique factorization domain.) For details, which are technical and difficult, see [2, 5, 6, 11].

15.4.5 Foundations of Geometry (1898–1902)

For over two millennia there was only one geometry – Euclidean. Since it was assumed to represent self-evident truths, namely those of the real world, it was believed to be the only possible geometry. The nineteenth century saw the rise of several new geometries: projective, non-Euclidean (both hyperbolic and elliptic), Riemannian, and algebraic. Some order in the nature of geometry, including an examination of the foundations of the various geometries, was called for. Klein's answer was given in his Erlangen Program of 1872 [8]. Another approach was to axiomatize some of the geometries. This was done by Pasch in 1882 for projective geometry and soon thereafter by Peano for Euclidean geometry. Both of their approaches, however, had important shortcomings as far as Hilbert was concerned: they tied their axioms to a physical reality, did not examine independence (in Peano's case) and consistency, and did not focus on the important issue of continuity in geometry. See below, as well as [1, pp. 90–91] and [10].

Euclid's presentation of geometry was found to be deficient in several ways. For one, it possessed deductive gaps in the reasoning because it lacked several types of axioms. Moreover, the notions of point, line, and plane were *defined,* but the definitions were logically unsustainable. In his classic *Grundlagen der Geometrie* (*Foundations of Geometry*) of 1899 Hilbert gave an axiomatization of Euclidean geometry which remedied these deficiencies: it contained 20 axioms (including axioms of order and continuity) as against Euclid's five, and it viewed "point," "line," and "plane" as *undefined concepts* – nowadays called "primitive terms." See [5, 8, 10].

But Hilbert's aim went beyond giving a more comprehensive set of axioms than Euclid's. Here is Weyl on the topic [11, p. 636]:

> It is one thing to build up geometry on sure foundations, another to inquire into the logical structure of the edifice thus erected. If I am not mistaken, Hilbert is the first who moves freely on this higher "metageometric" level: systematically he studies the mutual independence of his axioms [and their (relative) consistency]. . . . His method is the *construction of models.* [Models to establish the independence of Euclid's parallel axiom were given in the mid-nineteenth century.]

Hilbert's models were algebraic. Interesting ideas on the interplay of algebra and geometry were pursued [5]. He would deal with other "meta" issues, such as (absolute) consistency and completeness, in the broader context of mathematics as a whole. See Sect. 15.4.7 below.

The early twentieth century saw the vigorous reemergence, after more than 2,000 years of near dormancy, of the axiomatic method, this time also in areas beyond geometry. Hilbert was very influential in this movement toward abstraction and axiomatization. His *Foundations of Geometry* attracted worldwide attention as soon as it was published (in 1899); the ninth edition appeared in 1962.

It is important to note that Hilbert's axiomatics is not Euclid's. To Euclid axioms were self-evident truths, describing an existing reality. To Hilbert they were neither self-evident nor true. They were *assumptions* about *undefined terms*, which might be considered to be *implicitly* defined by the axioms. No truth-value was associated with the axioms. Hilbert's celebrated statement that "It must be possible to replace in all geometric statements the words *point, line, plane* by *table, chair, mug*" [3, p. 391] makes it starkly clear how his axiomatics differs from Euclid's. The Greeks would have been shocked! See [1, 8–11] for further details.

15.4.6 Analysis (1902–1912) and Physics (1910–1922)

The next two fields to which Hilbert made major contributions are analysis and physics. Since his work in these fields is rather technical, we will make only a very few remarks.

His focus in analysis was on two subfields: integral equations and calculus of variations. The major conceptual breakthrough in the former was his study of infinite-dimensional spaces and his introduction of what came to be known as Hilbert spaces. In the latter field, a major technical accomplishment was a rigorous proof of the Dirichlet Principle. Weierstrass was very critical of Riemann's use of the Dirichlet Principle since it was mathematically not well grounded, and subsequently produced a counterexample. This had a major impact on the development of complex analysis in the nineteenth century. Hilbert's rehabilitation of the Principle therefore legitimized Riemann's approach to the subject, which has since thrived. See [8, 11] and Sect. 15.5.

Using analytic tools, Hilbert achieved in 1908 a major breakthrough in number theory by solving Waring's Problem (proposed in 1770), using intricate ideas from analysis. The problem was to show, given a positive integer k, that the equation $n = x_1^k + x_2^k + \ldots + x_s^k$ holds for every integer n, where s depends on k but not on n. The eminent number-theorist G. H. Hardy was most impressed: "It would hardly be possible for me to exaggerate the admiration which I feel for the solution of this historic problem" [9, p. 114]. See also [8, 11].

As for Hilbert's work in physics, "It is generally acknowledged that Hilbert's achievements in this field lack the profundity and inventiveness of his mathematical work proper," noted Freudenthal [3, p. 392]. "But his application of integral equations to kinetic gas theory and to the elementary theory of radiation were notable contributions," according to Weyl [11, p. 653]. See also [1].

15.4.7 Foundations of Mathematics (1922–1930)

The roots of the crisis in foundations go back to Cantor's creation of set theory in the late nineteenth century, about which many mathematicians had reservations, but which Hilbert embraced wholeheartedly. "No one shall expel us from the paradise which Cantor created for us," he exclaimed [8, p. 1003]. Cracks, however, began to appear in Cantor's paradise when paradoxes were found in set theory – by, among others, Cantor and Russell. These were serious matters which demanded attention.

To this end, Zermelo and others gave various axiomatizations of set theory. Although these avoided the known paradoxes, they did not guarantee that others might not show up.

Hilbert determined to remedy the situation by undertaking to show the absolute consistency of various axiomatic systems, including of course that of set theory. The primitive terms and the axioms of such systems were considered to be strings of symbols to which no meaning was attached. They were to be manipulated according to established rules of inference to obtain the theorems of the system. The methods by which this was to be accomplished were finite, and so acceptable to all. Hilbert's ideas formed the essence of a school in the foundations of mathematics called *formalism*.

The formalists have been accused of removing all meaning from mathematics and reducing it to symbol manipulation. The charge is unfair. Hilbert's aim was to deal with the *foundations* of mathematics rather than with the daily practice of the mathematician. And to show that mathematics is free of inconsistencies one first needed to formalize the subject. This was formalism in the service of informality. See [1, 8, 9, 11].

Hilbert's ideas were fiercely opposed by those, headed by L.E.J. Brouwer, who came to be known as *intuitionists*. They claimed that no formal analysis of axiomatic systems is necessary. In fact, mathematics should not be founded on systems of axioms. The mathematicians' intuition, beginning with that of number, will guide them in avoiding contradictions. They must, however, pay special attention to definitions and methods of proof. These must be constructive and finitistic. In particular, the law of the excluded middle, completed infinities, the axiom of choice, and proof by contradiction are all outlawed.

The debate between the formalists and intuitionists has not been resolved. But Hilbert's grand design for proving consistency was laid to rest by Gödel's Incompleteness Theorems of 1931. These showed the inherent limitations of the axiomatic method: The consistency of a large class of axiomatic systems, including those for arithmetic and set theory, cannot be established within the systems. Moreover, if consistent, these systems are incomplete. See [8, 11, 12] for details.

These results did not, of course, invalidate the axiomatic method, which thrived in the first half of the twentieth century. Writing in 1944, Weyl claimed that "not a little of the attractiveness of modern mathematical research is due to a happy blending of axiomatic and genetic procedures" [11, p. 645]. And Tarski judged that, despite Gödel's results, "Hilbert will deservedly be called the father of metamathematics" [9, p. 218].

15.4.8 Mathematical Problems

In 1900, with very significant achievements behind him, Hilbert was asked to give a plenary lecture at the second International Congress of Mathematicians held in Paris. He chose the topic "Mathematical Problems," presenting 23 problems which he felt were important for mathematicians of the new century to consider. Indeed, the problems have served as inspiring guideposts for the development of important ideas. During the Congress "it became quite clear that David Hilbert had captured the imagination of the mathematical world with his list of problems for the twentieth century" [9, p. 84]. The solution of any one of them would entitle its solver to join what came to be known as "the honors class" [12].

The problems (as printed in [6]) are preceded by about eight pages of often enlightening comments, including historical illustrations, on the role of problems in general, and in particular on possible approaches to attacking some of the 23 problems which he proposed. We give a very brief selection, first of Hilbert's introductory comments [6]:

> As long as a branch of science offers an abundance of problems, so long is it alive ... (p. 438). It is by the solution of problems that the investigator tests the temper of his skill; he finds new methods and new outlooks and gains a wider and freer horizon... (p. 438). It is an error to believe that rigor in the proof is the enemy of simplicity. On the contrary, we find it confirmed by numerous examples that the rigorous method is at the same time the simpler and the more easily comprehended (p. 441). If we do not succeed in solving a mathematical problem, the reason frequently consists in our failure to recognize the more general standpoint from which the problem before us appears as a single link in a chain of related problems (p. 443). [It is a] conviction (which every mathematician shares ...) that every definite mathematical problem must necessarily be susceptible of an exact settlement, either in the form of an actual answer to the question asked, or by the proof of the impossibility of its solution ... (p. 444). We hear within us the perpetual call: There is the problem. Seek its solution. You can find it by pure reason, for in mathematics there is no *ignorabimus* (p. 445).

Now to some of the problems: To prove the continuum hypothesis (no. 1); to prove that the axioms for arithmetic are consistent (no. 2); to determine whether there is an "elementary" theory of volume for polyhedra (that is, one not using calculus) similar to the elementary theory of area for polygons (no. 3); to axiomatize physics (no. 6); to establish the transcendence of certain real numbers, for example, $2^{\sqrt{2}}$ (no. 7); to prove the Riemann hypothesis (no. 8); to establish the most general reciprocity law (no. 9); given a diophantine equation, to devise a procedure to determine if the equation is solvable using finitely many operations (no. 10); and to establish class field theory (no. 12).

Sixteen of Hilbert's problems have been solved (nos. 1, 2, 3, 4, 5, 7, 9, 10, 11, 13, 14, 15, 17, 18, 21, and 22; the first was no. 3), four have been essentially solved (nos. 12, 19, 20, and 23), and three have not (nos. 6, 8, and 16). Hilbert thought that no. 8 (the Riemann hypothesis) would be solved before no. 7 (the transcendence of certain real numbers, solved in the 1930s). The hazards of prognostication! See [1,4,6,12] for details.

15.4.9 Conclusion

"Like some mathematical Alexander [Hilbert] left his name written large across the map of mathematics. There [is] ... Hilbert space, Hilbert inequality, Hilbert transform, Hilbert invariant integral, Hilbert irreducibility theorem, Hilbert base theorem, Hilbert axiom, Hilbert subgroups, Hilbert class-field" [9, p. 216]. He founded a great mathematical center at Göttingen. In time, it attracted some of the foremost mathematicians to the University, which became the Mecca of mathematics. Among its permanent members or visitors were Artin, Courant, Dehn, Einstein, Feller, Neugebauer, von Neumann, Emmy Noether, Ore, Polya, Olga Taussky, Weyl, and Wigner. As for students – suffice it to say that Hilbert was the supervisor of sixty-nine theses!

In 1910, at the height of his mathematical power, the Hungarian Academy of Sciences awarded him the Bolyai Prize. Poincaré was chosen to make the presentation. Among his comments, he chose to emphasize for special mention the following qualities of Hilbert's work: "the variety of the investigations, the importance of the problems attacked, the elegance and the simplicity of the methods, the clarity of the exposition, and the care for absolute rigor" [9, p. 125]. "No mathematician of equal stature has risen from our generation," maintained Weyl in an address in 1944 [11, p. 612].

References

1. L. Corry, David Hilbert and the axiomatization of physics, *Arch. Hist. Exact Sciences* 51 (1997) 83–198.
2. D. A. Cox, *Galois Theory*, John Wiley and Sons, 2004.
3. H. Freudenthal, Hilbert, David. In *Dictionary of Scientific Biography*, ed. by C. C. Gillispie, Charles Scribner's Sons, 1981, Vol. 6, pp. 388–395.
4. J. Gray, *The Hilbert Challenge*, Oxford Univ. Press, 2000.
5. D. Hilbert, *Foundations of Geometry*, Open Court, 1902.
6. D. Hilbert, Mathematical problems, *Bull. Amer. Math. Soc.* 8 (1902) 437–479.
7. I. James, *Remarkable Mathematicians*, Math. Assoc. of Amer., 2002.
8. M. Kline, *Mathematical Thought from Ancient to Modern Times*, Oxford Univ. Press, 1972.
9. C. Reid, *Hilbert*, Springer-Verlag, 1970.
10. M. Toepell, On the origins of David Hilbert's "Grundlagen der Goemtrie," *Arch. Hist. Exact Sciences* 35 (1986) 329–344.
11. H. Weyl, David Hilbert and his mathematical work, *Bull. Amer. Math. Soc.* 50 (1944) 612–654.
12. B. Yandell, *The Honors Class,* A K Peters, 2002.

15.5 Karl Weierstrass (1815–1897)

15.5.1 Life

Weierstrass was born in Ostenfelde, Germany, the oldest of four children. His father was cultured and educated, but domineering. His mother died when Karl was eleven. When fourteen he entered the Catholic Gymnasium in Paderborn (near Münster), where his father became treasurer at the customs office. He was a brilliant student and won prizes in German, Greek, Latin, and mathematics. At this time he also read Crelle's *Journal für die reine und angewandte Mathematik* and worked part-time as a bookkeeper to help with family finances.

At age nineteen he entered the University of Bonn, studying – at his father's urging – finance and administration. But his real interest was mathematics. "The conflict between duty and inclination led to physical and mental strain" [2, p. 219]. He began to study mathematics on his own, starting with Laplace's *Mécanique céleste.* He left the university without completing his course of studies, greatly disappointing his father.

At twenty-four he moved to Münster to prepare for the teacher's examination. A year later he qualified to teach in gymnasia. He taught for fourteen years, the following subjects: mathematics, physics, German, botany, geography, history, gymnastics, and calligraphy. He devoted all his free time to studying mathematics.

The turning point in his life occurred in 1854, when – nearly forty years old – he published an important paper in *Crelle's* Journal on elliptic and abelian functions (for definitions see Sect. 15.5.4(b)). His interest in these functions was aroused by Gudermann, his mathematics teacher at Münster, to whom he was afterward eternally grateful.

Although that article was just a preliminary version of Weierstrass' forthcoming masterpiece, Liouville called it "one of those works that marks an epoch in science" [2, p. 221]. The complete version, "Theory of abelian functions," followed two years later (although it was conceived in 1844). In it, according to Hilbert, he had realized one of the greatest achievement of analysis, the solution of the Jacobian inversion problem for hyperelliptic integrals [2, p. 221]. In 1855 the University of Königsberg awarded him an honorary doctorate. In 1857 he became a professor at the University of Berlin. For more details see [1, 2, 10].

15.5.2 Foundations of Real Analysis

Weierstrass has been described as the "father of modern analysis." He contributed to all branches of the subject: calculus, differential and integral equations, calculus of variations, infinite series, elliptic and abelian functions, and real and complex analysis. His work is characterized by attention to foundations and by scrupulous logical reasoning. Klein commented on Weierstrass' overall approach to mathematics: "[He] is first of all a logician; he proceeds slowly, systematically, step-by-step. When he works, he strives for the definitive form" [4, p. 291]. Kline, it should be noted, was a critic of Weierstrass' approach.

The calculus giants of the seventeenth and eighteenth centuries – Newton, Leibniz, Euler, Lagrange, and others – introduced the basic concepts of the subject, conceived its algorithms, and derived many of its fundamental results. But the subject was largely heuristic, lacking logical foundations. The nineteenth century

Fig. 15.5 Karl Weierstrass (1815–1897)

ushered in a rigorous spirit in mathematics which included an examination of the foundation of various fields. Cauchy initiated this process in calculus in his *Cours d'analyse* of 1821. He selected several fundamental concepts – limit, continuity, convergence, derivative, and integral – highlighted "limit" as the one in terms of which all the others were defined, and derived by fairly rigorous means many of the important results of the calculus. But there were several major foundational problems with his approach: verbal definitions of limit and continuity, frequent use of infinitesimals, and intuitive appeal to geometry in proving the existence of various limits. See [5, 7].

Weierstrass and Dedekind (among others) determined to remedy this unsatisfactory situation, with the goal of establishing theorems in a "purely arithmetic" manner (as Dedekind put it). This came to be known as the "arithmetization of analysis" (a term coined by Felix Klein [4]). It meant establishing analysis rigorously on an arithmetic basis. Since the real numbers are in the foreground or background of much of analysis, and since from the inception of the calculus they were viewed geometrically, the goal became to establish them arithmetically, based on the rationals (and ultimately on the positive integers). This was accomplished independently in the1870s by several mathematicians, Weierstrass among them.

The remaining foundational task was to give a rigorous definition of the limit concept, to replace Cauchy's intuitive conception. This Weierstrass accomplished with his precise $\epsilon - \delta$ definitions of limit and continuity (those in use today). He thereby banished infinitesimals from analysis (until Robinson resurrected them some 100 years later). The foundations for the arithmetization of analysis were laid. To Plato "God geometrized" while to Weierstrass, Dedekind, and others "Man arithmetized." See [4, 5, 12], and Chap. 4 for further details.

15.5.3 Complex Analysis

Although complex numbers were used in analysis in the eighteenth century (by Euler among others), it was only in the nineteenth that complex analysis (also known as complex function theory) was founded as an independent subject – by Cauchy, who made important inroads in it in a series of papers beginning in the 1820s. The second stage in the evolution of complex analysis – the grand conception of the subject – was laid in the 1850s and 1860s by Riemann and Weierstrass. But their approaches to the subject were entirely different.

Riemann's was global and geometric, and was based on the notion of a Riemann surface and on the Dirichlet Principle, whereas Weierstrass' was local and algebraic, grounded in power series and analytic continuation. In a letter to H. A. Schwartz he asserted [4, p. 259]:

> The more I ponder the principles of [complex] function theory – and I do so incessantly – the more I am convinced that it must be founded on simple algebraic truths.

Weierstrass was opposed to Riemann's use of geometric intuition and physical arguments. In particular, he severely criticized the Dirichlet Principle for being mathematically not well grounded, and produced a counterexample. Thereafter, *his* approach to complex analysis became dominant [15, p. 98]:

> Only with the works of Klein and Lie and the rehabilitation of the Dirichlet Principle [in the early twentieth century] by Hilbert could the Riemann theory again gradually recover from the blow delivered to it by Weierstrass.

See [3, 4, 12, 16] for more details.

15.5.4 Other Work

We discuss briefly three of Weierstrass' other accomplishments.

(a) Continuity

The notion of continuity is subtle. Euler was the first to define it, but in a sense different from ours (see Chap. 5). Cauchy gave the first essentially modern and fairly rigorous definition in the 1820s. But he used infinitesimals, which were not defined rigorously at the time, and he viewed the real numbers geometrically, as no arithmetic treatment was available. For example, he claimed that "a remarkable property of continuous functions of a single variable is to be able to be represented geometrically by means of straight lines or continuous curves" [11, p. 261]. Little wonder that Cauchy and his contemporaries believed, and some of them "proved," that a continuous function is differentiable except possibly at isolated points. It was therefore "shocking" when Weierstrass (and independently Riemann) produced a function which is everywhere continuous and nowhere differentiable (see Chap. 6). This and similar examples showed that the notion of continuity is considerably broader than that of differentiability and established continuity as an important concept for investigation in its own right. It also showed the limitations of intuitive geometric reasoning in analysis and the need for careful analytic formulations of basic concepts.

Counterexamples to widely held notions have often played important roles in mathematics – clarifying concepts and pointing to significant departures (see [6, 13], and Chap. 8). We have described above two such examples due to Weierstrass – one having to do with the Dirichlet Principle, the other with continuity vs. differentiability. For a third, Cauchy "proved" that a convergent series of continuous functions is a continuous function. Soon thereafter Abel gave a counterexample, but it took another 20 years to discover where Cauchy went wrong. The discovery was made by Weierstrass (and independently by Seidel) who introduced the concept of uniform convergence and showed that a *uniformly* convergent series of continuous functions is indeed continuous. One was dealing with subtle concepts.

(b) Elliptic and abelian functions

An integral of the form $\int R[x, \sqrt{p(x)}]\mathrm{d}x$, where R is a rational function and p is a polynomial of degree 3 or 4 with distinct roots, is called an *elliptic integral*, because the first example of such an integral occurred in the formula for the arc length of an ellipse. Elliptic integrals were studied in the seventeenth and eighteenth centuries, but the crucial breakthrough occurred in the early nineteenth when Abel and Jacobi decided to "invert" these integrals, that is, to study their inverse functions, named *elliptic functions*. It is these functions that turned out to be the key objects to investigate. (Cf. the integral $\int[1/\sqrt{1-x^2}]\mathrm{d}x$ with its inverse function, $\sin x$.) A crucial early idea was to extend these functions to the complex domain.

Integrals of the form $\int R(z, w)\mathrm{d}z$, where R is an algebraic function (that is, a function $w = f(z)$ defined implicitly by the polynomial equation $R(z, w) = 0$) and z and w complex variables, are called *abelian integrals* (so named by Jacobi after Abel, who first studied them), and their inverses *abelian functions*. They are generalizations of elliptic integrals and functions, respectively. The study of elliptic and abelian functions became an important branch of analysis in the nineteenth century, with applications in number theory and beyond. See [4, 12, 16].

Weierstrass' first two major papers, on the basis of which he became Professor at Berlin, were (we recall) on elliptic and abelian functions. In his inaugural address in the late 1850s to the Berlin Academy of Sciences he confessed that

> A comparatively younger branch of mathematical analysis, the theory of elliptic functions, has, from the time in which I first became acquainted with it... exercised a strong attraction on me, and has retained a definite influence on the entire course of my mathematical development [4, p. 258].

Indeed, throughout his life he considered his work on elliptic and abelian functions to be his most important. According to Morris Kline, that work "completed, remodeled, and filled with elegance the theory of elliptic functions" [12, p. 651]. Since complex analysis is fundamental to the study of elliptic and abelian functions,

Weierstrass' determination to scrutinize its foundations was undertaken at least in part to gain a deeper insight into these functions. See [3, 4, 12, 16] for details.

(c) Linear algebra

Analysis was not the only area in which Weierstrass had significant success. Matrix theory was another. He made fundamental contributions in particular to the spectral theory of matrices, including such basic topics as eigenvalues and canonical forms. Historian Thomas Hawkins goes further [9, pp. 156, 157]:

> Insofar as anyone deserves the title of founder of the theory of matrices, it is WEIERSTRASS. ... He is the central figure in the developments occurring in the 19th century. His theory of elementary divisors provided a theoretical core, a substantial foundation, upon which to build. ... Most of the applications of the theory of matrices that were discovered in the 19th century were also discovered as applications of WEIERSTRASS' theory of elementary divisors.

15.5.5 Conclusion

We have only scratched the surface of Weierstrass' achievements. Here are several others to which his name has been permanently attached: The Weierstrass approximation theorem, which says that a continuous function can be uniformly approximated by polynomials; the Bolzano-Weierstrass theorem, which states that every infinite, bounded set of real numbers has a limit point; the Weierstrass factorization theorem, which gives the representation of an entire function in terms of an infinite product of "prime functions;" the Casorati–Weierstrass theorem, which says that in every neighborhood of an isolated essential singularity an analytic function takes values arbitrarily close to any assigned complex number; the Weierstrass M-test, which deals with the comparison of series for convergence; and the Weierstrass p-function, an example of an elliptic function of order 2. See [4, 12] for details.

Weierstrass demanded of himself the very strictest standards, with the result that he published little. His ideas and his reputation spread through his excellent lectures which drew both students and established mathematicians from around the world. The former included Frobenius, Killing, and Netto, the latter Hensel, Hölder, Klein, Lie, Minkowski, and Mittag-Leffler. He was kind to his students and generous in suggesting topics for dissertations. When Mittag-Leffler arrived in Paris to study analysis under Hermite, the latter told him: "You have made a mistake, sir, you should follow Weierstrass' course at Berlin. He is the master of all of us" [1, p. 422]. Hawkins describes Killing's appraisal of his teacher [8, p. 104]:

> Killing appreciated [Weierstrass'] openness with students, his willingness to engage in scientific discussion outside the lecture hall, his concern for the personal welfare of his students, and his generosity with mathematical ideas. Furthermore, although Weierstrass is nowadays thought of primarily as an analyst, his actual mathematical interests were much broader than his published opus might imply.

Especially noteworthy is Weierstrass' relationship with the brilliant Sonia Kovalevskaya, who became the leading female mathematician of the nineteenth century. Unable to convince the University Senate in Berlin to admit her as a student (this was in 1870, when women were as a rule not eligible for entrance to university), Weierstrass taught her privately for the next 4 years, and subsequently kept up a scientific correspondence with her until her premature death in 1891. He was instrumental in having her get a position as lecturer in mathematics at Stockholm in 1883 and a professorship for life in 1889. See [1, 12] for details.

It has often been said that mathematics is a young person's game, that the best work mathematicians do is when they are in their twenties or early thirties. Outstanding counterexamples to what Susan Landau claims is "the myth of the young mathematician" are Karl Weierstrass, Sophus Lie, and Emmy Noether; all three made their most outstanding contributions when nearing forty. See [14].

Weierstrass was most proud of his work on abelian functions, and much of his fame in the nineteenth century rested on it. His results in this field are, however, less significant today. For us, his main legacy is his unrelenting insistence on maintaining high standards of rigor and seeking the fundamental ideas underlying mathematical concepts and theories. "Weierstrassian rigor" has come to denote rigor of the strictest standard. According to historian of mathematics Kurt Biermann, "Weierstrass was the most important nineteenth-century German mathematician after Gauss and Riemann" [2, p. 224].

References

1. E. T. Bell, *Men of Mathematics*, Simon and Schuster, 1937.
2. K.- R. Biermann, Weierstrass, Karl Theodor Wilhelm. In *Dictionary of Scientific Biography*, ed. by C. C. Gillispie, Scribner's, 1970–1980, Vol 14, pp. 219–224.
3. U. Bottazzini, "Algebraic truths" vs "geometric fantasies": Weierstrass' response to Riemann, *Proceedings of the Intern. Congr. of Mathematicians, Beijing, 2002*, Vol. III, pp. 923–934.
4. U. Bottazzini, *The Higher Calculus: A History of Real and Complex Analysis from Euler to Weierstrass*, Springer-Verlag, 1986.
5. C. H. Edwards, *The Historical Development of the Calculus*, Springer-Verlag, 1979.
6. B. R. Gelbaum and J. M. H. Olmsted, *Counterexamples in Analysis*, Holden-Day, 1964.
7. J. W. Grabiner, *The Origins of Cauchy's Rigorous Calculus*, MIT Press, 1981.
8. T. Hawkins, *The Emergence of the Theory of Lie Groups: An Essay in the History of Mathematics, 1869–1926*, Springer-Verlag, 2000.
9. T. Hawkins, Weierstrass and the theory of matrices, *Arch. Hist. Exact Sciences* 17 (1977) 119–163.
10. I. James, *Remarkable Mathematicians, From Euler to Von Neumann*, Math. Assoc. of Amer., 2002.
11. P. Kitcher, *The Nature of Mathematical Knowledge*, Oxford Univ. Press, 1983.
12. M. Kline, *Mathematical Thought from Ancient to Modern Times*, Oxford Univ. Press, 1972.
13. S. Klymchuk, *Counterexamples in Calculus*, Math. Assoc. of Amer., 2010.
14. S. Landau, The myth of the young mathematician, *Notices of the Amer. Math. Soc.* 44 (1997) 1284.

15. E. Neuenschwander, Studies in the history of complex function theory II: Interactions among the French School, Riemann, and Weierstrass, *Bulletin of the Amer. Math. Soc.* 5 (1981) 87–105.
16. J. Stillwell, *Mathematics and its History*, 2nd ed., Springer-Verlag, 2002.

Index

I. Kleiner, *Excursions in the History of Mathematics*, DOI 10.1007/978-0-8176-8268-2,

MIX
Papier aus verantwortungsvollen Quellen
Paper from responsible sources
FSC® C105338

If you have any concerns about our products,
you can contact us on
ProductSafety@springernature.com

In case Publisher is established outside the EU,
the EU authorized representative is:
Springer Nature Customer Service Center GmbH
Europaplatz 3, 69115 Heidelberg, Germany

Printed by Libri Plureos GmbH
in Hamburg, Germany